THE MAXIMUM ENTROPY FORMALISM

A Conference Held at the Massachusetts Institute of Technology on May 2-4, 1978

Edited by Raphael D. Levine and Myron Tribus

The MIT Press
Cambridge, Massachusetts, and London, England

PUBLISHER'S NOTE This format is intended to reduce the cost of publishing certain works in book form and to shorten the gap between editorial preparation and final publication. The time and expense of detailed editing and composition in print have been avoided by photographing the text of the book directly from typescript provided by the authors.

This book was printed and bound in the United States of America

Library of Congress Cataloging in Publication Data

Maximum Entropy Formalism Conference, Massachusetts
Institute of Technology, 1978.
The maximum entropy formalism.

Includes bibliographical references.
1. Entropy (Information theory)--Congresses.
I. Levine, Raphael D. II. Tribus, Myron.
III. Massachusetts Institute of Technology.
IV. Title.
Q370.M38 1978 001.53'9 78-10799
ISBN 0-262-12080-1

CONTENTS

FOREWORD

This volume presents the lectures delivered at the Maximum Entropy Formalism Conference held at the Massachusetts Institute of Technology, May 2-4, 1978. The papers cover the background, the principles and the practice of the formalism as we understand it today. The wide range of applications in the natural sciences and engineering and the varied backgrounds of the participants at the conference bear witness to the considerable potential and scope of the formalism. Of particular interest however are the many cross correlations between the different papers. This very much validates our original contention in organizing the conference: there clearly is a common cultural heritage shared among the workers in this field. It also shows that a would-be practitioner would do well to examine applications outside of his immediate area of interest. It is with such a person in mind that the lectures have been collected together in this volume. It offers both an introduction to the method and to the literature and then proceeds, in a progress report fashion, to provide realistic examples of the methodology. It is our hope, however, that the more erudite reader will also find here new insights and novel applications.

The conference took place in 1978 because we felt it was an appropriate time to stop and take stock. The number of active workers in this field and the range of applications are constantly on the increase. The present conference may well be the last opportunity to span a wide range of applications in one volume. (Even so, we are guilty of an error of omission. Applications in Economics and the Social Sciences are not discussed in any detail.)

It turns out that 1978 also marks the anniversaries of the key papers that contributed to our understanding of why it makes sense to assign value to probabilities by maximizing the entropy. Some 100 years ago Boltzmann (1) presented the notion of the most probable distribution, the forerunner of the maximal entropy procedure. About 75 years ago, Planck (2) maximized the entropy so as to derive the distribution which bears his name and which changed the course of physics, and Gibbs (3) presented a concise and precise statement that is still a source of inspiration. Some 50 years ago von Neumann (4) introduced entropy into quantum mechanics and Hartley (5) sought to define the concept of information. Forty years ago Elsasser (6) suggested that the entropy of a quantal system should be maximized. Thirty years ago, Cox (7) discussed the algebra of probable inference, while Shannon (8) and Wiener (9) sought to refine the concept of information, with Wiener putting particular stress on physical applications. Finally, 20 years ago, Jaynes (10) invoked the Shannon axiomatic characterization of the 'uncertainty' or entropy function and the Cox interpretation of probability as reflecting a state of knowledge, to cast the principle in its general form, as a principle of inductive reasoning independent of any specific application. The more recent history will be found in the pages of this volume, which also includes new contributions by Professors Elsasser, Cox, and Jaynes.

In concluding, we thank the speakers for their contribution to the success of the meeting and to this volume; the participants for their keen interest which required that at night they be physically ejected from the conference room; and the sponsors of the conference: the Department of Chemistry, the Department of Mechanical Engineering, and the Center for Advanced Engineering Study of M.I.T., and the Air Force Office of Scientific Research.

M. Tribus
R. D. Levine

Cambridge, Massachusetts
June 1978

References

1. Boltzmann, L. 1877. *Uber die Beziehung zwischen dem zweiten Hauptsatze der mechanischen Warmetheorie und der Wahrscheinlichkeitsrechnung, respective den Satzen uber das Warmegleichgewicht.* Wein Ber. 76:373-95.

2. Planck, M. 1906. *Theorie der Warmestrahlung.* Leipzig: J. A. Barth.

3. Gibbs, J. W. 1902. *Elementary Principles in Statistical Mechanics.* New Haven: Yale University Press.

4. von Neumann, J. 1927. *Thermodynamik quantenmechanischer Gesamtheiten.* Gott. Nach. 273-91.

5. Hartley, R. V. L. 1928. *Transmission of Information.* Bell Syst. Tech. J. 7:535-63.

6. Elsasser, W. M. 1937. *On Quantum Measurements and the Role of Uncertainty Relations in Statistical Mechanics.* Phys. Rev. 52:987-999.

7. Cox, R. T. 1946. *Probability, Frequency and Reasonable Expectations.* Am. J. Phys. 14:1-14.

8. Shannon, C. E. 1948. *A Mathematical Theory of Communication.* Bell Syst. Tech. J. 27:379-423.

9. Wiener, N. 1948. *Cybernetics.* Cambridge, MIT Press.

10. Jaynes, E. T. 1956. *Information Theory and Statistical Mechanics.* Phys. Rev. 106:620-30.

Noam Agmon
Department of Chemistry
M.I.T.
Cambridge, MA 02139

Yoram Alhassid
Department of Physics
M.I.T.
Cambridge, MA 02139

John C. Allred
Physics Department
University of Houston
Houston, TX 77004

Bjarne Andresen
H. C. Ørsted Institutet
Universitetsparken 5
2100 Copenhagen Ø
Denmark

Jorge Barojas W.
Physics Department
Universidad Autonoma
Metropolitana-Iztapalapa
Aptdo Postal 55-534
Mexico 13, D.F.

Bruce Berne
Department of Chemistry
Columbia University
New York, NY 10027

Lesche Bernhard
Institut fur Quantenchemie
Holbeinstrasse 48
1 Berlin 45
Germany

Manuel Berrondo
Instituto de Fisica
University of Mexico
Aptdo Postal 20-364
Mexico 20, D.F.

R. Stephen Berry
Department of Chemistry
University of Chicago
Chicago, IL 60637

Edward A. Burke
Department of the
Air Force
Hanscom Air Force Base
MA 01731

Walter Carrington
Harvard School of Public
Health
14 Magnolia Avenue
Cambridge, MA 02138

Gregory J. Chaitin
IBM Corporation
P.O. Box 218
Yorktown Heights, NY 10598

Michael A. Collins
Department of Chemistry
M.I.T.
Cambridge, MA 02139

Richard T. Cox
300 Northfield Place
Baltimore, MD 21210

Roger Day
Department of Physiology
Harvard School of
Public Health
Boston, MA 02115

Walter M. Elsasser
Department of Earth and
Planetary Sciences
Johns Hopkins University
Baltimore, MD 21218

Robert B. Evans
School of Mechanical Engineering
Georgia Institute of Technology
Atlanta, GA 30332

Martin A. Garstens
913 Buckingham Drive
Silver Spring, MD 20901

Elias P. Gyftopoulos
Department of Nuclear Engineering
M.I.T.
Cambridge, MA 02139

Mechtild Heinlein
T.O. Berlin
1000 Berlin 45
Germany

David Hestenes
Physics Department
Arizona State University
Tempe, AZ 85281

E. T. Jaynes
Department of Physics
Washington University
St. Louis, MO 63130

James C. Keck
Department of Mechanical Engineering
M.I.T.
Cambridge, MA 02139

E. H. Kerner
Physics Department
University of Delaware
Newark, DE 19711

Marvin S. Keshner
Department of Electrical Engineering
M.I.T.
Cambridge, MA 02139

Robert M. Kiehn
Physics Department
University of Houston
Houston, TX 77004

James L. Kinsey
Department of Chemistry
M.I.T.
Cambridge, MA 02139

Bernard O. Koopman
Farrar Road
Lincoln, MA 01773

Rolf Landauer
T. J. Watson Research Center
IBM
P.O. Box 218
Yorktown Heights, NY 10598

Raphael D. Levine
Department of Chemistry
M.I.T.
Cambridge, MA 02139

C. Alden Mead
Chemistry Department
University of Minnesota
Minneapolis, MN 55455

Abraham Nitzan
Department of Chemistry
University of Tel Aviv
61390 Ramat-Aviv
Tel Aviv
Israel

Irwin Oppenheim
Department of Chemistry
M.I.T.
Cambridge, MA 02139

Philip Pechukas
Department of Chemistry
Columbia University
New York, NY 10027

Paul Penfield, Jr.
Department of Electrical Engineering
M.I.T.
Cambridge, MA 02139

John G. Pierce, Director
Exploratory Research Division
Center for Naval Analyses
1401 Wilson Boulevard
Arlington, VA 22209

Baldwin Robertson
National Bureau of Standards
Washington, D.C. 20234

John Ross
Department of Chemistry
M.I.T.
Cambridge, MA 02139

Jerome Rothstein
Department of Computer and Information Science
The Ohio State University
Columbus, OH 43210

Ernst Ruch
Institut fur Quantenchemie der freien Universitat
Holbeinstrasse 48
1000 Berlin 45
Germany

William M. Siebert
Department of Electrical Engineering
M.I.T.
Cambridge, MA 02139

Ray Solomonoff
Zator Company
140 Mt. Auburn Street
Cambridge, MA 02138

Laszlo Tisza
Department of Physics
M.I.T.
Cambridge, MA 02139

Myron Tribus
Center for Advanced Engineering Study
M.I.T.
Cambridge, MA 02139

THE MAXIMUM ENTROPY FORMALISM

THIRTY YEARS OF INFORMATION THEORY

Myron Tribus

This paper is concerned with the impact of Shannon's 1948 paper (1) "A Mathematical Theory of Communication" which was focused on the field of communications. I do not propose in this paper to deal with the influences exerted by Walter M. Elsasser, Richard T. Cox or Edwin T. Jaynes, even though the anniversaries of their three papers provides the impulse for this gathering. My objective is to trace the spread of Shannon's well publicised paper. I must say, from a strictly personal view, that the works of Cox and Jaynes have had an extraordinary impact on me and I acknowledge freely my indebtedness to them for all they have taught me. On another occasion I hope to have the opportunity to treat their papers and that of Elsasser.

My more limited objective is to comment upon the influence of Shannon's paper in fields other than communications. I approach the task with some hesitation. In 1961 Professor Shannon, in a private conversation, made it quite clear to me that he considered applications of his work to problems outside of communication theory to be suspect and he did not attach fundamental significance to them.

Despite Shannon's misgivings, as time goes on Shannon's great contribution to the literature has made its existence felt in an ever increasing number of fields. Shannon's clarification of the concepts of uncertainty and information and the ability to give a quantitative measure to these concepts has served as a powerful stimulant on the imagination of others. Of course there have been numerous unjustified excursions in the name of information theory. But my studies show many solid accomplishments clearly inspired by this famous paper.

To test the spread of Shannon's influence I once thumbed through the engineering index and found that the first entry under the heading "Information Theory" occurred in 1951, only three years after his paper appeared. In 1952 there were eight entries; in 1953 there were 13 entries, including three symposia and a reference to a bibliography containing over 1,000 references to information theory. By 1958, 10 years after Shannon's paper, the engineering index contained several pages of titles covering papers in the field of information. Many of the entries were references to symposia which themselves contained 15 or 20 papers on information theory. I can report also that the engineering index listing under the title "Information Theory" is incomplete. For example, in 1961 I prepared a paper on information theory and thermodynamics. It was listed by the index under "Thermodynamics" but not under "Information Theory." There is, therefore, no easy way to track down all of Shannon's influence. For example, under the heading "Entropy" the engineering index for 1961 merely says: "See Thermodynamics."

Once you know the field that was influenced, however, it is fairly easy to document this influence.

It is rather difficult to say where the field of information theory ends and other fields begin. Certainly information theory overlaps significantly with many other fields. For the purposes of this paper I shall take the view that any field of inquiry which uses the function

$$S = -K\sum_i p_i \ln p_i$$

(or its continuous analog) in one of the following three ways is using information theory:

a) As a criterion for the choice of probability distributions;
b) To determine the degree of uncertainty about a proposition; and
c) As a measure of the rate of information acquisition.

Claude Shannon was not the first to use the function $-\sum_i p_i \ln p_i$, for this function (or its continuous analog) had been used as long ago as 1872 by Boltzmann (2). What Shannon did was to give a universal meaning to the function $-\sum_i p_i \ln p_i$ and thereby make it possible for others to find applications. Warren Weaver, in his popularization of Shannon's work foresaw the widespread influence it was bound to have. Weaver understood that any paper which clarified our understanding of knowledge was certain to affect all fields which deal in knowledge.

In the 1961 interview with Shannon, to which I referred, I obtained one anecdote which it seems to me worth recording on this occasion. I had asked Dr. Shannon what his personal reaction had been when he realized he had identified a measure of uncertainty. Shannon said that he had been puzzled and wondered what to call his function. "Information" seemed to him to be a good candidate as a name, but "Information" was already badly overworked. Shannon said he sought the advice of John von Neumann, whose response was direct, "You should call it 'entropy' and for two reasons: first, the function is already in use in thermodynamics under that name; second, and more importantly, most people don't know what entropy really is, and if you use the word 'entropy' in an argument you will win every time!"

A field in which Shannon's interpretation of entropy has had a profound effect is classical thermodynamics. In this field, the entropy function has a long and involved history. Clausius coined the word "entropy" from the Greek language to mean "transformation" and defined it via the equation

$$dS = dQ/T$$

where

dQ = increment of energy added to a body as heat during a reversible process

T = absolute temperature during the reversible heat addition

dS = change in entropy.

The function as Clausius introduced it was at the time quite mysterious. It was an extensive property, like mass, energy, volume, momentum, charge and number of atoms of the chemical species. Unlike these quantities, however, entropy did not obey a conservation law. At a time so profoundly influenced by the success of the various conservation laws of physics, when determinism was at its peak, a measure which always increased was indeed mysterious and called for an "explanation."

Among scientists and engineers there are those who speak of "discovering" the "laws of nature" and then there are those who speak of "inventing" these laws. The former are the most numerous. They include the majority of teachers and, therefore, of students. The chief difference between the two types lies in differing interpretations of what constitutes an "explanation." The former would require that all mysteries be explained in terms of things already known or new axioms which square with the old. The latter are ready to rewrite and reorganize all human knowledge in their quest for consistency. To the latter there are no "explanations" - merely consistent ways to trace ideas to their common logical sources.

I have dwelt upon the word "explanation" because the arguments which have raged in thermodynamics hinge greatly upon this point. The information theory approach raises a question in the following way. Shannon's measure gives, by ordinary differentiation:

$$dS = -K\sum_i (\ln p_i + 1)\, d\, p_i$$

But since $\sum_i p_i = 1$, we have $\sum d\,p_i = 0$ and, therefore:

$$dS = -K \sum_i \ln p_i \, d\,p_i .$$

On the other hand, from classical thermodynamics, we have:

$$dS = dQ/T.$$

How then are we to "explain" these two equations for dS?

In 1953 Brillouin formally stated the case for a close connection between these two entropies (4). Brillouin followed his early paper with several others and with his famous book on science and information theory (5). In these publications Brillouin showed there was an intimate connection between entropy and information but Brillouin did not think it necessary to define one in terms of the other. Brillouin treated the field of thermodynamics as correct and pre-existing and he treated Shannon's information theory as a correct theory. He then proceeded to demonstrate a consistency between the two entropies.

As early as 1911 Van der Waals had proposed that there ought to be a connection between Bayes' equation in probability theory and the second law of thermodynamics (6). As we now know, Shannon's measure can be derived through the use of Bayes' equation (7,8). In 1932 G. N. Lewis, in a discussion of irreversibility, had written: "Gain in entropy means loss of information -- nothing more."(9) As I mentioned earlier, Boltzmann had used the "H-Function" as early as 1872.

In 1938 Slater based his book on the definition

$$S = -k \sum f_i \ln f_i$$

choosing this expression for entropy over the more conventional one

$$S = k \ln W$$

which is used by most authors in statistical thermodynamics. (f_i = fraction in microstate i; W = number of ways to realize a given macrostate.)

With these historical antecedents Brillouin evidently felt no need to take one concept as more primitive and felt no need to "explain" the others in terms of it. But he pointed the way.

It was E. T. Jaynes who first saw the complete answer and presented it in a sequence of brilliant and historical papers (10,11). Jaynes proposed that Shannon's measure of uncertainty (entropy) be used to define the values for probabilities.

Prior to Jaynes' contribution, workers in statistical mechanics were forced to rely upon classical thermodynamics for their ideas about entropy and to develop statistical mechanics as "an analog" (12,13). Prior to Jaynes' paper there had been no clear and unequivocal reason to define entropy in statistical mechanics other than the after-the-fact conclusion that since "statistical mechanics works," it must be all right. In 1961 after a careful study of Jaynes' paper, I demonstrated that all of the laws of classical thermodynamics, and in particular, the concepts of heat and temperature, could be defined from Shannon's entropy using Jaynes' principle of maximum entropy (14,15). The debates engendered by this approach have been extensive and, on occasion, bitter. I have concluded from them that thermodynamics is as much a branch of theology as it is a branch of science! It is very comforting to be able to show that although Gibbs did not explain how or why he decided in 1901 to set the logarithm of the probability of a system linearly proportional to the energy and numbers of particles, thereby producing his "statistical analogs," he does say that the expected value of ln p is a minimum, that is, $-\int p \ln p \, dv$ is a maximum (16). So we may conclude that the Gibbs' and Jaynes' formalisms are essentially the same.

Using the information measure as a primitive idea, my colleague Robert Evans has been able to demonstrate that thermodynamic information is proportional to what has been called "availability" in thermodynamics (17). Acually Evans has shown that a new function which he calls "essergy" (and which is related to the function used in Germany under the name "exergie") is at once a measure of information and a measure of the potential work of systems. We have used the essergy function in the optimization of thermal systems such as sea water demineralization plants and steam turbine generators (18,19,20). The original impetus for these developments came from Shannon's work.

Jaynes has converted Shannon's measure to a powerful instrument for the generation of statistical hypotheses and he has applied it as a tool in statistical inference. The mathematical aspects of Jaynes' use of Shannon's measure are straightforward. If the given information is in the form

$$\sum p_i = 1$$

$$\sum p_i \, g_r(x_i) = \langle g_r \rangle \qquad r = 1, 2, .. m$$

The minimally prejudiced (i.e., least presumptive) probability distribution is the set of p_i which obeys the (m+1) equations above and maximizes

$$S = -K \sum_i p_i \ln p_i .$$

The resulting distribution is:

$$p_i = \exp [-\lambda_0 - \lambda_1 g_1(x_i) - \lambda_2 g_2(x_i) \ . \ . \ .]$$

where the λ's are Lagrangian multipliers.

The entropy is:

$$S = \lambda_0 + \sum_{r=1}^{m} \lambda_r \langle g_r \rangle$$

And the potential function, λ_0, is:

$$\lambda_0 = \ln \sum_i \exp [-\sum_r \lambda_r g_r(x_i)].$$

The Lagrange multipliers are related to the given data by the equation:

$$-\frac{\partial \lambda_0}{\partial \lambda_r} = \langle g_r \rangle .$$

The higher moments are given by:

$$\frac{\partial^2 \lambda_0}{\partial \lambda_r^2} = \text{var } (g_r) \qquad \frac{\partial^2 \lambda_0}{\partial \lambda_r \partial \lambda_s} = \text{covar } (g_r g_s).$$

For other relations see Jaynes' (11) or (25). Jaynes' principle shows that if Shannon's measure is taken to be the measure of uncertainty not just a measure, the formal results of statistical mechanical reasoning can be carried over to other fields.

Jaynes proposed, therefore, that in problems of statistical inference (which means problems for which the given data are inadequate for a deterministic prediction of what will [or did] happen) the probabilities should be assigned so as to maximize:

$$S = -K \sum p_i \ln p_i$$

subject to:

$$\sum_i p_i = 1$$

and any other given information. Jaynes thus essentially

defines probabilities via Shannon's measure. From this basis one may see that any field which uses probability theory is a candidate to be influenced by Shannon's work.

I must confess in 1957 (3) I felt there was no "explanation" to be had and complained publicly that Shannon had confused things by calling two different ideas by the same name! I was forced to retract that ill considered judgment in a footnote to reference 3 and wish now that I had retracted the whole paper!

An "explanation" of the two entropies can be had only if we clarify our ideas about what we mean by:

a) Heat, dQ
b) Temperature, T
c) Probability, p_i.

What Shannon's work did, as interpreted by Jaynes, was make it possible to take the information based entropy as the primitive concept and explain dQ, T and p_i in terms of S.

Jaynes' initial publication demonstrated that this approach permitted a derivation of statistical mechanics. He later went on to give illustration of the use of this principle in decision theory (21), communication theory (22) and transport theory (23).

It was also straightforward to use the same approach in reliability engineering (24). In this case the maximum entropy principle was used with Bayes' equation to show how to form an initial estimate of reliability and then incorporate field test and laboratory data as they become available.

Shannon's measure occurs in statistics but unless one is attuned to its meaning, the significance of Shannon's function in any one case is apt to go unnoticed. For example, in hypothesis testing we have demonstrated that the empirical fitting of experimental data to a probability distribution is expedited and rendered more accurate via maximum entropy methods (25).

We have also demonstrated that in the analysis of the contingency tables the first approximation to an exact Bayesian test for statistical dependence between attributes A_i and B_j is given by the measure:

$$\Delta S^* = -N \sum_i f_i \ln f_i - N \sum_j f_j \ln f_j + N \sum_i \sum_j f_{ij} \ln f_{ij}$$

which simply tells us to compare the sum of entropies in the margins of the contingency table with the entropy of the center and see where the information resides. It is especially satisfying to be able to report that the first order approximation to the right hand side of the above equation is the familiar chi-square statistic (26).

Once the measure of uncertainty had been established psychologists were quick to recognize its utility. In 1949, only a year after Shannon's article appeared, Miller and Frick (31) published an article illustrating the relevance of the theory to psychology.

The first class of psychological problems to attract attention quite naturally centered on man as a communicator or receiver of information. Shannon had already given some impetus to the field by his analysis of the information content of symbols in the English language. Shannon's clever techniques provided brand new tools for psychologists already attracted to engineering literature by Wiener's provocative publications on cybernetics, which had just appeared. Indeed, the appearance of popular treatments of feedback theory, computers and information theory has had a profound effect on all of science. (The impact of engineering upon society is not entirely due to engineering hardware. That observation could form the basis for several essays.) By 1954 McGill was using the differences in entropy between the margins and the center of a contingency table to measure a subject's response to stimuli (32). This work was unknown to me until I began to prepare for this assignment. It is interesting to note that one may <u>derive</u> Shannon's measure from a Bayesian analysis of contingency tables (26) and then show the chi-square measure as its first approximation. Attneave uses the chi-square measure to see how significant the entropy difference is! Attneave lists 87 primary references to information theory in psychology. The range of topics considered is most impressive. Included are such titles as "Information Theory and Immediate Recall," "Information in Absolute Judgments," "Informational Aspects of Visual Perception," "Uncertainty and Conflict," "Model for Learning," "Information from Dot and Matrix Patterns," "Relation Between Error Variance and Information Transmitted in a Simple Pointing Task," "Information in Auditory Displays."

Pollack investigated the information transmission when nine different musical tones were used and found 2.67 bits with and 2.19 bits without objective standard tones (33). Attneave reports that Rogers tested a concert master of a symphony orchestra having "absolute pitch" and found he could transmit 5.5 bits!

Measuring the ability of man and animals to transmit and receive information through various sensory channels is an obvious way to apply information theory and Shannon's work could be taken over almost unchanged for this purpose. Miller has pointed out (34), however, that the conceptual foundations of information theory have as much importance as the numerical measure. They provide a methodology for organization and patterning. Miller points out that concepts of gestalt psychology may be put to test via information theoretic concepts.

The view that psychoneural activity is the economical encoding of experience may be attributed directly to Shannon's influence.

It may properly be said that the penetration of field X by field Y is complete when the concepts of field Y are used to prepare a textbook for students in field X. In this sense, the appearance of Garner's book in 1962 explaining psychological concepts via information theory was a milestone (35).

An economist, Kuhn (36), has used concepts of uncertainty, feedback and decision analysis to produce a unified treatment of learning, motivation, language, culture personality, personal transactions, organizations, social systems, ecomomics, public policy and the political process. It is Kuhn's thesis that the concepts of engineering are to be subsumed as special cases the concepts of these social sciences.

New information theory applications in turbulence theory are emerging from the work of Silver and his colleague Tyldesly in Glasgow (39).

John P. Dowds has developed his own method of applying the entropy concept to an examination of the data from oil fields. Based upon such an analysis, he purchased the rights to an abandoned oil well in Oklahoma and today this well, which he christened "Rock Entropy #1", is a producer sufficiently generous to support Mr. Dowds' further researches into the uses of entropy (40).

My colleague A. O. Converse has also applied the entropy concept in a search technique, using the entropy as a measure of the information obtained from each sample point in the search for the maximum of a function in a bounded interval (41).

The maximum entropy principle has been used in land use planning and in the prediction of travel between different communities (38).

There seems to be a growing use of information theory in medicine. I have searched, via machine techniques, for articles referring to Shannon's 1948 article and Jaynes' 1957 article and turned up papers on rheumatism, bacterial populations in the Beaufort sea (42), information processing of schizophrenics (43,44), molecular biology (45), diagnostic value of clinical tests and the nature of living systems (46).

The ideas have also spread over the globe. J. Sonuga in Lagos, Nigeria uses the entropy principle to analyze runoff (47). In the Netherlands Van Marlen and Dijkstra use information theory to select peaks for retrieval of mass spectra (48). In the Free University of Berlin, Eckhorn and Popel apply information theory to the visual system of a cat (49). In Canada Reilly and Blau use entropy in building mathematical models of chemical reacting systems, (50) as does Levine in Israel (51).

And I suppose we can say the subject has really arrived when it is referenced by the U.S. Congress Office of Technology Assessment! (52)

The future we can perceive but dimly. But surely Shannon's paper deserves to be read by future generations with the same sense of intellectual adventure that we associate with the great works of the past. That famous publication takes its rightful place alongside the works of Newton, Carnot, Gibbs, Einstein, Mendeleev and the other giants of science on whose shoulders we all stand.

1. Shannon, C. E., "A Mathematical Theory of Communication, The Bell System Technical Journal, 27, 379-623, 1948.

2. Boltzmann, L., "Weitere Studien uber das Warmegleichgewicht unter Gasmolekulen", K. Acad. Wiss. (Wein) Sitzb., II Abt. 66, 275.

3. Tribus, M., Thermodynamics, A Survey of the Field, Recent Advances in the Engineering Sciences, McGraw Hill, New York, 1958.

4. Brillouin, L., "Negentropy Principle of Information", J. Appl. Phys., 24, N9, 1152-63, September 1953.

5. Brillouin, L., Science and Information Theory, Academic Press, New York, 1962.

6. Van der Walls, J., "Uber die Erklarung der Naturgesetze auf Statisch-Mechanischer Grundlage", Physik. Zeitschr. XII, 547-549, 1911.

7. Tribus, M., Evans, R. and Crellin, G., "The Use of Entropy in Hypothesis Testing", Tenth National Symposium on Reliability and Quality Control, January 7-9 1964.

8. Good, I. J., Probability and the Weighing of Evidence, Ch. 6, Griffin, London/Hafner, New York, 1950.

9. Lewis, G. N., "The Symmetry of Time in Physics", Science, 71, 569, June 1930.

10. Jaynes, E. T., "Information Theory and Statistical Mechanics", Phys. Rev., 106, 620, 1957.

11. Jaynes, E. T., "Information Theory and Statistical Mechanics", Phys. Rev., 108, 171, 1957.

12. Gibbs, J. W., Elementary Principles in Statistical Mechanics, Yale University Press, New Haven, 1948.

13. Denbigh, K. G., The Principles of Chemical Equilibrium, Cambridge University Press, Cambridge, 1955.

14. Tribus, M., "Information Theory as the Basis for Thermostatics and Thermodynamics", Jour. Appl. Mech. 28, 1-8, March 1961.

15. Tribus, M., Thermostatics and Thermodynamics, D. Van Nostrand, New York, 1961.

16. Gibbs, J. W., The Collected Works of J. Willard Gibbs, 2 Chapter II, Theorem 5, Yale University Press, New Haven, 1948. (Original publication 1902)

17. Evans, R. B., "Basic Relationships Among Entorpy, Exergy, Energy and Availability", Chapter 2, Appendix A in Principles of Desalination, ed: K. S. Speigler, Academic Press, New York, 1966.

18. El-Sayed, Y. M. and Evans, R. B., "On the Use of Exergy and Thermoeconomics in the Design of Desalination Plants", Thayer School of Engineering, Dartmouth College, Hanover, New Hampshire, January 1968.

19. El-Sayed, Y. M. and Aplenc, A. J., "Application of the Thermoeconomic Approach to the Analysis and Optimization of a Vapor-Compression Desalting System", Journal of Engineering for Power, 17-26, January 1970.

20. El-Sayed, Y. M. and Evans, R. B., "Thermoeconomics and the Design of Heat Systems", Journal of Engineering for Power, 27-35, January 1970.

21. Jaynes, E. T., "New Engineering Applications of Information Theory", Proceedings of the First Symposium on Engineering Applications of Random Function Theory, John Wiley & Sons, New York, 1963.

22. Jaynes, E. T., "Note on Unique Decipherability", IEEE Trans. on Infor. Theory, IT-5, 98, September 1959.

23. Jaynes, E. T. and Scalapino, D., "Non-Local Transport Theory" (Paper presented at the first meeting of the Society for Natural Philosophy, Baltimore, Maryland, 25 March 1963).

24. Tribus, M., "The Use of the Maximum Entropy Estimate in Reliability Engineering", Recent Developments in Information and Decision Processes, MacMillan, New York, 1962.

25. Tribus, M., Rational Descriptions, Decisions and Designs, Chapter VII, Pergamon Press, Oxford, 1969.

26. Tribus, M., ibid., Chapter VI.

27. Lees, S., "Uncertainty and Imprecision", J. Basic Engineering, ASME 88, 369-378, 1966.

28. Lees, S., "On Measurement", to be published by Applied Mechanics Reviews.

29. Attneave, F., Applications of Information Theory to Psychology, Henry Holt and Co., New York, 1959.

30. Shannon, C. E. and Weaver, W., The Mathematical Theory of Communication, University of Illinois Press, Urbana, 1949.

31. Miller, G. A. and Frick, R. C., "Statistical Behavioristics and Sequences of Responses", Psychol. Rev., 56, 311-324, 1949.

32. McGill, W. J., "Multivariate Information Transmission", Psychometrika, 19, 97-116, 1954.

33. Pollack, I., "The Information of Elementary Auditory Displays", J. Acoust. Soc. Am., 25, 765-769, 1953.

34. Miller, G. A., "What is Information Measurement?", American Psychologist, 8, 3-11, 1953.

35. Garner, W. R., Uncertainty and Structure as Psychological Concepts, John Wiley & Sons, New York, 1962.

36. Kuhn, A., The Study of Society--A Unified Approach, Irwin-Dorsey Press, Homewood, Illinois, 1963.

37. Wilson, A. G., A Statistical Theory of Spatial Distribution Models, Transportation Research, 1, 253-269, Pergamon Press, Oxford, 1967.

38. Wilson, A. G., "The Use of Entropy Maximizing Models in the Theory of Trip Distribution, Mode Split and Route Split", Journal of Transport Economics and Policy, January 1969.

39. Tyldesley, J. R., "A Thermodynamic Approach to Turbulence Phenomena", Private Communication, Department of Mechanical Engineering, Glasgow University.

40. Dowds, J. P., "Application of Information Theory in Establishing Oil Field Trends", Reprinted from Computers in the Mineral Industries, ed: George A. Parks, School of Earth Sciences, Stanford University, 1964.

41. Converse, A. O., "The Use of Uncertainty in a Simultaneous Search", Operations Research, 15, N6, 1088-95, November-December 1967.

42. Kaneko, T., Atlas, R. M., Krichevsky, M., "Diversity of Bacterial Populations in Beaufort Sea", Nature, 270, N5638, 569-599, 1977.

43. Wijesinghe, O., "Effect of Varying Rate of Presentation on Information - Transmission of Schizophrenic and Control Group", Brit. Jour. Psychiatry, 130, 509-513, May 1977.

44. Neufeld, R. W. J., "Response Selection Processes in Paranoid and Non-Paranoid Schizophrenia", Perceptual and Motor Skills, 44, N2, 499-505, 1977.

45. Berger, J., "Information Theory and Central Dogma of Molecular Biology", Jour. Theor. Biol., 65 , N2, 393-345, 1977.

46. Miller, J. G., "Nature of Living Systems", Behavioral Science, 21, N5, 295-319, 1976.

47. Sonuga, J. O., "Engropy Principle Applied to Rainfall Runoff Process", Jour. Hydrology, 30, N1-2, 81-94, 1976.

48. Van Marlen, G. and Dijkstra, A., "Information Theory Applied to Selection of Peaks for Retrieval of Mass Spectra", Analytical Chem., 48, N3, 595-598, 1976.

49. Eckhorn, R., and Popel, B., "Rigorous and Extended Application of Information Theory to Afferent Visual System of a Cat", Kybernetic, 16, N4, 191-200, 1974.

50. Reilly, P. M. and Blau, G. E., "Use of Statistical Methods to Build Mathematical Models of Clinical Reacting Systems", Canadian Journal of Chemical Engineering, 52, N3, 289-299, 1974.

51. Levine, R. D., "Entropy and Microscopic Disequilibrium. Isothermal Time Evolution With Application to Vibrational Relaxation", Jour. Chem. Phys., 65, N8, 3284-3301, 1976.

52. Coates, J. F., "Technology Assessment--Tool Kit", Chemical Technology, 6, N6, 372-383, 1976.

WHERE DO WE STAND ON MAXIMUM ENTROPY?

Edwin T. Jaynes

A. Historical Background

B. Present Features

C. Speculations for the Future

D. An Application: Irreversible Statistical Mechanics

Summary. In Part A we place the Principle of Maximum Entropy in its historical perspective as a natural extension and unification of two separate lines of development, both of which had long used special cases of it. The first line is identified with the names Bernoulli, Laplace, Jeffreys, Cox; the second with Maxwell, Boltzmann, Gibbs, Shannon.

Part B considers some general properties of the present maximum entropy formalism, stressing its consistency and interderivability with the other principles of probability theory. In this connection we answer some published criticisms of the principle.

In part C we try to view the principle in the wider context of Statistical Decision Theory in general, and speculate on possible future applications and further theoretical developments. The Principle of Maximum Entropy, together with the seemingly disparate principles of Group Invariance and Marginalization, may in time be seen as special cases of a still more general principle for translating information into a probability assignment.

Part D, which should logically precede C, is relegated to the end because it is of a more technical nature, requiring also the full formalism of quantum mechanics. Readers not familiar with this will find the first three Sections a self-contained exposition.

In Part D we present some of the details and results of what is at present the most highly developed application of the Principle of Maximum Entropy; the extension of Gibbs' formalism to irreversible processes. Here we consider the most general application of the principle, without taking advantage of any special features (such as interest in only a subspace of states, or a subset of operators) that might be found in particular problems. An alternative formulation, which does take such advantage--and is thus closer to the spirit of previous "kinetic equation" approaches at the cost of some generality, appears in the presentation of Dr. Baldwin Robertson.

A. Historical Background

The ideas to be discussed at this Symposium are found clearly expressed already in ancient sources, particularly the Old Testament, Herodotus, and Ovennus. All note the virtue of making wise decisions by taking into account all possibilities, i.e., by not presuming more information than we possess. But probability theory, in the form which goes beyond these moral exhortations and considers actual numerical values of probabilities and expectations, begins with the Ludo aleae of Gerolamo Cardano, some time in the mid-sixteenth century. Wilks (1961) places this "around 1520," although Cardano's Section "On Luck in Play" contains internal evidence that shows the date of its writing to be 1564, still 90 years before the Pascal-Fermat correspondence.

Already in these earliest works, special cases of the Principle of Maximum Entropy are recognized intuitively and, of necessity, used. For there is no application of probability theory in which one can evade that all-important first step: assigning some initial numerical values of probabilities so that the calculation can get started. Even in the most elementary homework problems, such as "Find the probability of getting at least two heads in four tosses of a coin," we have no basis for the calculation until we make some initial judgment, usually that "heads" shall have the probability 1/2 independently at each toss. But by what reasoning does one arrive at this initial assignment? If it is questioned, how shall we defend it?

The basis underlying such initial assignments was stated as an explicit formal principle in the Ars Conjectandi of James (= Jacob) Bernoulli (1713). Unfortunately, it was given the curious name: Principle of Insufficient Reason which has had, ever since, a psychologically repellant quality that prevents many from seeing the positive merit of the idea itself. Keynes (1921) helped somewhat by renaming it the Principle of Indifference; but by then the damage had been done. Had Bernoulli called his principle, more appropriately, the Desideratum of Consistency, nobody would have ventured to deprecate it, and today statistical theory would be in considerably better shape than it is.

The essence of the principle is just: (1) we recognize that a probability assignment is a means of describing a certain state of knowledge. (2) if the available evidence gives us no reason to consider proposition A_1 either more or less likely than A_2, then the only honest way we can describe that state of knowledge is to assign them equal probabilities: $p_1 = p_2$. Any other procedure would be inconsistent in the sense that, by a mere interchange of the labels (1, 2) we could then generate a new problem in which our state of knowledge is the same but in which we are assigning different probabilities. (3) Extending this reasoning, one arrives at the rule

$$p(A) = \frac{M}{N} = \frac{\text{(Number of cases favorable to A)}}{\text{(Total number of equally possible cases)}} \qquad \text{(A1)}$$

which served as the basic definition of probability for the next 150 years.

The only valid criticism of this principle, it seems to me, is that in the original form (enumeration of the "equally possible" cases) it cannot be applied to all problems. Indeed, nobody could have emphasized this more strongly than Bernoulli himself. After noting its use where applicable, he adds, "But here, finally, we seem to have met our problem, since this may be done only in a very few cases and almost nowhere other than in games of chance the inventors of which, in order to provide equal chances for the players, took pains to set up so that the numbers of cases would be known and --- so that all these cases could happen with equal ease." After citing some examples, Bernoulli continues in the next paragraph, "But what mortal will ever determine, for example, the number of diseases --- these and other such things depend upon causes completely hidden from us ---."

It was for the explicitly stated purpose of finding probabilities when the number of "equally possible" cases is infinite or beyond our powers to determine, that Bernoulli turns next to his celebrated theorem, today called the weak law of large numbers. His idea was that, if a probability p cannot be calculated in the manner $p = M/N$ by direct application of the Principle of Insufficient Reason, then in some cases we may still reason backwards and estimate the ratio M/N approximately by observing frequencies in many trials.

That there ought to be some kind of connection between a theoretical probability and an observable frequency was a vaguely seen intuition in the earlier works; but Bernoulli, seeing clearly the distinction between the concepts, recognized that the existence of a connection between them cannot be merely postulated; it requires mathematical demonstration. If in a binary experiment we assign a constant probability of success p, independently at each trial, then we find for the probability of seeing m successes in n trials the binomial distribution

$$P(m|n,p) = \binom{n}{m} p^m (1-p)^{n-m} \quad . \tag{A2}$$

Bernoulli then shows that as $n \to \infty$, the observed frequency $f = m/n$ of successes tends to the probability p in the sense that for all $\varepsilon > 0$,

$$P(p - \varepsilon < f < p + \varepsilon | p,n) \to 1 \tag{A3}$$

and thus (in a sense made precise only in the later work of Bayes and Laplace) for sufficiently large n, the observed frequency is practically certain to be close to the number p sought.

But Bernoulli's result does not tell us how large n must be for a given accuracy. For this, one needs the more detailed limit theorem; as n increases, f may be considered a continuous variable, and the probability that $(f < m/n < f + df)$ goes into a gaussian, or normal, distribution:

$$P(df|n,p) \sim \left[\frac{n}{2\pi p(1-p)}\right]^{\frac{1}{2}} \exp\left[-\frac{n(f-p)^2}{2p(1-p)}\right] df \tag{A4}$$

in the sense of the leading term of an asymptotic expansion. For example, if $p = 2/3$, then from (A4), in $n = 1000$ trials, there is a 99% probability that the observed f will lie in the interval 0.667 ± 0.038, and an even chance that it will fall in 0.667 ± 0.010. The result (A4) was first given in this generality by Laplace; it had been found earlier by de Moivre for the case $p = \frac{1}{2}$. And in turn, the de Moivre-Laplace theorem (A4) became the ancestor of our present Central Limit Theorem.

Since these limit theorems are sometimes held to be the most important and sophisticated fruits of probability theory, we note that they depend crucially on the assumption of independence of different trials. The slightest positive correlation between trials i and j, if it persists for arbitrarily large $|i - j|$, will render these theorems qualitatively incorrect.

Laplace's contributions to probability theory go rather far beyond mere analytical refinements of other peoples' results. Most important for statistical theory today, he saw the general principle needed to solve problems of the type formulated by Bernoulli, but left unfinished by the Bernoulli and de Moivre-Laplace limit theorems. These results concern only the so-called "sampling distribution." That is, given $p = M/N$, what is the probability that we shall see particular sample numbers (m,n)? The results (A1)-(A4) describe a state of knowledge in which the "population numbers" (M,N) are known, the sample number unknown. But in the problem Bernoulli tried to solve, the sample is known and the population is not only unknown--

its very existence is only a tentative hypothesis (what mortal will ever determine the number of diseases, etc.).

We have, therefore, an inversion problem. The above theorems show that, given (M,N) and the correctness of the whole conceptual model, then it is likely that in many trials the observed frequency f will be close to the probability p. Presumably, then, given the observed f in many trials, it is likely that p is close to f. But can this be made into a precise theorem like (A4)? The binomial law (A2) gives the probability of m, given (M,N,n). Can we turn this around and find a formula for the probability of M, given (m,N,n)? This is the problem of inverse probabilities.

A particular inversion of the binomial distribution was offered by a British clergyman and amateur mathematician, Thomas Bayes (1763) in what has become perhaps the most famous and controversial work in probability theory. His reasoning was obscure and hard to describe; but his actual result is easy to state. Given the data (m,n), he finds for the probability that M/N lies in the interval $p < (M/N) < p + dp$,

$$P(dp|m,n) = \frac{(n+1)!}{m!(n-m)!} p^m (1-p)^{n-m} dp \quad , \tag{A5}$$

today called a Beta distribution. It is not a binomial distribution because the variable is p rather than m and the numerical coefficient is different, but it is a trivial mathematical exercise [expand the logarithm of (A5) in a power series about its peak] to show that, for large n, (A5) goes asymptotically into just (A4) with f and p everywhere interchanged. Thus, if in $n = 1000$ trials we observe $m = 667$ successes, then on this evidence there would be a 99% probability that p lies in (0.667 ± 0.038), etc.

In the gaussian approximation, according to Bayes' solution, there is complete mathematical symmetry between the probability of f given p, and of p given f. This would certainly seem to be the neatest and simplest imaginable solution to Bernoulli's inversion problem.

Laplace, in his famous memoir of 1774 on the "probabilities of causes," perceived the principle underlying inverse probabilities in far greater generality. Let E stand for some observable event and $\{C_1 \ldots C_N\}$ the set of its conceivable causes. Suppose that we have found, according to some conceptual model, the "sampling distribution" or "direct" probabilities of E for each cause: $P(E|C_i)$, $i = 1,2,\ldots,N$. Then, says Laplace, if initially the causes C_i are considered equally likely, then having seen the event E, the different causes are indicated with probability proportional to $P(E|C_i)$. That is, with uniform prior probabilities, the posterior probabilities of the C_i are

$$P(C_i|E) = \left[\sum_{j=1}^{N} P(E|C_j) \right]^{-1} P(E|C_i) \quad . \qquad \text{(A6)}$$

This is a tremendous generalization of the Bernoulli-Bayes results (A2), (A5). If the event E consists in finding m successes in n trials, and the causes C_i correspond to the possible values of M in the Bernoulli model, then $P(E|C_i)$ is the binomial distribution (A2); and in the limit $N \to \infty$ (A6) goes into Bayes' result (A5).

Later, Laplace generalized (A6) further by noting that, if initially the C_i are not considered equally likely, but have prior probabilities $P(C_i|I)$, where I stands for the prior information, then the terms in (A6) should be weighted according to $P(C_i|I)$:

$$P(C_i|E,I) = \frac{P(E|C_i)P(C_i|I)}{\sum_j P(E|C_j)P(C_j|I)} \qquad \text{(A7)}$$

but, following long-established custom, it is Laplace's result (A7) that is always called, in the modern literature, "Bayes' theorem."

Laplace proceeded to apply (A6) to a variety of problems that arise in astronomy, meteorology, geodesy, population statistics, etc. He would use it typically as follows. Comparing experimental observations with some existing theory, or calculation, one will never find perfect agreement. Are the discrepancies so small that they might reasonably be attributed to measurement errors, or are they so large that they indicate, with high probability, the existence of some new systematic cause? If so, Laplace would undertake to find that cause. Such uses of inverse probability--what would be called today "significance tests" by statisticians, and "detection of signals in noise" by electrical engineers--led him to some of the most important discoveries in celestial mechanics.

Yet there were difficulties that prevented others from following Laplace's path, in spite of its demonstrated usefulness. In the first place, Laplace simply stated the results (A6), (A7) as intuitive, ad hoc recipes without any derivation from compelling desiderata; and this left room for much agonizing over their logical justification and uniqueness. For an account of this, see Keynes (1921). However, we now know that Laplace's result (A7) is, in fact, the entirely correct and unique solution to the inversion problem.

More importantly, it became apparent that, in spite of first appearances, the results of Bayes and Laplace did not, after all, solve the problem that Bernoulli had set out to deal with. Recall, Bernoulli's original motivation was that the Principle of Insufficient Reason is inapplicable in so many real problems, because we are unable to break things down into an enumeration of "equally possible" cases. His hope--left unrealized at his death in 1705--had been that, by inversion of his theorem one could avoid having to use Insufficient Reason. Yet when the inversion problem was finally solved by Bayes and Laplace, the prior probabilities $P(C_i|I)$ that Bernoulli had sought to avoid, intruded themselves inevitably right back into the picture!

The only useful results Laplace got came from (A6), based on the uniform prior probabilities $P(C_i|I) = 1/N$ from the Principle of Insufficient Reason. That is, of course, not because Laplace failed to understand the generalization (A7) as some have charged--it was Laplace who, in his Essai Philosophique, pointed out the need for that generalization. Rather, Laplace did not have any principle for finding prior probabilities in cases where the prior information fails to render the possibilities "equally likely."

At this point, the history of statistical theory takes a sharp 90° turn away from the original goal, and we are only slowly straightening out again today. One might have thought, particularly in view of the great pragmatic success achieved by Laplace with (A6), that the next workers would try to build constructively on the foundations laid down by him. The next order of business should have been seeking new and more general principles for determining prior probabilities, thus extending the range of problems where probability theory is useful to (A7). Instead, only fifteen years after Laplace's death, there started a series of increasingly violent attacks on his work. Totally ignoring the successful results they had yielded, Laplace's methods based on (A6) were rejected and ridiculed, along with the whole conception of probability theory expounded by Bernoulli and Laplace. The main early references to this counter-stream of thought are Ellis (1842), Boole (1854), Venn (1866), and von Mises (1928).

As already emphasized, Bernoulli's definition of probability (A1) was developed for the purpose of representing mathematically a particular state of knowledge; and the equations of probability theory then represent the process of plausible, or inductive, reasoning in cases where there is not enough information at hand to permit deductive reasoning. In particular, Laplace's result (A7) represents the process of "learning by experience," the prior probability $P(C|I)$ changing to the posterior probability $P(C|E,I)$ as a result of obtaining new evidence E.

This counter-stream of thought, however, rejected the notion of probability as describing a state of knowledge, and insisted that by "probability" one must mean only "frequency in a random experiment." For a time this viewpoint dominated the field so completely that those who were students in the period 1930-1960 were hardly aware that any other conception had ever existed.

If anyone wishes to study the properties of frequencies in random experiments he is, of course, perfectly free to do so; and we wish him every success. But if he wants to talk about frequencies, why can't he just use the word "frequency?" Why does he insist on appropriating the word "probability," which had already a long-established and very different technical meaning?

Most of the debate that has been in progress for over a century on "frequency vs. non-frequency definitions of probability" seems to me not concerned with any substantive issue at all; but merely arguing over who has the right to use a word. Now the historical priority belongs clearly to Bernoulli and Laplace. Therefore, in the interests not only of responsible scholarship, but also of clear exposition and to avoid becoming entangled in semantic irrelevancies, we ought to use the word "probability" in the original sense of Bernoulli and Laplace; and if we mean something else, call it something else.

With the usage just recommended, the term "frequency theory of probability" is a pure incongruity; just as much so as "theory of square circles." One might speak properly of a "frequency theory of inference," or the better term "sampling theory," now in general use among statisticians (because the only distributions admitted are the ones we have called sampling distributions). This stands in contrast to the "Bayesian theory" developed by Laplace, which admits the notion of probability of an hypothesis.

Having two opposed schools of thought about how to handle problems of inference, the stage is set for an interesting contest. The sampling theorists, forbidden by their ideology to use Bayes' theorem as Laplace did in the form (A6), must seek other methods for dealing with Laplace's problems. What methods, then, did they invent? How do their procedures and results compare with Laplace's?

The sampling theory developed slowly over the first half of this Century by the labors of many, prominent names being Fisher, "Student," Pearson, Neyman, Kendall, Cramér, Wald. They proceeded through a variety of ad hoc intuitive principles, each appearing reasonable at first glance, but for which defects or limitations on generality always appeared. For example, the Chi-squared test, maximum likelihood, unbiased and/or efficient estimators, confidence intervals, fiducial distributions, conditioning on ancillary statistics, power functions and sequential methods for hypothesis testing. Certain technical difficulties ("nuisance" parameters, non-existence of sufficient or ancillary statistics, inability to take prior information into account) remained behind as isolated pockets of resistance which sampling theory has never been able to overcome. Nevertheless, there was discernible progress over the years, accompanied by an unending stream of attacks on Laplace's ideas and methods, sometimes degenerating into personal attacks on Laplace himself [see, for example, the biographical sketch by E. T. Bell (1937), entitled "From Peasant to Snob"].

Enter Jeffreys. After 1939, the sampling theorists had another target for their scorn. Sir Harold Jeffreys, finding in geophysics some problems of "extracting signals from noise" very much like those treated by Laplace, found himself unconvinced by Fisher's arguments, and produced a book in which the methods of Laplace were reinstated and applied, in the precise, compact modern notation that did not exist in the time of Laplace, to a mass of current scientific problems. The result was a vastly more comprehensive treatment of inference than Laplace's, but with two points in common: (A) the applications worked out beautifully, encountering no such technical difficulties as the "nuisance parameters" noted above; and yielding the same or demonstrably better results than those found by sampling theory methods. For many specific examples, see Jaynes (1976). (B) Unfortunately, like Laplace, Jeffreys did not derive his principles as necessary consequences of any compelling desiderata; and thus left room to continue the same old arguments over their justification.

The sampling theorists, seizing eagerly upon point (B) while again totalling ignoring point (A), proceeded to give Jeffreys the same treatment as Laplace, which he had to endure for some thirty years before the tide began to turn.

As a student in the mid-1940's, I discovered the book of Jeffreys (1939) and was enormously impressed by the smooth, effortless way he was able to derive the useful results of the theory, as well as the sensible philosophy he expressed. But I too felt that something was missing in the exposition of fundamentals in the first Chapter and, learning about the attacks on Jeffreys' methods by virtually every other writer on statistics, felt some mental reservations.

But just at the right moment there appeared a work that removed all doubts and set the direction of my own life's work. An unpretentious little article by Professor R. T. Cox (1946) turned the problem under debate around and, for the first time, looked at it in a constructive way. Instead of making dogmatic assertions that it is or is not legitimate to use probability in the sense of degree of plausibility rather than frequency, he had the good sense to ask a question: Is it possible to construct a consistent set of mathematical rules for carrying out plausible, rather than deductive, reasoning? He found that, if we try to represent degrees of plausibility by real numbers, then the conditions of consistency can be stated in the form of functional equations, whose general solutions can be found. The results were: out of all possible monotonic functions which might in principle serve our purpose, there exists a particular scale on which to measure degrees of plausibility which we henceforth call probability, with particularly simple properties. Denoting various propositions by A, B, etc., and using the notation, $AB \equiv$ "Both A and B are true," $\bar{A} \equiv$ "A is false," $p(A|B) \equiv$ probability of A given B, the consistent rules of combination take the form of the familiar product rule and sum rule:

$$p(AB|C) = p(A|BC)\, p(B|C) \quad , \tag{A8}$$

$$p(A|B) + p(\bar{A}|B) = 1 \quad . \tag{A9}$$

By mathematical transformations we can, of course, alter the form of these rules; but what Cox proved was that any alteration of their content will enable us to exhibit inconsistencies (in the sense that two methods of calculation, each permitted by the rules, will yield different results). But (A8), (A9) are, in fact, the basic rules of probability theory; all other equations needed for applications can be derived from them. Thus, Cox proved that any method of inference in which we represent degrees of plausibility by real numbers, is necessarily either equivalent to Laplace's, or inconsistent.

For me, this was exactly the argument needed to clinch matters; for Cox's analysis makes no reference whatsoever to frequencies or random experiments. From the day I first read Cox's article I have never for a moment doubted the basic soundness and inevitability of the Laplace-Jeffreys methods, while recognizing that the theory needs further development to extend its range of applicability.

Indeed, such further development was started by Jeffreys. Recall, in our narrative we left Laplace (or rather, Laplace left us) at Eq. (A6), seeing the need but not the means to make the transition to (A7), which would open up an enormously wider range of applications for Bayesian inference. Since the function of the prior probabilities is to describe the prior information, we need to develop new or more general principles for determination of those priors by logical analysis of prior information when it does not consist of frequencies; just what should have been the next order of business after Laplace.

Recognizing this, Jeffreys resumed the constructive development of this theory at the point where Laplace had left off. If we need to convert prior information into a prior probability assignment, perhaps we should start at the beginning and learn first how to express "complete ignorance" of a continuously variable parameter, where Bernoulli's principle will not apply.

Bayes and Laplace had used uniform prior densities, as the most obvious analog of the Bernoulli uniform discrete assignment. But it was clear, even in the time of Laplace, that this rule is ambiguous because it is not invariant under a change of parameters. A uniform density for θ does not correspond to a uniform density for $\alpha = \theta^3$; or $\beta = \log \theta$; so for which choice of parameters should the uniform density apply?

In the first (1939) Edition of his book, Jeffreys made a tentative start on this problem, in which he found his now famous rule: to express ignorance of a scale parameter σ, whose possible domain is $0 < \sigma < \infty$, assign uniform prior density to its logarithm: $P(d\sigma|I) = d\sigma/\sigma$. The first arguments advanced in support of this rule were not particularly clear or convincing to others (including this writer). But other desiderata were found; and we have now succeeded in proving via the integral equations of marginalization theory (Jaynes, 1979) that Jeffreys' prior $d\sigma/\sigma$ is, in fact, uniquely determined as the only prior for a scale parameter that is "completely uninformative" in the sense that it leads us to the same conclusions about other parameters θ as if the parameter σ had been removed from the model [see Eq. (C33) below].

In the second (1948) Edition, Jeffreys gave a much more general "Invariance Theory" for determining ignorance priors, which showed amazing prevision by coming within a hair's breadth of discovering both the principles of Maximum Entropy and Transformation Groups. He wrote down the actual entropy expression (note the date!), but then used it only to generate a quadratic form by expansion about its peak. Jeffreys' invariance theory is still of great importance today, and the question of its relation to other methods that have been proposed is still under study.

In the meantime, what had been happening in the sampling theory camp? The culmination of this approach came in the late 1940's when for the first time, Abraham Wald succeeded in removing all <u>ad hockeries</u> and presenting general rules of conduct for making decisions in the face of uncertainty, that he proved to be uniquely optimal by certain very simple and compelling desiderata of reasonable behavior. But quickly a number of people--including I. J. Good (1950), L. J. Savage (1954), and the present writer--realized independently that, if we just ignore Wald's entirely different vocabulary and diametrically opposed philosophy, and look only at the specific mathematical steps that were now to be used in solving specific problems, they were identical with the rules given by Laplace in the eighteenth century, which generations of statisticians had rejected as metaphysical nonsense!

It is one of those ironies that make the history of science so interesting, that the missing Bayes-optimality proofs, which Laplace and Jeffreys had failed to supply, were at last found inadvertently, while trying to prove the opposite, by an early ardent disciple of the von Mises "collective" approach. It is also a tribute to Wald's intellectual honesty that he was able to recognize this, and in his final work (Wald, 1950) he called these optimal rules, "Bayes strategies."

Thus came the "Bayesian Revolution" in statistics, which is now all but over. This writer's recent polemics (Jaynes, 1976) will probably be one of the last battles waged. Today, most active research in statistics is Bayesian, a good deal of it directed to the above problem of determining priors by logical analysis; and the parts of sampling theory which do not lie in ruins are just the ones (such as sufficient statistics and sequential analysis) that can be justified in Bayesian terms.

This history of basic statistical theory, showing how developments over more than two centuries set the stage naturally for the Principle of Maximum Entropy, has been recounted at some length because it is unfamiliar to most scientists and engineers. Although the second line converging on this principle is much better known to this audience, our account can be no briefer because there is so much to be unlearned.

The Second Line: Maxwell, Boltzmann, Gibbs, Shannon. Over the past 120 years another line of development was taking place, which had astonishingly little contact with the "statistical inference" line just described. In the 1850's James Clerk Maxwell started the first serious work on the application of probability analysis to the kinetic theory of gases. He was confronted immediately with the problem of assigning initial probabilities to various positions and velocities of molecules. To see how he dealt with it, we quote his first (1859) words on the problem of finding the probability distribution for velocity direction of a spherical molecules after an impact: "In order that a collision may take place, the line of motion of one of the balls must pass the center of the other at a distance less than the sum of their radii; that is, it must pass through a circle whose centre is that of the other ball, and radius the sum of the radii of the balls. Within this circle every position is equally probable, and therefore --- ."

Here again, as that necessary first step in a probability analysis, Maxwell had to apply the Principle of Indifference; in this case to a two-dimensional continuous variable. But already at this point we see a new feature. As long as we talk about some abstract quantity θ without specifying its physical meaning, we see no reason why we could not as well work with $\alpha = \theta^3$, or $\beta = \log \theta$; and there is an unresolved ambiguity. But as soon as we learn that our quantity has the physical meaning of position within the circular collision cross-section, our intuition takes over with a compelling force and tells us that the probability of impinging on any particular region should be taken proportional to the area of that region; and not to the cube of the area, or the logarithm of the area. If we toss pennies onto a wooden floor, something inside us convinces us that the probability of landing on any one plank should be taken proportional to the width of the plank; and not to the cube of the width, or the logarithm of the width.

In other words, merely knowing the physical meaning of our parameters, already constitutes highly relevant prior information which our intuition is able to use at once; in favorable cases its effect is to give us an inner conviction that there is no ambiguity after all in applying the Principle of Indifference. Can we analyze how our intuition does this, extract the essence, and express it as a formal mathematical principle that might apply in cases where our intuition fails us? This problem is not completely solved today, although I believe we have made a good start on it in the principle of transformation groups (Jaynes, 1968, 1973, 1979). Perhaps these remarks will encourage others to try their hand at resolving these puzzles; this is an area where important new results might turn up with comparatively little effort, given the right inspiration on how to approach them.

Maxwell built a lengthy, highly non-trivial, and needless to say, successful analysis on the foundation just quoted. He was able to predict such things as the equation of state, velocity distribution law, diffusion coefficient, viscosity, and thermal conductivity of the gas. The case of viscosity was particularly interesting because Maxwell's theory led to the prediction that viscosity is independent of density, which seemed to contradict common sense. But when the experiments were performed, they confirmed Maxwell's prediction; and what had seemed a difficulty with his theory became its greatest triumph.

Enter Boltzmann. So far we have considered only the problem of expressing initial ignorance by a probability assignment. This is the first fundamental problem, since "complete initial ignorance" is the natural and inevitable starting point from which to measure our positive knowledge; just as zero is the natural and inevitable starting point when we add a column of numbers. But in most real problems we do not have initial ignorance about the questions to be answered. Indeed, unless we had some definite prior knowledge about the parameters to be measured or the hypotheses to be tested, we would seldom have either the means or the motivation to plan an experiment to get more knowledge. But to express positive initial knowledge by a probability assignment is just the problem of getting from (A6) to (A7), bequeathed to us by Laplace.

The first step toward finding an explicit solution to this problem was made by Boltzmann, although it was stated in very different terms at the time. He wanted to find how molecules will distribute themselves in a conservative force field (say, a gravitational or centrifugal field; or an electric field acting on ions). The force acting on a molecule at position x is then $F = -\text{grad}\ \phi$, where $\phi(x)$ is its potential energy. A molecule with mass m, position x, velocity v thus has energy $E = \frac{1}{2} mv^2 + \phi(x)$. We neglect the interaction energy of molecules with each other and suppose they are enclosed in a container of volume V, whose walls are rigid and impermeable to both molecules and heat. But Boltzmann was not completely ignorant about how the molecules are distributed, because he knew that however they move, the total number N of molecules present cannot change, and the total energy

$$E = \sum_{i=1}^{N} \left[\frac{1}{2} m\, v_i^2 + \phi(x_i) \right] \tag{A10}$$

must remain constant. Because of the energy constraint, evidently, all positions and velocities are not equally likely.

At this point, Boltzmann found it easier to think about discrete distributions than continuous ones (a kind of prevision of quantum theory); and so he divided the phase space (position-momentum space) available to the molecules into discrete cells. In principle, these could be defined in any way; but let us think of the k'th cell as being a region R_k so small that the energy E_k of a molecule does not vary appreciably within it; but also so large that it can accommodate a large number, $N_k >> 1$, of molecules. The cells $\{R_k,\ 1 \leq k \leq s\}$ are to fill up the accessible phase space (which because of the energy constraint has a finite volume) without overlapping.

The problem is then: given N, E, and $\phi(x)$, what is the best prediction we can make of the number of N_k of molecules in R_k? In Boltzmann's reasoning at this point, we have the beginning of the Principle of Maximum Entropy. He asked first: In how many ways could a given set of occupation numbers N_k be realized? The answer is the multinomial coefficient

$$W(N_k) = \frac{N!}{N_1!\, N_2! \, \ldots \, N_s!} \quad . \tag{A11}$$

This particular distribution will have total energy

$$E = \sum_{k=1}^{s} N_k \, E_k \tag{A12}$$

and of course, the N_k are also constrained by

$$N = \sum_{k=1}^{s} N_k \quad . \tag{A13}$$

Now any set $\{N_k\}$ of occupation numbers for which E, N agree with the given information, represents a possible distribution, compatible with all that is specified. Out of the millions of such possible distributions, which is most likely to be realized? Boltzmann's answer was that the "most probable" distribution is the one that can be realized in the greatest number of ways; i.e., the one that maximizes (A11) subject to the constraints (A12), (A13), if the cells are equally large (phase volume).

Since the N_k are large, we may use the Stirling approximation for the factorials, whereupon (A11) can be written

$$\log W = -N \sum_{k=1}^{s} \left(\frac{N_k}{N}\right) \log\left(\frac{N_k}{N}\right) \quad . \tag{A14}$$

The mathematical solution by Lagrange multipliers is straightforward, and the result is: the "most probable" value of N_k is

$$\hat{N}_k = \frac{N}{Z(\beta)} \exp\left(-\beta E_k\right) \tag{A15}$$

where

$$Z(\beta) \equiv \sum_{k=1}^{s} \exp\left(-\beta E_k\right) \tag{A16}$$

and the parameter β is to be chosen so that the energy constraint (A12) is satisfied.

This simple result contains a great deal of physical information. Let us choose a particular set of cells R_k as follows. Divide up the coordinate space V and the velocity space into cells X_a, Y_b respectively, such that the potential and kinetic energies $\phi(x)$, $\frac{1}{2} m v^2$ do not vary appreciably within them, and take $R_k = X_a \otimes Y_b$. Then, writing $N_k = N_{ab}$, Boltzmann's prediction of the number of molecules in X_a irrespective of their velocity, is from (A15)

$$\hat{N}_a = \sum_b \hat{N}_{ab} = A(\beta)\ \exp\left(-\beta\phi_a\right) \qquad \text{(A17)}$$

where the normalization constant $A(\beta)$ is determined from $\sum N_a = N$. This is the famous Boltzmann distribution law. In a gravitational field, $\phi(x) = mgz$, it gives the usual "barometric formula" for decrease of the atmospheric density with height:

$$\rho(z) = \rho(0)\ \exp(-\beta mgz) \quad . \qquad \text{(A18)}$$

Now this can be deduced also from the macroscopic equation of state: for one mole, $PV = RT$, or $P(z) = (RT/mN_o)\rho(z)$, where N_o is Avogadro's number. But hydrostatic equilibrium requires $-dP/dz = g\rho(z)$, which gives on integration, for uniform temperature, $\rho(z) = \rho(0)\ \exp(-N_o\, mgz/RT)$. Comparing with (A18), we find the meaning of the parameter: $\beta = (kT)^{-1}$, where T is the Kelvin temperature and $k \equiv R/N_o$ is Boltzmann's constant.

We can, equally well, sum (A15) over the space cells X_a and find the predicted number of molecules with velocity in the cell Y_b, irrespective of their position in space; but a far more interesting result is contained already in (A15) without this summation. Let us ask, instead; What fraction of the molecules in the space cell X_a are predicted to have velocity in the cell Y_b? This is, from (A15) and (A17),

$$f_b = \hat{N}_{ab}/\hat{N}_a = B(\beta)\ \exp\left(-\beta m\, v_b^2/2\right) \qquad \text{(A20)}$$

This is, of course, just the Maxwellian velocity distribution law; but with the new and at first sight astonishing feature that it is independent of position in space. Even though the force field is accelerating and decelerating molecules as they move from one region to another, when they arrive at their new location they have exactly the same mean square velocity as when they started! If this result is correct (as indeed it proved to be) it means that a Maxwellian velocity distribution, once established, is maintained automatically, without any help from collisions, as the molecules move about in any conservative force field.

From Boltzmann's reasoning, then, we get a very unexpected and nontrivial dynamical prediction by an analysis that, seemingly, ignores the dynamics altogether! This is only the first of many such examples where it appears that we are "getting something for nothing," the answer coming too easily to believe. Poincaré, in his essays on "Science and Method," felt this paradox very keenly, and wondered how by exploiting our ignorance we can make correct predictions in a few lines of calculation, that would be quite impossible to obtain if we attempted a detailed calculation of the 10^{23} individual trajectories.

It requires very deep thought to understand why we are not, in this argument and others to come, getting something for nothing. In fact, Boltzmann's argument does take the dynamics into account, but in a very efficient manner. Information about the dynamics entered his equations at two places: (1) the conservation of total energy; and (2) the fact that he defined his cells in terms of phase volume, which is conserved in the dynamical motion (Liouville's theorem). The fact that this was enough to predict the correct spatial and velocity distribution of the molecules shows that the millions of intricate dynamical details that were not taken into account, were actually irrelevant to the predictions, and would have cancelled out anyway if he had taken the trouble to calculate them.

Boltzmann's reasoning was super-efficient; far more so than he ever realized. Whether by luck or inspiration, he put into his equations only the dynamical information that happened to be relevant to the questions he was asking. Obviously, it would be of some importance to discover the secret of how this come about, and to understand it so well that we can exploit it in other problems.

If we can learn how to recognize and remove irrelevant information at the beginning of a problem, we shall be spared having to carry out immense calculations, only to discover at the end that practically everything we calculated was irrelevant to the question we were asking. And that is exactly what we are after by applying Information Theory [actually, the secret was revealed in my second paper (Jaynes, 1957b); but to the best of my knowledge no other person has yet noticed it there; so I will explain it again in Section D below. The point is that Boltzmann was asking only questions about experimentally reproducible equilibrium properties].

In Boltzmann's "method of the most probable distribution," we have already the essential mathematical content of the Principle of Maximum Entropy. But in spite of the conventional name, it did not really involve probability. Boltzmann was not trying to calculate a probability distribution; he was estimating some physically real occupation numbers N_k, by a criterion (value of W) that counts the number of real physical possibilities; a definite number that has nothing to do with anybody's state of knowledge. The transition from this to our present more abstract Principle of Maximum Entropy, although mathematically trivial, was so difficult conceptually that it required almost another Century to bring about. In fact, this required three more steps and even today the development of irreversible Statistical Mechanics is being held up as much by conceptual difficulties as by mathematical ones.

Enter Gibbs. Curiously, the ideas that we associate today with the name of Gibbs were stated briefly in an early work of Boltzmann (1871); but were not pursued as Boltzmann became occupied with his more specialized H-theorem. Further development of the general theory was therefore left to Gibbs (1902). The Boltzmann argument just given will not work when the molecules have appreciable interactions, since then the total energy cannot be written in the additive form (A12). So we go to a much more abstract picture. Whereas the preceding argument was applied to an actually existing large collection of molecules, we now let the entire macroscopic system of interest become, in effect, a "molecule," and imagine a large collection of copies of it.

This idea, and even the term "phase" to stand for the collection of all coordinates and momenta, appears also in a work of Maxwell (1876). Therefore, when Gibbs adopted this notion, which he called an "ensemble," it was not, as is apparently thought by those who use the term "Gibbs ensemble," an innovation on his part. He used ensemble language rather as a concession to an already established custom. The idea became associated later with the von Mises "Kollektiv" but was actually much older, dating back to Venn (1866); and Fechner's book *Kollektivmasslehre* appeared in 1897.

It is important for our purposes to appreciate this little historical fact and to note that, far from having invented the notion of an ensemble, Gibbs himself (*loc cit*., p. 17) deemphasized its importance. We can detect a hint of cynicism in his words when he states: "It is in fact customary in the discussion of probabilities to describe anything which is imperfectly known as something taken at random from a great number of things which are completely described." He continues that, if we prefer to avoid any reference to an ensemble of systems, we may recognize that we are merely talking about "the probability that the phase of a system falls within certain limits at a certain time ---."

In other words, even in 1902 it was customary to talk about a probability as if it were a frequency; even if it is a frequency only in an imaginary *ad hoc* collection invented just for that purpose. Of course, any probability whatsoever can be thought of in this way if one wishes to; but Gibbs recognized that in fact we are only describing our imperfect knowledge about a single system.

The reason it is important to appreciate this is that we then understand Gibbs' later treatment of several topics, one of which had been thought to be a serious omission on his part. If we are describing only a state of knowledge about a single system, then clearly there can be nothing physically real about frequencies in the ensemble; and it makes no sense to ask, "which ensemble is the correct one?" In other words: different ensembles are not in 1:1 correspondence with different physical situations; they correspond only to different states of knowledge about a single physical situation. Gibbs understood this clearly; and that, I suggest, is the reason why he does not say a word about ergodic theorems, or hypotheses, but instead gives a totally different reason for his choice of the canonical ensembles.

Technical details of Gibbs' work will be deferred to Sec. D below, where we generalize his algorithm. Suffice it to say here that Gibbs introduces his canonical ensemble, and works out its properties, without explaining why he chooses that particular distribution. Only in Chap. XII, after its properties--including its maximum entropy property--have been set forth, does he note that the distribution with the minimum expectation of log p (i.e., maximum entropy) for a prescribed distribution of the constants of the motion has certain desirable properties. In fact, this criterion suffices to generate all the ensembles--canonical, grand canonical, microcanonical, and rotational--discussed by Gibbs.

This is, clearly, just a generalized form of the Principle of Indifference. The possibility of a different justification in the frequency sense, via ergodic theorems, had been discussed by Maxwell, Boltzmann, and others for some thirty years; as noted in more detail before (Jaynes, 1967) if Gibbs thought that any such further justification was needed, it is certainly curious that he neglected to mention it.

After Gibbs' work, however, the frequency view of probability took such absolute control over mens' minds that the ensemble became something physically real, to the extent that the following phraseology appears. Thermal equilibrium is defined as the situation where the system is "in a canonical distribution." Assignment of uniform prior probabilities was considered to be not a mere description of a state of knowledge, but a basic postulate of physical fact, justified by the agreement of our predictions with experiment.

In my student days this was the kind of language always used, although it seemed to me absurd; the individual system is not "in a distribution;" it is in a <u>state</u>. The experiments, moreover, do not verify "equal <u>a priori</u> probabilities" or "random <u>a priori</u> phases;" they verify only the predicted macroscopic equation of state, heat capacity, etc., and the predictions for these would have been the same for many ensembles, uniform or nonuniform microscopically. Therefore, the reason for the success of Statistical Mechanics must be altogether different from our having found the "correct" ensemble.

Intuitively, it must be true that use of the canonical ensemble, while sufficient to predict thermal equilibrium properties, is very far from necessary; in some sense, "almost every" member of a very wide class of ensembles would all lead to the same predictions for the particular macroscopic quantities actually observed. But I did not have any hint as to exactly what that class is; and needless to say, had not the faintest success in persuading anyone else of such heretical views.

We stress that, on this matter of the exact status of ensembles, you have to read Gibbs' own words in order to know accurately what his position was. For example, Ter Haar (1954, p. 128) tells us that "Gibbs introduced ensembles in order to use them for statistical considerations rather than to illustrate the behavior of physical systems ---." But Gibbs himself (_loc_. _cit_. p. 150) says, "--- our ensembles are chosen to illustrate the probabilities of events in the real world ---."

It might be thought that such questions are only matters of personal taste, and a scientist ought to occupy himself with more serious things. But one's personal taste determines which research problems he believes to be the important ones in need of attention; and the total domination by the frequency view caused all attention to be directed instead to the aforementioned "ergodic" problems; to justify the methods of Statistical Mechanics by proving from the dynamic equations of motion that the canonical ensemble correctly represents the _frequencies_ with which, over a long time, an individual system coupled to a heat bath, finds itself in various states.

This problem metamorphosed from the original conception of Boltzmann and Maxwell that the phase point of an isolated (system + heat bath) ultimately passes through every state compatible with the total energy, to the statement that the time average of any phase function f(p,q) for a single system is equal to the ensemble average of f; and this statement in turn was reduced (by von Neumann and Birkhoff in the 1930's) to the condition of metric transitivity (i.e., the full phase space shall have no subspace of positive measure that is invariant under the motion). But here things become extremely complicated, and there is little further progress. For example, even if one proves that in a certain sense "almost every" continuous flow is metrically transitive, one would still have to prove that the particular flows generated by a Hamiltonian are not exceptions.

Such a proof certainly cannot be given in generality, since counter-examples are known. One such is worth noting: in the writer's "Neoclassical Theory" of electrodynamics (Jaynes, 1973) we write a complete classical Hamiltonian system of equations for an atom (represented as a set of harmonic oscillators) interacting with light. But we find [loc. cit. Eq. (52)] that not only is the total energy a constant of the motion, the quantity $\Sigma_n W_n/\nu_n$ is conserved, where W_n, ν_n are the energy and frequency of the n'th normal mode of oscillation of the atom.

Setting this new constant of the motion equal to Planck's constant h, we have a classical derivation of the $E = h\nu$ law usually associated with quantum theory! Indeed, quantum theory simply takes this as a basic empirically justified postulate; and never makes any attempt to explain why such a relation exists. In Neoclassical Theory it is explained as a consequence of a new uniform integral of the motion, of a type never suspected in classical Statistical Mechanics. Because of it, for example, there is no Liouville theorem in the "action shell" subspace of states actually accessible to the system, and statistical properties of the motion are qualitatively different from those of the usual classical Statistical Mechanics. But all this emerges from a simple, innocent-looking classical Hamiltonian, involving only harmonic oscillators with a particular coupling law (linear in the field oscillators, bilinear in the atom oscillators). Having seen this example, who can be sure that the same thing is not happening more generally?

This was recognized by Truesdell (1960) in a work that I recommend as by far the clearest exposition, carried to the most far-reaching physical results, of any discussion of ergodic theory. He comes up against, "--- an old problem, one of the ugliest which the student of statistical mechanics must face: What can be said about the integrals of a dynamical system?" The answer is, "Practically nothing." In view of such simple counter-examples as that provided by Neoclassical theory, confident statements to the effect that real systems are almost certainly ergodic, seem like so much whistling in the dark.

Nevertheless, ergodic theory considered as a topic in its own right, does contain some important results. Unlike some others, Truesdell does not confuse the issue by trying to mix up probability notions and dynamical ones. Instead, he states unequivocally that his purpose is to calculate time averages. This is a definite, well posed dynamical problem having nothing to do with any probability considerations; and Truesdell proceeds to show, in greater depth than any other writer known to me, exactly what implications the Birkhoff theorem has for this question. Since we cannot prove, and in view of counter-examples have no valid reason to expect, that the flow is metrically transitive over the entire phase space S, the original hopes of Boltzmann and Maxwell must remain unrealized; but in return for this we get something far more valuable, which just misses being noticed.

The flow will be metrically transitive on some (unknown) sub-space S' determined by the (unknown) uniform integrals of the motion; and the time average of any phase function f(p,q) will, by the Birkhoff theorem, be equal to its phase space average over that subspace. Furthermore, the fraction of time that the system spends in any particular region s in S' is equal to the ratio of phase volumes: $\sigma(s)/\sigma(S')$.

These are just the properties that Boltzmann and Maxwell wanted; but they apply only to some subspace S' which cannot be known until we have determined all the uniform integrals of the motion. That is the purely dynamical theorem; and I think that if today we could resurrect Maxwell and tell it to him, his reaction would be: "Of course, that is obviously right and it is just what I was trying to say. The trouble was that I was groping for words, because in my day we did not have the mathematical vocabulary, arising out of measure theory and the theory of transformation groups, that is needed to state it precisely."

That more valuable result is tantalizingly close when Truesdell considers "--- the idea that however many integrals a system has, generally we shall not know the value of any but the energy, so we should assign equal *a priori* probability to the possible values of the rest, which amounts to disregarding the rest of them. Now an idea of this sort, by itself, is just unsound." It is indeed unsound, in the context of Truesdell's purpose to calculate correct time averages from the dynamics; for those time averages must in general depend on all the integrals of the motion, whether or not we happen to know about them.

The point that he just fails to see is that if, nevertheless, we only have the courage to go ahead and do the calculation he rejects as unsound, we can then compare its results with experimental time averages. If they disagree, then we have obtained experimental evidence of the existence of new integrals of the motion, and the nature of the deviation gives a clue as to what they may be. So, if our calculation should indeed prove to be "unsound," the result would be far more valuable to physics than a "successful" calculation!

To all this, however, one proviso must be added. Even if one could prove transitivity for the entire phase space, this result would not explain the success of equilibrium statistical mechanics, for reasons expounded in great detail before (Jaynes, 1967). These theorems apply only to time averages over enormous (strictly, infinite) time; and an average over a finite time T will approach its limiting value for $T \to \infty$ only if T is so long that the phase point of the system has explored a "representative sample" of the accessible phase volume. But the very existence of time-dependent irreversible processes shows that the "representative sampling time" must be very long compared to the time in which our measurements are made. So the equality of phase space averages with infinite time averages fails, on two counts, to explain the equality of canonical ensemble averages and experimental values. We can conclude only that the "ergodic" attempts to justify Gibbs' statistical mechanics foundered not only on impossibly difficult technical problems of integrals of the motion; but also on a basic logical defect arising from the impossibly long averaging times.

Enter Shannon. It was the work of Claude Shannon (1948) on Information Theory which showed us the way out of this dilemma. Like all major advances, it had many precursors, whose full significance could be seen only later. One finds them not only in the work of Boltzmann and Gibbs just noted, but also in that of G. N. Lewis, L. Szilard, J. von Neumann, and W. Elsasser, to mention only the most obvious examples.

Shannon's articles appeared just at the time when I was taking a course in Statistical Mechanics from Professor Eugene Wigner; and my mind was occupied with the difficulties, which he always took care to stress, faced by the theory at that time; the short sketch above notes only a few of them. Reading Shannon filled me with the same admiration that all readers felt, for the beauty and importance of the material; but also with a growing uneasiness about its meaning. In a communication process, the message M_i is assigned probability p_i, and the entropy $H = -\Sigma p_i \log p_i$ is a measure of "information." But whose information? It seems at first that if information is being "sent," it must be possessed by the sender. But the sender knows perfectly well which message he wants to send; what could it possibly mean to speak of the probability that he will send message M_i?

We take a step in the direction of making sense out of this if we suppose that H measures, not the information of the sender, but the ignorance of the receiver, that is removed by receipt of the message. Indeed, many subsequent commentators appear to adopt this interpretation. Shannon, however, proceeds to use H to determine the channel capacity C required to transmit the message at a given rate. But whether a channel can or cannot transmit message M in time T obviously depends only on properties of the message and the channel--and not at all on the prior ignorance of the receiver! So this interpretation will not work either.

Agonizing over this, I was driven to conclude that the different messages considered must be the set of all those that will, or might be, sent over the channel during its useful life; and therefore Shannon's H measures the degree of ignorance of the communication engineer when he designs the technical equipment in the channel. Such a viewpoint would, to say the least, seem natural to an engineer employed by the Bell Telephone Laboratories--yet it is curious that nowhere does Shannon see fit to tell the reader explicitly whose state of knowledge he is considering, although the whole content of the theory depends crucially on this.

It is the obvious importance of Shannon's theorems that first commands our attention and respect; but as I realized only later, it was just his vagueness on these conceptual questions--allowing every reader to interpret the work in his own way--that made Shannon's writings, like those of Niels Bohr, so eminently suited to become the Scriptures of a new Religion, as they so quickly did in both cases.

Of course, we do not for a moment suggest that Shannon was deliberately vague; indeed, on other matters few writers have achieved such clarity and precision. Rather, I think, a certain amount of caution was forced on him by a growing paradox that Information Theory generates within the milieu of probability theory as it was then conceived--a paradox only vaguely sensed by those who had been taught only the strict frequency definition of probability, and clearly visible only to those familiar with the work of Jeffreys and Cox. What do the probabilities p_i mean? Do they stand for the frequencies with which the different messages are sent?

Think, for a moment, about the last telegram you sent or received. If the Western Union Company remains in business for another ten thousand years, how many times do you think it will be asked to transmit that identical message?

The situation here is not really different from that in statistical mechanics, where our first job is to assign probabilities to the various possible quantum states of a system. In both cases the number of possibilities is so great that a time millions of times the age of the universe would not suffice to realize all of them. But it seems to be much easier to think clearly about messages than quantum states. Here at last, it seemed to me, was an example where the absurdity of a frequency interpretation is so obvious that no one can fail to see it; but the usefulness of the probability approach was equally clear. The probabilities assigned to individual messages are not measurable frequencies; they are only a means of describing a state of knowledge; just the original sense in which Laplace and Jeffreys interpreted a probability distribution.

The reason for the vagueness is then apparent; to a person who has been trained to think of probability only in the sense of frequency in a random experiment (as was surely the case for anyone educated at M.I.T. in the 1930's!), the idea that a probability distribution represents a mere state of knowledge is strictly taboo. A probability distribution would not be "objective" unless it represents a real physical situation. The question: "Whose information are we describing?" doesn't make sense, because the notion of a probability for a person with a certain state of knowledge just doesn't exist. So Shannon is forced to do the most careful egg-walking, speaking of a probability as if it were a real, measurable frequency, while using it in a way that shows clearly that it is not.

For example, Shannon considers the entropies H_1 calculated from single letter frequencies, H_2 from digram frequencies, H_3 from trigram frequencies, etc., as a sequence of successive approximations to the "true" entropy of the source, which is $H = \lim H_n$ for $n \to \infty$. Application of his theorems presupposes that all this is known. But suppose we try to determine the "true" ten-gram frequencies of English text. The number of different ten-grams is about 1.4×10^{14}; to determine them all to something like five percent accuracy, we should need a sample of English text containing about 10^{17} ten-grams. That is thousands of times greater than all the English text in the Library of Congress, and indeed much greater than all the English text recorded since the invention of printing.

If we had overcome that difficulty, and could measure those ten-gram frequencies (by scanning the entire text) at the rate of 1000 per second, it would require about 4400 years to take the data; and to record it on paper at a rate of 1000 entries per sheet, would require a stack of paper about 7000 miles high. Evidently, then, we are destined never to know the "true" entropy of the English language; and in the application of Shannon's theorems to real communication systems we shall have to accept some compromise.

Now, our story reaches its climax. Shannon discusses the problem of encoding a message, say English text, into binary digits in the most efficient way. The essential step is to assign probabilities to each of the conceivable messages in a way which incorporates the prior knowledge we have about the structure of English. Having this probability assignment, a construction found independently by Shannon and R. M. Fano yields the encoding rules which minimize the expected transmission time of a message.

But, as noted, we shall never know the "true" probabilities of English messages; and so Shannon suggests the principle by which we may construct the distribution p_i actually used for applications: "--- we may choose to use some of our statistical knowledge of English in constructing a code, but not all of it. In such a case we consider the source with the maximum entropy subject to the statistical conditions we wish to retain. The entropy of this source determines the channel capacity which is necessary and sufficient." [emphasis mine].

Shannon does not follow up this suggestion with the equations, but turns at this point to other matters. But if you start to solve this problem of maximizing the entropy subject to certain constraints, you will soon discover that you are writing down some very familiar equations. The probability distribution over messages is just the Gibbs canonical distribution with certain parameters. To find the values of the parameters, you must evaluate a certain partition function, etc.

Here was a problem of statistical inference--or what is the same thing, statistical decision theory--in which we are to decide on the best way of encoding a message, making use of certain partial information about the message. The solution turns out to be mathematically identical with the Gibbs formalism of statistical mechanics, which physicists had been trying, long and unsuccessfully, to justify in an entirely different way.

The conclusion, it seemed to me, was inescapable. We can have our justification for the rules of statistical mechanics, in a way that is incomparably simpler than anyone had thought possible, if we are willing to pay the price. The price is simply that we must loosen the connections between probability and frequency, by returning to the original viewpoint of Bernoulli and Laplace. The only new feature is that their Principle of Insufficient Reason is now generalized to the Principle of Maximum Entropy. Once this is accepted, the general formalism of statistical mechanics--partition functions, grand canonical ensemble, laws of thermodynamics, fluctuation laws--can be derived in a few lines without wasting a minute on ergodic theory. The pedagogical implications are clear.

The price we have paid for this simplification is that we cannot interpret the canonical distribution as giving the frequencies with which a system goes into the various states. But nobody had ever justified or needed that interpretation anyway. In recognizing that the canonical distribution represents only our state of knowledge when we have certain partial information derived from macroscopic measurements, we are not losing anything we had before, but only frankly admitting the situation that has always existed; and indeed, which Gibbs had recognized.

On the other hand, what we have gained by this change in interpretation is far more than we bargained for. Even if one had been completely successful in proving ergodic theorems, and had continued to ignore the difficulty about length of time over which the averages have to be taken, this still would have given a justification for the methods of Gibbs only in the equilibrium case. But the principle of maximum entropy, being entirely independent of the equations of motion, contains no such restriction. If one grants that it represents a valid method of reasoning at all, one must grant that it gives us also the long-hoped-for general formalism for treatment of irreversible processes!

The last statement above breaks into new ground, and claims for statistical mechanics based on Information Theory, a far wider range of validity and applicability than was ever claimed for conventional statistical mechanics. Just for that reason, the issue is no longer one of mere philosophical preference for one viewpoint or another; the issue is now one of definite mathematical fact. For the assertion just made can be put to the test by carrying out specific calculations, and will prove to be either right or wrong.

Some Personal Recollections. All this was clear to me by 1951; nevertheless, no attempt at publication was made for another five years. There were technical problems of extending the formalism to continuous distributions and the density matrix, that were not solved for many years; but the reason for the initial delay was quite different.

In the Summer of 1951, Professor G. Uhlenbeck gave his famous course on Statistical Mechanics at Stanford, and following the lectures I had many conversations with him, over lunch, about the foundations of the theory and current progress on it. I had expected, naively, that he would be enthusiastic about Shannon's work, and as eager as I to exploit these ideas for Statistical Mechanics. Instead, he seemed to think that the basic problems were, in principle, solved by the then recent work of Bogoliubov and van Hove (which seemed to me filling in details, but not touching at all on the real basic problems)--and adamantly rejected all suggestions that there is any connection between entropy and information.

His initial reaction to my remarks was exactly like my initial reaction to Shannon's: "Whose information?" His position, which I never succeeded in shaking one iota, was: "Entropy cannot be a measure of 'amount of ignorance,' because different people have different amounts of ignorance; entropy is a definite physical quantity that can be measured in the laboratory with thermometers and calorimeters." Although the answer to this was clear in my own mind, I was unable, at the time, to convey that answer to him. In trying to explain a new idea I was, like Maxwell, groping for words because the way of thinking and habits of language then current had to be broken before I could express a different way of thinking.

Today, it seems trivially easy to answer Professor Uhlenbeck's objection as follows: "Certainly, different people have different amounts of ignorance. The entropy of a thermodynamic system is a measure of the degree of ignorance of a person *whose sole knowledge about its microstate consists of the values of the macroscopic quantities* X_i *which define its thermodynamic state*. This is a completely 'objective' quantity, in the sense that it is a function only of the X_i, and does not depend on anybody's personality. There is then no reason why it cannot be measured in the laboratory."

It was my total inability to communicate this argument to Professor Uhlenbeck that caused me to spend another five years thinking over these matters, trying to write down my thoughts more clearly and explicitly, and making sure in my own mind that I could answer all the objections that Uhlenbeck and others had raised. Finally, in the Summer of 1956 I collected this into two papers, sending the first off to the Physical Review on August 29.

Now another irony takes place; it is left to the Reader to guess to whom the Editor (S. Goudsmit) sent it for refereeing. That Unknown Referee's comments (now framed on my office wall as an encouragement to young men who today have to fight for new ideas against an Establishment that wants only new mathematics) opine that the work is clearly written, but since it expounds only a certain philosophy of interpretation and has no application whatsoever in Physics, it is out of place in a Physics journal. But a second referee thought differently, and so the papers were accepted after all, appearing in 1957. Within a year there were over 2000 requests for reprints.

Needless to say, my own understanding of the technical problems continued to evolve for many years afterward. A schoolboy, having just learned the rules of arithmetic, does not see immediately how to apply them to the extraction of cube roots, although he has in his grasp all the principles needed for this. Similarly, I did not see how to set down the explicit equations for irreversible processes because I simply *could not believe* that the solution to such a complicated problem could be as simple as the Maximum Entropy Principle was giving; and spent six more years (1956-1962) trying to mutilate the principle by grafting new and more complicated rococo embellishments onto it. In my Brandeis lectures of 1962, tongue and pen somehow managed to state the right rule [Eq. (50)]; but the inner mind did not fully assent; it still seemed like getting something for nothing.

The final breakthrough came in the Christmas vacation period of 1962 when, after all else had failed, I finally had the courage to sit down and work out all the details of the calculations that result from using only the Maximum Entropy Principle; _and nothing else_. Within three days the new formalism was in hand, masses of the known correct results of Onsager, Wiener, Kirkwood, Callen, Kubo, Mori, MacLennon, were pouring out as special cases, just as fast as I could write them down; and it was clear that this was it. Two months later, my students were the first to have assigned homework problems to predict irreversible processes by solving Wiener-Hopf integral equations.

As it turned out, no more principles were needed beyond those stated in my first paper; one has merely to take them absolutely literally and _apply_ them, putting into the equations the macroscopic information that one does, in fact, have about a nonequilibrium state; and all else follows inevitably.

From this the reader will understand why I have considerable sympathy for those who today have difficulty in accepting the Principle of Maximum Entropy, because (1) the results seem to come too easily to believe; and (2) it seems at first glance as if the dynamics has been ignored. In fact, I struggled for eleven years with exactly the same feeling, before seeing clearly not only _why_, but also in detail _how_ the formalism is able to function so efficiently.

The point is that we are not ignoring the dynamics, and we are not getting something for nothing, because we are asking of the formalism only some extremely simple questions; we are asking only for predictions of _experimentally reproducible_ things; and for these all circumstances that are not under the experimenter's control must, of necessity, be irrelevant.

If certain macroscopically controlled conditions are found, in the laboratory, to be sufficient to determine a reproducible outcome, then it must follow that _information_ about those macroscopic conditions tells us everything about the microscopic state that is relevant for theoretical _prediction_ of that outcome. It may seem at first "unsound" to assign equal _a priori_ probabilities to all other details, as the Maximum Entropy Principle does; but in fact we are assigning uniform probabilities only to details that are irrelevant for questions about reproducible phenomena.

To assume further information by putting some additional fine-grained structure into our ensembles would, in all probability, not lead to incorrect predictions; it would only force us to calculate intricate details that would, in the end, cancel out of our final predictions. Solution by the Maximum Entropy Principle is so unbelievably simple just because it eliminates those irrelevant details right at the beginning of the calculation by averaging over them.

To discover this argument requires only that one think, very carefully, about why Boltzmann's method of the most probable distribution was able to predict the correct spatial and velocity distribution of the molecules; and this could have been done at any time in the past 100 years. Whether or not one wishes to recognize it, this--and not ergodic properties--is the real reason why all Statistical Mechanics works. But once the argument is understood, it is clear that it applies equally well whether the macroscopic state is equilibrium or nonequilibrium, and whether the observed phenomenon is reversible or irreversible.

I hope that this historical account will also convey to the reader that the Principle of Maximum Entropy, although a powerful tool, is hardly a radical innovation. Its philosophy was clearly foreshadowed by Laplace and Jeffreys; its mathematics by Boltzmann and Gibbs.

B. Present Features and Applications.

Let us set down, for reference, a bit of the basic Maximum Entropy formalism for the finite discrete case, putting off generalizations until they are needed. There are n different possibilities, which would be distinguished adequately by a single index $(i = 1,2,\ldots,n)$. Nevertheless we find it helpful, both for notation and for the applications we have in mind, to introduce in addition a real variable x, which can take on the discrete values $(x_i, 1 \le i \le n)$, defined in any way and not necessarily all distinct. If we have certain information I about x, the problem is to represent this by a probability distribution $\{p_i\}$ which has maximum entropy while agreeing with I.

Clearly, such a problem cannot be well-posed for arbitrary information; I must be such that, given any proposed distribution $\{p_i\}$, we can determine unambiguously whether I does or does not agree with $\{p_i\}$. Such information will be called testable. For example, consider:

$I_1 \equiv$ "It is certain that $\tanh x < 0.7$."

$I_2 \equiv$ "There is at least a 90% probability that $\tanh x < 0.7$."

$I_3 \equiv$ "The mean value of $\tanh x$ is 0.675."

I_4 ≡ "The mean value of tanh x is probably less than 0.7."

I_5 ≡ "There is some reason to believe that tanh x = 0.675."

Statements I_1, I_2, I_3 are testable, and may be used as constraints in maximizing the entropy. I_4 and I_5, although clearly relevant to inference about x, are too vague to be testable, and we have at present no formal principle by which such information can be used in a mathematical theory. However, the fact that our intuitive common sense does make use of nontestable information suggests that new principles for this, as yet undiscovered, must exist.

Since n is finite, the entropy has an absolute maximum value log n, and any constraint can only lower this. If we think of the $\{p_i\}$ as cartesian coordinates of a point P in an n-dimensional space, P is constrained by $p_i \geq 0$, $\Sigma p_i = 1$ to lie on a domain D which is a "triangular" segment of an (n-1)-dimensional hyperplane. On D the entropy varies continuously, taking on all values in $0 \leq H \leq \log n$ and reaching its absolute maximum at the center. Any testable information will restrict P to some subregion D' of D, and clearly the entropy has some least upper bound $H \leq \log n$ on D'. So the maximum entropy problem must have a solution if D' is a closed set.

There may be more than one solution: for example, the information I_6 ≡ "The entropy of the distribution $\{p_i\}$ is not greater than log(n-1)" is clearly testable, and if $n > 2$ it yields an infinite number of solutions. Furthermore, strictly speaking, if D' is an open set there may not be any solution, the upper bound being approached but not actually reached on D'. Such a case is generated by I_7 ≡ "$p_1^2 + p_2^2 < n^{-2}$". However, since we are concerned with physical problems where the distinction between open and closed sets cannot matter, we would accept a point on the closure of D' (in this example, on its boundary) as a valid solution, although corresponding strictly only to I_8 ≡ "$p_1^2 + p_2^2 \leq n^{-2}$."

But these considerations are mathematical niceties that one has to mention only because he will be criticized if he does not. In the real applications that matter, we have not yet found a case which does not have a unique solution.

In principle, every different kind of testable information will generate a different kind of mathematical problem. But there is one important class of problems for which the general solution was given once and for all, by Gibbs. If the constraints consist of specifying mean values of certain functions $\{f_1(x), f_2(x), \ldots, f_m(x)\}$:

$$\sum_{i=1}^{n} p_i \, f_k(x_i) = F_k \quad , \qquad 1 \le k \le m \qquad \text{(B1)}$$

where $\{F_k\}$ are numbers given in the statement of the problem, then if $m < n$, entropy maximization is a standard variational problem solvable by stationarity using the Lagrange multiplier technique. It has the formal solution:

$$p_i = \frac{1}{Z(\lambda_1 \ldots \lambda_m)} \exp\left[-\lambda_1 f_1(x_i) - \ldots - \lambda_m f_m(x_i)\right] \qquad \text{(B2)}$$

where

$$Z(\lambda_1, \ldots, \lambda_m) \equiv \sum_{i=1}^{n} \exp\left[-\lambda_1 f_1(x_i) - \ldots - \lambda_m f_m(x_i)\right] \qquad \text{(B3)}$$

is the partition function and $\{\lambda_k\}$ are the Lagrange multipliers, which are chosen so as to satisfy the constraints (B1). This is the case if

$$F_k = -\frac{\partial}{\partial \lambda_k} \log Z \quad , \qquad 1 \le k \le m \qquad \text{(B4)}$$

a set of m simultaneous equations for m unknowns. The value of the entropy maximum then attained is, as noted in my reminiscences, a function only of the given data:

$$S(F_1 \ldots F_m) = \log Z + \sum_k \lambda_k F_k \qquad \text{(B5)}$$

and if this function were known, the explicit solution of (B4) would be

$$\lambda_k = \frac{\partial S}{\partial F_k} \quad , \qquad 1 \le k \le m \quad . \qquad \text{(B6)}$$

Given this distribution, the best prediction we can make (in the sense of minimizing the expected square of the error) of any quantity $q(x)$, is then

$$\langle q(x) \rangle = \sum_{i=1}^{n} p_i \, q(x_i)$$

and numerous covariance and reciprocity rules are contained in the identity

$$\langle qf_k\rangle - \langle q\rangle\langle f_k\rangle = -\frac{\partial\langle q\rangle}{\partial\lambda_k} \tag{B7}$$

[note the special cases $q(x) = f_j(x)$, and $j = k$]. The functions $f_k(x)$ may contain also some parameters α_j:

$$f_k = f_k(x;\alpha_1 \dots \alpha_s)$$

(which in physical applications might have the meaning of volume, magnetic field intensity, angular velocity, etc.); and we have an important variational property; if we make an arbitrary small change in all the data of the problem $\{\delta F_k, \delta\alpha_r\}$, we may compare two slightly different maximum-entropy solutions. The difference in their entropies is found, after some calculation, to be

$$\delta S = \sum_k \lambda_k\ \delta Q_k \tag{B8}$$

where

$$\delta Q_k \equiv \delta\langle f_k\rangle - \langle\delta f_k\rangle \quad . \tag{B9}$$

The meaning of this identity has a familiar ring: there is no such function as $Q_k(F_1 \dots F_m;\alpha_1 \dots \alpha_s)$ because δQ_k is not an exact differential. However, the Lagrange multiplier λ_k is an integrating factor such that $\Sigma\lambda_k\ \delta Q_k$ is the exact differential of a "state function" $S(F_1 \dots F_m;\alpha_1 \dots\alpha_s)$.

I believe that Clausius would recognize here an interesting echo of his work, although we have only stated some general rules for plausible reasoning, making no necessary reference to physics. This is enough of the bare skeleton of the formalism to serve as the basis for some examples and discussion.

The Brandeis Dice Problem. First, we illustrate the formalism by working out the numerical solution to a problem which was used in the Introduction to my 1962 Brandeis lectures merely as a qualitative illustration of the ideas, but has since become a cause célèbre as some papers have been written attacking the Principle of Maximum Entropy on the grounds of this very example. So a close look at it will take us straight to the heart of some of the most common misconceptions and, I hope, give us some appreciation of what the Principle of Maximum Entropy does and does not (indeed, should not) accomplish for us.

When a die is tossed, the number of spots up can have any value i in $1 \leq i \leq 6$. Suppose a die has been tossed N times and we are told only that the average number of spots up was not 3.5 as we might expect from an "honest" die but 4.5. Given this information, and nothing else, what probability should we assign to i spots on the next toss? The Brandeis lectures started with a qualitative graphical discussion of this problem, which showed (or so I thought) how ordinary common sense forces us to a result with the qualitative properties of the maximum-entropy solution.

Let us see what solution the Principle of Maximum Entropy gives for this problem, if we interpret the data as imposing the mean value constraint

$$\sum_{i=1}^{6} i\, p_i = 4.5 \quad . \tag{B10}$$

The partition function is

$$Z(\lambda) = \sum_i e^{-\lambda i} = x(1-x)^{-1}(1-x^6) \tag{B11}$$

where $x \equiv e^{-\lambda}$. The constraint (B10) then becomes

$$-\frac{\partial}{\partial\lambda} \log Z = \frac{1 - 7x^6 + 6x^7}{(1-x)(1-x^6)} = 4.5$$

or

$$3x^7 - 5x^6 + 9x - 7 = 0 \quad . \tag{B12}$$

By computer, the desired root of this is $x = 1.44925$, which yields $\lambda = -0.37105$, $Z = 26.66365$, $\log Z = 3.28330$. The maximum-entropy probabilities are $p_i = Z^{-1} x^i$, or

$$\{p_1 \cdots p_6\} = \{0.05435,\ 0.07877,\ 0.11416,\ 0.16545,\ 0.23977,\ 0.34749\} \tag{B13}$$

From (B5), the entropy of this distribution is

$$S = 1.61358 \text{ natural units} \tag{B14}$$

as compared to the maximum of $\log_e 6 = 1.79176$, corresponding to no constraints and a uniform distribution.

Now, what does this result mean? In the first place, it is a distribution $\{p_r,\ 1 \leq r \leq 6\}$ on a space of only six points; the sample space S of a single trial. Therefore, our result as it stands is only a mean of describing a state of knowledge about the outcome of a single trial. It represents a state of knowledge in which one has only (1) the enumeration of the six possibilities; and (2) the mean value constraint (B10); and no other information. The distribution is "maximally noncommittal" with respect to all other matters; it is as uniform (by the criterion of the Shannon information measure) as it can get without violating the given constraint.

Any probability distribution over some sample space S enables us to make statements about (i.e., assign probabilities to) propositions or events defined within that space. It does not--and by its very nature cannot--make statements about any event lying outside that space. Therefore, our maximum-entropy distribution does not, and cannot, make any statement about frequencies.

Anything one says about a frequency in n tosses is a statement about an event in the n-fold extension space $S^n = S \otimes S \otimes \ldots \otimes S$ of n tosses, containing 6^n points (and of course, in any higher space which has S^n as a subspace).

It may be common practice to jump to the conclusion that a probability in one space is the same as a frequency in a different space; and indeed, the level of many expositions is such that the distinction is not recognized at all. But the first thing one has to learn about using the Principle of Maximum Entropy in real problems is that the mathematical rules of probability theory must be obeyed strictly; all conceptual sloppiness of this sort must be recognized and expunged.

There is, indeed, a connection between a probability p_i in space S and a frequency g_i in S^n; but we are justified in using only those connections which are deducible from the mathematical rules of probability theory. As we shall see in connection with fluctuation theory, some common attempts to identify probability and frequency actually stand in conflict with the rules of probability theory.

Probability and Frequency. To derive the simplest and most general connection, the sample space S^n of n trials may be labeled by $\{r_1, r_2, \ldots, r_n\}$, where $1 \leq r_k \leq 6$, and r_k is the number of spots up on the k'th toss. The most general probability assignment on S^n is a set of non-negative real numbers $P(r_1 \ldots r_n)$ such that

$$\sum_{r_1=1}^{6} \cdots \sum_{r_n=1}^{6} P(r_1 \ldots r_n) = 1 \quad . \tag{B15}$$

In any given sequence $\{r_1 \ldots r_n\}$ of results, the frequency with which i spots occurs is

$$g_i(r_1 \ldots r_n) = n^{-1} \sum_{k=1}^{n} \delta(r_k, i) \quad . \tag{B16}$$

This can take on (n+1) discrete values, and its expectation is

$$\langle g_i \rangle = \frac{1}{n} \sum_{k=1}^{n} \sum_{r_1=1}^{6} \cdots \sum_{r_n=1}^{6} P(r_1 \ldots r_n) \delta(r_k, i)$$

$$= \frac{1}{n}\left[p_1(i) + p_2(i) + \ldots + p_n(i)\right] \tag{B17}$$

where $p_k(i)$ is the probability of getting i spots on the k'th toss, regardless of what happens in other tosses. The expected frequency of an event is always equal to its average probability over the different trials.

Many experiments fall into the category of exchangeable sequences; i.e., it is clear that the underlying "mechanism" of the experiment, although unknown, is not changing from one trial to another. The probability of any particular sequence of results $\{r_1 \ldots r_n\}$ should then depend only on how many times a particular outcome $r = i$ happened; and not on which particular trials. Then the probability distribution $P\{r_k\}$ is invariant under permutations of the labels k. In this case, the probability of i spots is the same at each trial: $p_1(i) = p_2(i) = \ldots = p_n(i) = p_i$, and (B17) becomes

$$\langle g_i \rangle = p_i \quad . \tag{B18}$$

In an exchangeable sequence, the probability of an event at one trial is not the same as its frequency in many trials; but it is numerically equal to the expectation of that frequency; and this connection holds whatever correlations may exist between different trials.

The probability is therefore the "best" estimate of the frequency, in the sense that it minimizes the expected square of the error. But the result (B18) tells us nothing whatsoever about whether this is a reliable estimate; and indeed nothing in the space S of a single trial can tell us anything about the reliability of (B18).

To investigate this, note that by a similar calculation, the expected product of two frequencies is

$$\langle g_i g_j \rangle = n^{-2} \sum_{k,m=1}^{n} p_k(i)\ p(j,m|i,k) \tag{B19}$$

where $p(j,m|i,k)$ is the conditional probability that the m'th trial gives the result j, given that the k'th trial had the outcome i. Of course, if $m = k$ we have simply $p(jk|ik) = \delta_{ij}$.

In an exchangeable sequence $p(jm|ik)$ is independent of m,k for $m \neq k$; and so $p_k(i)\ p(jm|ik) = p_{ij}$, the probability of getting the outcomes i,j respectively at any two different tosses. The covariance of g_i, g_j then reduces to

$$\langle g_i g_j\rangle - \langle g_i\rangle\langle g_j\rangle = (p_{ij} - p_i p_j) + \frac{1}{n}(\delta_{ij}\ p_i - p_{ij}) \quad . \tag{B20}$$

If the probabilities are not independent, $p_{ij} \neq p_i p_j$, this does not go to zero for large n.

Let us examine the case $i = j$ more closely. Writing $p_{ii} = \alpha_i\ p_i$, α_i is the conditional probability that, having obtained the result i on one toss, we shall get it at some other specified toss. The variance of g_i is, from (B20), dropping the index i,

$$\langle g^2\rangle - \langle g\rangle^2 = p(\alpha - p) + \frac{1}{n}\ p(1-\alpha) \quad . \tag{B21}$$

Two extreme cases of inter-trial correlations are contained in (B21). For complete independence, $\alpha = p$, the variance reduces to $n^{-1}p(1-p)$, just the result of the de Moivre-Laplace limit theorem (A4). But as cautioned before, in any other case the variance does not tend to zero at all; there is no "law of large numbers." For complete dependence, $\alpha = 1$ (i.e., having seen the result of one toss, the die is certain to give the same result at all others), (B21) reduces to $p(1-p)$ which again makes excellent sense; in this case our uncertainty about the frequency in any number of tosses must be just our uncertainty about the first toss.

Note that the variance (B21) becomes zero for a slight negative correlation:

$$\alpha = p - \frac{1-p}{n-1} \tag{B22}$$

Due to the permutation invariance of $P(r_1 \ldots r_n)$ it is not possible to have a negative correlation stronger than this; as $n \to \infty$ it is not possible to have any negative correlation in an exchangeable sequence. This corresponds to the famous de Finetti (1937) representation theorem; in the literature of pure mathematics it is called the Hausdorff moment problem. An almost unbelievably simple proof has just been found by Heath and Sudderth (1976).

To summarize: given any probability assignment $P(r_1 \ldots r_n)$ on the space S^n, we can determine the probability distribution $W_i(t)$ for the frequency g_i to take on any of its possible values $g_i = (t/n)$, $0 \leq t \leq n$. The (mean) ± (standard deviation) over this distribution then provide a reasonable statement of our "best" estimate of g_i and its accuracy. In the case of an exchangeable sequence, this estimate is

$$(g_i)_{est} = p_i \pm \sqrt{p_i(1-p_i)}\left[R_i + \frac{1-R_i}{n}\right]^{1/2} \tag{B23}$$

where $R_i \equiv (\alpha_i - p_i)/(1-p_i)$ is a measure of the inter-trial correlation, ranging from $R=0$ for complete independence to $R=1$ for complete dependence.

Evidently, then, to suppose that a probability assignment at a single trial is also an assertion about a frequency in many trials in the sense of the Bernoulli and de Moivre-Laplace limit theorems, is in general unjustified unless (1) the successive trials form an exchangeable sequence, and (2) the correlation of different trials is strictly zero. However, there are other kinds of connections between probability and frequency; and maximum-entropy distributions have an exact and close relation to frequencies after all, as we shall see presently.

Relation to Bayes' Theorem. To prepare us to deal with some objections to the maximum-entropy solution (B13) we turn back to the basic product and sum rules of probability theory (A8), (A9) derived by Cox from requirements of consistency. Just as any argument of deductive logic can be resolved ultimately into many syllogisms, so any calculation of inductive logic (i.e., probability theory) is reducible to many applications of these rules.

We stress that these rules make no reference to frequencies; or to any random experiment. The numbers $p(A|B)$ are simply a convenient numerical scale on which to represent degrees of plausibility. As noted at the beginning of this work, it is the problem of determining initial numerical values by logical analysis of the prior information in more general cases than solved by Bernoulli and Laplace, that underlies our study.

Furthermore, in neither the statement nor the derivation of these rules is there any reference to the notion of a sample space. In a formally qualitative sense, therefore, they may be applied to any propositions A, B, C, ... with unambiguous meanings. Their complete qualitative correspondence with ordinary common sense was demonstrated in exhaustive detail by Polya (1954).

But in quantitative applications we find at once that merely defining two propositions, A, B is not sufficient to determine any numerical value for $p(A|B)$. This numerical value depends not only on A, B, but also on which alternative propositions A', A", etc. are to be considered if A should be false; and the problem is mathematically indeterminate until those alternatives are fully specified. In other words, we must define our "sample space" or "hypothesis space" before we have any mathematically well-posed problem.

In statistical applications (parameter estimation, hypothesis testing), the most important constructive rule is just the statement that the product rule is consistent; i.e., $p(AB|C)$ is symmetric in A and B, so $p(A|BC)p(B|C) = p(B|AC)p(A|C)$. If $p(B|C) \neq 0$, we thus obtain

$$p(A|BC) = p(A|C) \frac{p(B|AC)}{p(B|C)} \tag{B24}$$

in which we may call C the prior information, B the conditioning information. In typical applications, C represents the general background knowledge or assumptions used to formulate the problem, B is the new data of some experiment, and A is some hypothesis being tested. For example, in the Millikan oil-drop experiment, we might take A as the hypothesis: "the electronic charge lies in the interval $4.802 < e < 4.803$," while C represents the general assumed known laws of electrostatics and viscous hydrodynamics and the results of previous measurements, while B stands for the new data being used to find a revised "best" value of e. Equation (B24) then shows how the prior probability $p(A|C)$ is changed to the posterior probability $p(A|BC)$ as a result of acquiring the new information B.

In this kind of application, $p(B|AC)$ is a "direct" or "sampling" probability, since we reason in the direction of the causal influence, from an assumed cause A to a presumed observable result B: and $p(A|BC)$ is an "inverse" probability, in which we reason from an observed result B to an assumed cause A. On comparing with (A7) we see that (B24) is a more general form of Laplace's rule, in which we need not have an exhaustive set of possible causes. Therefore, since (A7) is always called "Bayes' theorem," we may as well apply the same name to (B24).

At the risk--or rather the certainty--of belaboring it, we stress again that we are concerned here with inductive reasoning of any kind, not necessarily related to random experiments or any repetitive process. On the other hand, nothing prevents us from applying the theory to a repetitive situation (i.e., n tosses of a die); and propositions about frequencies g_i are then just as legitimate pieces of data or objects of inquiry as any other propositions. Various kinds of connection between probability and frequency then appear, as mathematical consequences of (A8), (A9). We have just seen one of them.

But now, could we have solved the Brandeis dice problem by applying Bayes' theorem instead of maximum entropy? If so, how do the results compare? Friedman and Shimony (1971) (hereafter denoted FS) claimed to exhibit an inconsistency in the Principle of Maximum Entropy (hereafter denoted PME) by an argument which introduced a proposition d_ε, so ill-defined that they tried to use it as (1) a constraint in PME, (2) a conditioning statement in Bayes' theorem; and (3) an hypothesis whose posterior probability is calculated. Therefore, let us note the following.

If a statement d referring to a probability distribution in space S is testable (for example, if it specifies a mean value $\langle f \rangle$ for some function f(i) defined on S), then it can be used as a constraint in PME; but it cannot be used as a conditioning statement in Bayes' theorem because it is not a statement about any event in S or any other space.

Conversely, a statement D about an event in the space S^n (for example, an observed frequency) can be used as a conditioning statement in applying Bayes' theorem, whereupon it yields a posterior distribution on S^n which may be contracted to a marginal distribution on S; but D cannot be used as a constraint in applying PME in space S, because it is not a statement about any event in S, or about any probability distribution over S; i.e., it is not testable information in S.

At this point, informed students of statistical mechanics will be astonished at the suggestion that there is any inconsistency between application of PME in space S and of Bayes' theorem in S^n, since the former yields a canonical distribution, while the latter is just the Darwin-Fowler method, originally introduced as a rigorous way of justifying the canonical distribution! The mathematical fact shown by this well-known calculation (Schrödinger, 1948) is that, whether we use maximum entropy in space S with a constraint fixing an average $\langle f \rangle$ over a <u>probability distribution</u>, or apply Bayes' theorem in S^n with a conditioning statement fixing a numerically equal average $\overline{f}$ over <u>sample values</u>, we obtain for large n identical distributions in the space S. The result generalizes at once to the case of several simultaneous mean-value constraints.

This not only illustrates--contrary to the claims of FS--the consistency of PME with the other principles of probability theory, but it shows what a powerful tool PME is; i.e., how much simpler and more convenient mathematically it is to use PME in statistical calculations if the distribution on S is what we are seeking. PME leads us directly to the same final result, without any need to go into a higher space S^n and carry out passage to the limit $n \to \infty$ by saddle-point integration.

Of course, it is as true in probability theory as in carpentry that introduction of more powerful tools brings with it the obligation to exercise a higher level of understanding and judgment in using them. If you give a carpenter a fancy new power tool, he may use it to turn out more precise work in greater quantity; or he may just cut off his thumb with it. It depends on the carpenter.

The FS article led to considerably more discussion (see the references collected with the FS one) in which severed thumbs proliferated like hydras; but the level of confusion about the points already noted is such that it would be futile to attempt any analysis of the FS arguments.

FS suggest that a possible way of resolving all this is to deny that the probability of d_ε can be well-defined. Of course it cannot be; however, to understand the situation we need no "deep and systematic analysis of the concept of reasonable degree of belief." We need only raise our standards of exposition to the same level that is required in any other application of probability theory; i.e., we must define our propositions and sample spaces with enough precision to make a determinate mathematical problem.

There is a more serious difficulty in trying to reply to these criticisms. If FS dislike the maximum-entropy solution (B13) to this problem strongly enough to write three articles attacking it, then it would seem to follow that they prefer a different solution. But what different solution? One cannot form any clear idea of what is really troubling them, because in all these publications FS give no hint as to how, in their view, a more acceptable solution ought to differ from (B13).

The Rowlinson Criticism. In sharp contrast to the FS criticisms is that of J. S. Rowlinson (1970), who considers the same dice problem but does offer an alternative solution. For this reason, it is easy to give a precise quantitative reply to his criticism.

He starts with the all too familiar line: "Most scientists would say that the probability of an event is (or represents) the frequency with which it occurs in a given situation." Likewise, a critic of Columbus could have written (after he had returned from his first voyage): "Most geographers would say that the earth is flat."

Clarification of the centuries-old confusion about probability and frequency will not be achieved by taking votes; much less by quoting the philosophical writings of Leslie Ellis (1842). Rather, we must examine the mathematical facts concerning the rules of probability theory and the different sample spaces in which probabilities and frequencies are defined. We have seen, in the discussion following (B14) above, that anyone who glibly supposes that a probability in one space can be equated to a frequency in another, is assuming something which is not only not generally deducible from the principles of probability theory; it may stand in conflict with those principles.

There is no stranger experience than seeing printed criticisms which accuse one of saying the exact opposite of what he *has* said, explicitly and repeatedly. Thus my bewilderment at Rowlinson's statement that I reject "the methods used by Gibbs to establish the rules of statistical mechanics." I believe I can lay some claim to being the foremost current advocate and defender of Gibbs' methods! Anyone who takes the trouble to read Gibbs will see that, far from rejecting Gibbs' methods, I have adopted them enthusiastically and (thanks to the deeper understanding from Shannon) extended their range of application.

One of the major unsolved riddles of probability theory is: how to explain to another person exactly *what is the problem being solved?* It is well established that merely stating this in words does not suffice; repeatedly, starting with Laplace, writers have given the correct solution to a problem, only to have it attacked on the grounds that it is not the solution to some entirely different problem. This is at least the tenth time it has happened to me. As I tried to stress, the maximum-entropy solution (B13) describes the state of knowledge in which we are given the enumeration of the six possibilities, the mean value $\langle i \rangle = 4.5$, and nothing else. But Rowlinson proceeds to introduce models with an urn containing seven white and three black balls (or a population of urns with varying contents) from which one makes various numbers of random draws with replacement. One expects that different problems will have different solutions.

In Rowlinson's Urn model, we perform Bernoulli trials five times, with constant probability of success $p = 0.7$. Then the numbers s of successes is in $0 \leq s \leq 5$, and the expected number is $\langle s \rangle = 5 \times 0.7 = 3.5$. Setting $i \equiv s + 1$, we have $1 \leq i \leq 6$, $\langle i \rangle = 4.5$, the conditions stated in my dice problem. Thus he offers as a counter-proposal the binomial distribution

$$p'_i = \binom{5}{i-1} p^{i-1} (1-p)^{6-i} , \qquad 1 \leq i \leq 6 \quad . \tag{B25}$$

These numbers are

$$\{p'_1 \ldots p'_6\} = \{0.00243, 0.02835, 0.1323, 0.3087, 0.36015, 0.16807\} . \tag{B26}$$

and they yield an entropy $S' = 1.413615$, 0.2 unit lower than that of (B13). This lower entropy indicates that the urn model puts further constraints on the solution beyond that used in (B13). We see that these consist in the extreme values ($i = 1,6$) receiving less probability than before (only one of

$2^5 = 32$ possible outcomes can lead to $i = 1$, while ten of them yield $i = 3$, etc.).

Now if we knew that the experiment consisted of drawing five times from an urn with just the composition specified by Rowlinson, the result (B25) would indeed be the correct solution. But by what right does one assume this elaborate model structure when it is not given in the statement of the problem? One could, with equal right, assume any one of a hundred other specific models, leading to a hundred other counter-proposals. But it is just the point of the maximum-entropy principle that it achieves "objectivity" of our inferences, in the sense that we base our predictions only on the information that we do, in fact, have; and carefully avoid introducing any such gratuitous assumptions not warranted by our data. Any such assumption is far more likely to impose false constraints than to happen, by luck, onto an unknown correct one (which would be like guessing the combination to a safe).

At this point, Rowlinson says, "Those who favour the automatic use of the principle of maximum entropy would observe that the entropy of [our Eq. (B25)], 1.4136, is smaller than that of [B13], and so say that in proposing [B25] as a solution, 'information' has been assumed for which there is no justification." We do indeed say this, although Rowlinson simply rejects it out of hand without giving a reason. So to sustain our claim, let us calculate explicitly just how much Rowlinson's solution assumes without justification.

To clarify what is meant by "assuming information," suppose that an economist, Mr. A, is trying to forecast future price trends for some commodity. The condition of next week's market cannot be known with certainty, because it depends on intentions to buy or sell hidden in the minds of many different individuals. Evidently, a rational method of forecasting must somehow take account of all these unknown possibilities. Suppose that Mr. A's data are found to be equally compatible with 100 different possibilities. If he arbitrarily picked out 10 of these which happened to suit his fancy, and based his forecast only on them, ignoring the other 90, we should certainly consider that Mr. A was guilty of an egregious case of assuming information without justification. Our present problem is similar in concept, but quite different in numerical values.

We have stressed that, fundamentally, the maximum-entropy solution (B13) describes only a state of knowledge about a single trial, and is not an assertion about frequencies. But Rowlinson, as noted, also rejects this distinction and wants to judge the issue on the grounds of frequencies. Very well; let us now bring out the frequency connection that a maximum-entropy distribution does, after all, have (and which, incidentally, was pointed out in my Brandeis lectures, from which Rowlinson got this dice problem).

In N tosses, a set of observed frequencies $\{g_i\} = \{N_i/N\}$ (called g to avoid collision with previous notation) can be realized in

$$W = \frac{N!}{(Ng_1)!\ (Ng_2)!\ \ldots\ (Ng_6)!} \qquad \text{(B27)}$$

different ways. As we noted from Boltzmann's work, Eq. (A14), the Stirling approximation to the factorials yields an asymptotic formula

$$\log W \sim NS \qquad \text{(B28)}$$

where

$$S = -\sum_{i=1}^{6} g_i \log g_i \qquad \text{(B29)}$$

is the entropy of the observed <u>frequency</u> distribution. Given two different sets of frequencies $\{g_i\}$ and $\{g_i'\}$, the ratio: (number of ways g_i can be realized)/(number of ways g_i' can be realized) is given by an asymptotic formula

$$\frac{W}{W'} \sim A \exp[N(S - S')]\left\{1 + \frac{B}{N} + O(N^{-2})\right\} \qquad \text{(B30)}$$

where

$$A \equiv \prod_i (g_i'/g_i)^{1/2} \qquad \text{(B31)}$$

$$B \equiv \frac{1}{12}\sum_i \left(\frac{1}{g_i'} - \frac{1}{g_i}\right) \qquad \text{(B32)}$$

are independent of N, and represent corrections from the higher terms in the Stirling approximation. We write them down only to allay any doubts about the accuracy of the numbers to follow. In all cases considered here it is easily seen that they have no effect on our conclusions, and only the exponential factor matters.

Rowlinson mentions an experiment involving 20,000 throws of a die, to which we shall return later; but in the present comparison this leads to numbers beyond human comprehension. To keep the results more modest, let us assume only $N = 1000$ throws. If we take $\{g_i\}$ as the maximum-entropy distribution (B13) and $\{g_i'\}$ as Rowlinson's solution (B26), we find $A = 0.159$, $B = 34$, $S - S' = 0.200$; and thus, with $N = 1000$,

$$\frac{W}{W'} = 1.19 \times 10^{86} \quad . \tag{B34}$$

Both distributions agree with the datum $\langle i \rangle = 4.5$; but for every way in which Rowlinson's distribution can be realized, there are over 10^{86} ways in which the maximum entropy distribution can be realized (the age of the universe is less than 10^{18} seconds). It appears that information was indeed "assumed for which there is no justification."

This example should help to give us a proper respect for just what we are accomplishing when we maximize entropy. It shows the magnitude of the indiscretion we commit if we accept a distribution whose entropy is 0.2 unit less than the maximum value compatible with our data. In this example, to accept any distribution whose entropy is as much as 0.005 below the maximum value, would be to ignore over 99 percent of all possible ways in which the average $\langle i \rangle = 4.5$ could be realized.

For reasons unexplained, Rowlinson seizes upon the particular value $p_1 = 0.05435$ from the maximum-entropy solution (B13), and asks: "But what basis is there for trusting in this last number?" but fails to ask the same question about his own very different result $p_1' = 0.00243$. Since it is so seldom that one is able to give a quantitative reply to a rhetorical question, we should not pass up this opportunity.

Answer to the Rhetorical Question. Let us, as before, count up the number of possibilities compatible with the given data. In the original problem we were to find $\{p_1 \ldots p_n\}$ so as to maximize $H = -\Sigma p_i \log p_i$ subject to the constraints $\Sigma p_i = 1$, $\langle i \rangle = \Sigma i p_i$, a specified numerical value. If now we impose the additional constraint that p_1 is specified, we can define conditional probabilities

$$p_i' = \frac{p_i}{1 - p_1} \ , \qquad i = 2,3,\ldots n \tag{B35}$$

with entropy

$$H' = -\sum_{i=2}^{n} p_i' \log p_i' \ . \tag{B36}$$

These quantities are related by Shannon's basic functional equation

$$H(p_1 \ldots p_n) = H(p_1, 1 - p_1) + (1 - p_1) H'(p_2' \ldots p_n') \tag{B37}$$

and so, maximizing H with p_1 held fixed is equivalent to maximizing H'. We have the reduced maximum entropy problem:

maximize H' subject to

$$\sum_{i=2}^{n} p_i' = 1 \qquad \text{(B38)}$$

$$\langle i \rangle' = \sum_{i=2}^{n} i p_i' = 1 + \frac{\langle i \rangle - 1}{1 - p_1} \quad . \qquad \text{(B39)}$$

The solution proceeds as before, but now the maximum attainable entropy is a function $H_{max} = S(p_1, \langle i \rangle)$ of the specified value of p_1, as well as $\langle i \rangle$. The maximum of $S(p_1, 4.5)$ is of course the previous value (B14) of 1.61358, attained at the maximum-entropy value $p_1 = 0.05435$. Evaluating this also for Rowlinson's p_1', I find $S(p_1', 4.5) = 1.55716$, lower by 0.05642 units. By (B30) this means that, in 1000 tosses, for every way in which Rowlinson's value could be realized, regardless of all other frequencies except for the constraint $\langle i \rangle = 4.5$, there are over 10^{24} ways in which the maximum-entropy frequency could be realized.

We may give a more detailed answer: expanding $S(p_1, 4.5)$ about its peak, we find that as we depart from 0.05435, the number of ways in which the frequency g_1 could be realized drops off like

$$\exp[-14{,}200(g_1 - 0.05435)^2] \qquad \text{(B40)}$$

and so, for example, for 99% of all possible ways in which the average $\langle i \rangle = 4.5$ can be realized, g_1 lies in the interval (0.05435 ± 0.0153).

This would seem to be an adequate answer to the question, "But what basis is there for trusting in this number?" I stress that the numerical results just given are theorems, involving only a straightforward counting of the possibilities allowed by the given data. Therefore they stand independently of anybody's personal opinions about either dice or probability theory.

However, it is necessary that we understand very clearly the meaning of these frequency connections. They concern only the number of possible ways in which certain frequencies $\{g_i\}$ could be realized, compatible with our constraints. They do not assert that the maximum-entropy frequencies will be observed in a real experiment; indeed, neither the Principle of Maximum Entropy nor any other principle of probability theory can predict with certainty what will happen in a real experiment. The correct statement is rather: the frequency distribution $\{g_i\}$ with maximum entropy calculated from certain constraints is overwhelmingly the most likely one to be observed in a real experiment, provided that the physical constraints operative in the experiment are the same as those assumed in the calculation.

In our mathematical formalism, a "constraint" is some piece of information that leads us to modify a probability distribution; in the case of a mean value constraint, by inserting an exponential factor $\exp[-\lambda f(x)]$ with an adjustable Lagrange multiplier λ. It is perhaps not yet clear just what we mean by "constraints" in a physical experiment. Of course, by these we do not mean the gross constraining linkages by levers, cables, and gears of a mechanics textbook, but something more subtle. In our applications, a "physical constraint" is any physical influence that exerts a systematic tendency--however slight--on the outcome of an experiment. We give some specific examples of physical constraints in die tossing below.

From the above numbers we can understand the success of the work of J. C. Keck and R. D. Levine reported here. I am sure that their results must seem like pure magic to those who have not understood the maximum-entropy formalism. To find a distribution of populations over 20 molecular energy levels might seem to require 19 independent pieces of data. But if one knows, from approximate rate coefficients or from past experience, which constraints exist (in practice, even if only the one or two most important ones are taken into account), one can make quite confident predictions of distributions over many levels simply by maximizing the entropy.

In fact, most frequency distributions produced in real experiments are maximum-entropy distributions, simply because these can be realized in so many more ways than can any other. As $N \to \infty$, the combinatorial factors become so sharply peaked at the maximum entropy point that to produce any appreciably different distribution would require very effective physical constraints. Any statistically significant departure from a maximum-entropy prediction then constitutes strong--and if it persists, conclusive--evidence of the existence of new constraints that were not taken into account in the calculation. Thus the maximum-entropy formalism has the further "magical" property that it provides the most efficient procedure by which, if unknown constraints exist, they can be discovered. But this is only an updated version of the process noted in Section A by which Laplace discovered new systematic effects.

It is, perhaps, sufficiently clear from this how much a Physical Chemist has to gain by understanding, rather than attacking, maximum entropy methods.

But we still have not dealt with the most fundamental misunderstandings in the Rowlinson article. He turns next to the shape of the maximum-entropy distribution (B13), with another rhetorical question: "--- is there anything in the mechanics of throwing dice which suggests that if a die is not true the probabilities of scores 1,2,...6, should form the geometrical progression [our Eq. (B13)]?" He then cites some data of Wolf on 20,000 throws of a die which gave an average $\langle i \rangle = 3.5983$, plots the observed frequencies against the maximum-entropy distribution based on that constraint, and concludes that "departures from the random value of 1/6 bear no resemblance to those calculated from the rule of maximum entropy. What is clearly wrong with the indiscriminate use of this rule, and of the older rules from which it stems, is that they ignore the physics of the problem."

We have here a total, absolute misconception about every point I have been trying to explain above. If Wolf's data depart significantly from the maximum-entropy distribution based only on the constraint $\langle i \rangle = 3.5983$, then the proper conclusion is not that maximum entropy methods "ignore the physics" but rather that the maximum entropy method *brings out* the physics by showing us that another physical constraint exists beyond that used in the calculation. Unable to see the new physical information here revealed, he lashes out blindly against the principle that has revealed it.

Therefore, let us now give an analysis of Wolf's dice data showing just what things maximum entropy can give us here, if we only open our eyes to them.

Wolf's Dice Data. In the period roughly 1850-1890, the Zurich astronomer R. Wolf conducted and reported a mass of "random experiments." An account is given by Czuber (1908). Our present concern is with a particular die (identified as "Weiszer Würfel" in Czuber's two-way table, *loc. cit* p. 149) that was tossed 20,000 times and yielded the aforementioned mean value $\langle i \rangle = 3.5983$. We shall look at all details of the data presently, but first let us note a few elementary things about that "ignored" physics.

We all feel intuitively that a perfectly symmetrical die, fairly tossed, ought to show all faces equally often (but that statement is really circular, since there is no other way to define a "fair" method of tossing; so, suppose that by experimenting on a die known to be true, we have found such a fair method, and we continue to use it). The uniform frequency distribution $\{g_i = 1/6,\ 1 < i < 6\}$ then represents the nominal "unconstrained" situation of maximum possible entropy $S = \log 6$. Any imperfection in the die may then give rise to a "physical constraint" as we have defined that term. A little physical common sense can anticipate what these imperfections are likely to be.

The most obvious imperfection is that different faces have different numbers of spots. This affects the center of gravity, because the weight of ivory removed from a spot is obviously not (in any die I have seen) compensated by the paint then applied. Now the numbers of spots on opposite faces add up to seven. Thus the center of gravity is moved toward the "3" face, away from "4", by a small distance ε corresponding to the one spot discrepancy. The effect of this must be a slight frequency difference which is surely, for very small ε, proportional to ε:

$$g_4 - g_3 = \alpha\varepsilon \tag{B41}$$

where the coefficient α would be very difficult to calculate, but could be measured by experiments on dies with known ε. But the (2-5) face direction has a discrepancy of three spots, and (1-6) of five. Therefore, we anticipate the ratios:

$$(g_4 - g_3):(g_5 - g_2):(g_6 - g_1) = 1:3:5 \quad . \tag{B42}$$

But this says only that the spot frequencies vary linearly with i:

$$g_i = \frac{1}{6} + \alpha\varepsilon\, f_1(i) \quad , \qquad i \le i \le 6 \tag{B43}$$

where

$$f_1(i) \equiv (i - 3.5) \quad . \tag{B44}$$

The spot imperfections should then lead to a small linear skewing favoring the "6." This is the most obvious "physical constraint," and it changes the expected number of spots to

$$\langle i\rangle = \sum i\, g_i = 3.5 + 17.5\,\alpha\varepsilon \tag{B45}$$

or, to state it more suggestively, the function $f_1(i)$ acquires a non-zero expectation

$$\langle f_1\rangle = 17.5\,\alpha\varepsilon \quad . \tag{B46}$$

Now, what is the next most obvious imperfection to be expected? Evidently, it will involve departure from a perfect cube, the specific kind depending on the manufacturing methods; but let us consider only the highest quality die that a factory would be likely to make. If you were assigned the job of making a perfect cube of ivory, how would you do it with equipment likely to be available in a Physics Department shop or a small factory?

I think you would head for the milling machine, and mount your lump of ivory on a dividing head clamped to the work table with axis vertical. The first cut would be with an end mill, making the "top" face of the die. The construction of the machine guarantees that this will be accurately plane. Then you use side cutters to make the four side faces. For the finish cuts you will move the work table only in the direction of the cut, rotating the dividing head 90° from one face to the next. The accuracy of the equipment guarantees that you now have five of the faces of your cube, all very accurately plane and all angles accurately 90°, the top face accurately square.

But now the trouble begins; to make the final "bottom" face you have to remove the work from its mount, place it upside down on the table, and go over it with the end mill. Again, the construction of the machine guarantees that this final face will be accurately plane and parallel to the "top;" but it will be practically impossible to adjust the work table height so accurately that the final dimension is exactly equal to the other two. Of course, a skilled artisan with a great deal more time and equipment could do better; but this would run up the cost of manufacture for something that would never be detected in use. For factory production, there would be no motivation to do better than we have described.

Thus, the most likely geometrical imperfection in a high quality die is not lack of parallelism or of true 90° angles, but rather that one dimension will be slightly different from the other two.

Again, it is clear what kind of effect this will have on frequencies. Suppose the die comes out slightly "oblate," the (1-6) dimension being shorter than the (2-5) and (3-4) by some small amount δ. If the die were otherwise perfect, this would evidently increase the frequencies g_1, g_6 by some small amount $\beta\delta$, and decrease the other four to keep the sum equal to unity, where β is another coefficient hard to calculate but measurable. The result can be stated thus: the function

$$f_3(i) \equiv \begin{Bmatrix} +2, & i = 1,6 \\ -1, & i = 2,3,4,5 \end{Bmatrix} \tag{B47}$$

defined on the sample space, acquires a non-zero expectation

$$\langle f_3 \rangle = 6\beta\delta \tag{B48}$$

and the frequencies are

$$\begin{aligned} g_i &= \frac{1}{6} + \frac{1}{2}\beta\delta\, f_3(i) \\ &= \frac{1}{6}[1 + 3\beta\delta\, f_3(i)] \quad . \end{aligned} \tag{B49}$$

If now both imperfections are present, since the perturbations are so small we can in first approximation just superpose their effects:

$$g_i \simeq \frac{1}{6}[1 + 6\alpha\varepsilon\ f_1(i)][1 + 3\beta\delta\ f_3(i)] \quad . \tag{B50}$$

But this is hardly different from

$$g_i = \frac{1}{6} \exp[6\alpha\varepsilon\ f_1(i) + 3\beta\delta\ f_3(i)] \tag{B51}$$

and so a few elementary physical common-sense arguments have led us to something which begins to look familiar.

If we had done maximum entropy using the constraints (B46), (B48), we would find a distribution proportional to $\exp[-\lambda_1 f_1(i) - \lambda_3 f_3(i)]$, so that (B51) is a maximum-entropy distribution based on those constraints. We see that the Lagrange multiplier by which any information constraint is coupled into our probability distribution, is just a measure of the strength of the physical constraint required to realize a numerically equal frequency distribution:

$$\lambda_1 = -6\alpha\varepsilon \tag{B52}$$

$$\lambda_3 = -3\beta\delta \tag{B53}$$

and if our die has no other imperfections beyond the two noted, then it is overwhelmingly more likely to produce the distribution (B51) than any other.

If the observed frequencies show any statistically significant departure from (B51), then we have extracted from the data evidence of a third imperfection, which probably would have been totally invisible in the raw data; i.e., only when we have used the maximum entropy principle to "subtract off" the effect of the stronger influences, can we hope to detect a weaker one.

Our program for the maximum-entropy analysis of the die--or any other random experiment--is now defined except for the final step; how we decide whether a discrepancy is "statistically significant?"

The reader is cautioned that in all this discussion relating to Rowlinson we are being careless about distinctions between probability and frequency, because Rowlinson himself makes no distinction between them, and trying to correct this at every point quickly became tedious. The following analysis should be restated much more carefully to bring out the fact that it is only a very special case, although to the "frequentist" it appears to be the general case.

We have some "null hypothesis" H_o about our die, that leads us to assign the probabilities $\{p_1 \dots p_6\}$. We obtain data from N tosses, in which the observed frequencies are $\{g_i = N_i/N,\ 1 \le i \le 6\}$. If the numbers $\{g_1 \dots g_6\}$ are sufficiently close to $\{p_1 \dots p_6\}$ we shall say the fit is satisfactory; the null hypothesis is consistent with our data, and so there is no need, as far as this experiment indicates, to seek a better hypothesis. But how close is "close?" How do we measure the "distance" between the two distributions; and how large may that distance be before we begin to doubt the null hypothesis?

Early in this Century, Karl Pearson invented an intuitive, ad hoc procedure, called the Chi-squared test, to deal with this problem, which has been since widely adopted. Here we calculate the quantity

$$\chi^2 = N \sum_{i=1}^{6} \frac{(g_i - p_i)^2}{p_i} \qquad \text{(B54)}$$

and if it is greater than a certain "critical value" given in Tables, we reject the null hypothesis. In the present case (six categories, five "degrees of freedom" after normalization), the critical value at the conventional 5% significance level is

$$\chi_c^2 = 11.07 \qquad \text{(B55)}$$

which means that, if the null hypothesis is true there is only a 5% chance of seeing a value greater than χ_c^2. The critical value is independent of N, because for a frequentist who believes that p_i is an assertion of a limiting frequency in the sense of the de Moivre-Laplace limit theorem (A4), if H_o is true, then the deviations should fall off as $|g_i - p_i| = O(N^{-\frac{1}{2}})$. A more careful approach shows that this holds only if our model is an exchangeable sequence with zero correlations; and even in this case the χ^2 criterion of "closeness" has no theoretical justification (i.e., no uniqueness property) in the basic principles of probability theory.

In fact, for the case of independent exchangeable trials, there is a criterion with a direct information-theory justification (Kullback, 1959) in the "minimum discrimination information statistic"

$$\psi \equiv N \sum_{i=1}^{6} g_i \log(g_i/p_i) \qquad \text{(B56)}$$

and the numerical value of ψ, rather than χ^2, will lead us to inferences directly justifiable by Bayes' theorem. If the deviations $(g_i - p_i)$ are large, these criteria can be very different.

However, by a lucky mathematical accident, if the deviations are small (as we already know them to be for our dice problem) an expansion in powers of $(g_i - p_i)$ [in the logarithm, write $g/p = 1 + (g-p)/g + (g-p)^2/gp$] yields

$$\psi = \frac{1}{2}\chi^2 + o\left(N^{-\frac{1}{2}}\right) \tag{B57}$$

the neglected terms falling off as indicated, provided that $|g_i - p_i| = O\left(N^{-\frac{1}{2}}\right)$. The result is that in our problem, from a pragmatic standpoint it doesn't matter whether we use χ^2 or ψ. So I shall apply the χ^2 test to Wolf's data, because it is so much more familiar to most people.

Wolf's empirical frequencies $\{g_i\}$ are given in the second column of Table 1. As a first orientation, let us test them against the null hypothesis $\{H_o : p_i = 1/6,\ 1 < i < 6\}$ of a uniform die. We find the result

$$\chi_o^2 = 271 \tag{B58}$$

over twenty times the critical value (B55). The hypothesis H_o is decisively rejected.

Next, let us follow Rowlinson by considering a new hypothesis H_1 which prescribes the maximum-entropy solution based on Wolf's average $<i> = 3.5983$, or,

$$<f_1(i)> = 0.0983 \quad . \tag{B59}$$

This will give us a distribution $p_i \sim \exp[-\lambda f_1(i)]$. From the partition function (B11) with this new datum we find $\lambda = 0.03373$ and the probabilities given in the third column of Table 1. The fourth column gives the differences $\Delta_i = g_i - p_i$, while in the fifth we list the partial contributions to Chi-squared:

$$c_i \equiv 20{,}000 \frac{(g_i - p_i)^2}{p_i}$$

which add up to the value

$$\chi_1^2 = 199.4 \quad . \tag{B60}$$

The fit is improved only slightly; and H_1 is also decisively rejected.

Table 1. One Constraint

i	g_i	p_i	Δ_i	c_i
1	0.16230	0.15294	+ 0.0094	11.46
2	0.17245	0.15818	+ 0.0143	25.75
3	0.14485	0.16361	- 0.0188	43.02
4	0.14205	0.16922	- 0.0272	87.25
5	0.18175	0.17502	+ 0.0067	5.18
6	0.19660	0.18103	+ 0.0156	26.78
				199.43

At this point, Rowlinson wants to reject not only H_1, but also the whole principle of maximum entropy. But now I stress still another time what the principle is really telling us: a statistically significant deviation is evidence of a new physical constraint; and the nature of the deviation gives us a clue as to what that constraint is. After subtracting off, by maximum entropy, the deviation attributable to the first constraint, the nature of the most important remaining one is revealed. Indeed, from a glance at the deviations $\Delta_i = g_i - p_i$ the answer leaps out at us; Wolf's die was slightly "prolate," the (3-4) dimension being greater than the (2-5) and (1-6) ones. So, instead of (B47), the new constraint is

$$f_2(i) \equiv \begin{cases} +1, & i = 1,2,5,6 \\ -2, & i = 3,4 \end{cases} \tag{B61}$$

and Wolf's data yield the result

$$\langle f_2 \rangle = 0.1393 \; . \tag{B62}$$

So now let us subtract off, by maximum entropy, the effect of both of these constraints; and thus discover whether Wolf's die had a third imperfection.

With the two constraints (B59), (B62) we have two Lagrange multipliers and a partition function

$$Z(\lambda_1,\lambda_2) = \sum_{i=1}^{6} \exp\left[-\lambda_1 f_1(i) - \lambda_2 f_2(i)\right]$$

$$= x^{-5/2}\, y(1+x)(1+x^4+x^2y^{-3}) \tag{B63}$$

where $x \equiv \exp(-\lambda_1)$, $y \equiv \exp(-\lambda_2)$. The maximum-entropy probabilities are then

$$\{p_1 \cdots p_6\} = Z^{-1}\, x^{-5/2}\, y\left\{1, x, x^2y^{-3}, x^3y^{-3}, x^4, x^5\right\} \tag{B64}$$

Writing out the constraint equations (B4) and eliminating y from them, we find that x is determined by

$$(6F_1 - 4F_2 - 11)x^5 + (6F_1 - 4F_2 - 5)x^4 + (6F_1 + 4F_2 + 5)x + (6F_1 + 4F_2 + 11) = 0 \tag{B65}$$

or, with Wolf's numerical values (B59), (B62),

$$5.4837x^5 + 2.4837x^4 - 3.0735x - 6.0735 = 0 \tag{B66}$$

This has only one real root, at $x = 1.032233$, from which we have $\lambda_1 = -0.0317244$, $y = 1.074415$, $\lambda_2 = -0.0717764$. The new maximum-entropy probabilities are given in Table 2, which contains the same information as Table 1, but for the new hypothesis H_2.

Table 2. Two Constraints

i	g_i	p_i	Δ_i	c_i
1	0.16230	0.16433	- 0.0020	0.502
2	0.17245	0.16963	+ 0.0028	0.938
3	0.14485	0.14117	+ 0.0037	1.919
4	0.14205	0.14573	- 0.0037	1.859
5	0.18175	0.18656	- 0.0048	2.480
6	0.19660	0.19258	+ 0.0040	1.678
				9.375

We see that the second constraint has greatly improved the fit. Chi-squared has been reduced to

$$\chi_2^2 = 9.375 \quad . \tag{B67}$$

This is less than the critical value 11.07, so there is now no statistically significant evidence for any further imperfections; i.e., if the given p_i were the "exact" values, it is reasonably likely that the distribution g_i would deviate from p_i by the observed amount, by chance alone. Or, to put it in a way perhaps more appropriate to this problem, if the die were tossed another 20,000 times, we would not expect the frequencies g_i to be repeated exactly; the new frequencies g_i', might reasonably be expected to deviate from the first set g_i by about as much as the distributions g_i, p_i differ.

That this is reasonable can be seen directly without calculating Chi-squared. For if the result i is obtained n_i times in N tosses, we might expect this to fluctuate in successive repetitions of the whole experiment by about $\pm\sqrt{n_i}$. Thus the observed frequencies $g_i = n_i/N$ should fluctuate by about $\Delta g_i = \pm\sqrt{n_i}/N$; for $g_i = 1/6$, $N = 20{,}000$, this gives $\Delta g_i \simeq 0.0029$. But this is just of the order of the observed deviations Δ_i. Therefore, it would be futile to search the $\{\Delta_i\}$ of Table 2 for a third imperfection. Not only does their distribution fail to suggest any simple hypothesis; if the die were tossed another 20,000 times, in all probability the new Δ_i' would be entirely different. With our two-parameter hypothesis H_2 we are down "in the noise" of random variations, and any further systematic influences are too small to be seen unless we go up to a million tosses, by which time the die will be changed anyway by wear.

A technical point might be raised by Statisticians: "You have estimated two parameters λ_1, λ_2 from the data; therefore you should use the test for three degrees of freedom rather than five." This reduction is appropriate if the parameters are chosen by the criterion of minimizing χ^2. That is, if we choose them for the express purpose of making χ^2 small and still fail to do so, it does not speak well for the hypothesis and a penalty is in order. But our parameters were chosen by a criterion that took no note of χ^2; and therefore the proper question is only; "How well does the result fit the data?" and not: "How did you find the parameters?" Had we chosen our parameters to minimize χ^2, we would have found a still lower value; but one that is not relevant to the point being made here, which is the performance of the maximum entropy criterion, as advocated long before this die problem was thought of.

The maximum entropy method with two Lagrange multipliers thus successfully determines a distribution with five independent quantities. The "ensemble" canonical with respect to the constraints $f_1(i)$, $f_2(i)$ describing the two imperfections that common sense leads us to expect in a die, agrees with Wolf's data about as well as can be hoped for in a statistical problem.

It was stressed above that in this theory the connections between probability and frequency are loosened and we noted, in the discussion following (B40), that the connections remaining are now theorems rather than conjectures. As we now see, they are not loosened enough to hamper us in dealing with real random experiments. If we had been given only the two constraints (B59), (B62) we could have reproduced, by maximum entropy, all of Wolf's frequency data.

This is an interesting caricature of the results of Keck and Levine, and shows again how much our critics would gain by understanding, rather than attacking, this principle. Far from "ignoring the physics," it leads us to concentrate our attention on the part of the physics that is relevant. Success in using it does not require that we take into account all dynamical details; it is enough if we can recognize, whether by common-sense analysis or by inspection of data, what are the systematic influences at work, that represent the "physical constraints?" If by any means we can recognize these, maximum entropy then takes over and supplies the rest of the solution, which does not depend on dynamical details but only on counting the possibilities.

In effect, then, by subtracting off the systematic effects we reduce the problem to Bernoulli's "equally possible" cases; the deviations Δ_i from the canonical distribution that remain in Table 2 are the same as the deviations from $p_i = 1/6$ that we would expect if the die had no imperfections.

Success of our predictions is not guaranteed in advance, as Rowlinson supposed it should be when he wanted to reject the entire principle at the stage of Table 1. But this supposition merely reflects his rejection, at the very outset, of the distinction between probability and frequency that I keep stressing. If one is not moved by theoretical arguments for that distinction, we now see a pragmatic reason for it. The probabilities p_i in Table 1 are an entirely correct description of our state of knowledge about a single toss, when we know about only the constraint $f_1(i)$. It is a theorem that they are also numerically equal to the frequencies which could happen in the greatest number of ways if no other physical constraint existed. But our probabilities will agree with measured frequencies only when we have recognized and put into our equations the constraints representing all the systematic influences at work in the real experiment.

This, I submit, is exactly as it should be in a statistical theory; at no point are we ever justified in claiming that our predictions must be right; only that, in order to make any better ones we should need more information than was given. It is when a theory purports to do more than this (by failing to recognize the distinction between probability and frequency) that it may be charged with promising us something for nothing.

Since the fit is now satisfactory, the above values of λ_1, λ_2 give us the numerical values of the systematic influences in Wolf's experiment: from (B52), (B53) we have

$$\alpha\varepsilon = \frac{0.03172}{6} = 0.0053 , \tag{B68}$$

$$\beta\delta = \frac{0.07178}{3} = 0.024 . \tag{B69}$$

So, if today some enterprising person at Monte Carlo or Las Vegas will undertake to measure for us the coefficients α, β, then we can determine--100 years after the fact--just how far (in terms of its nominal dimensions) the center of gravity of Wolf's die was displaced (presumably by excavation of the spots), and how much longer it was in the (3-4) direction than in the (2-5) or (1-6). We can also certify that it had no other significant imperfections (at least, none that affected its frequencies). Note, however, that α, β are not, strictly speaking, physical constants only of the die; a little further common-sense reasoning makes it clear that they must depend also on how the die was tossed; for example, tossing it with a large angular momentum about a (3-4) axis will decrease the effect of the $f_1(i)$ constraint, while if it spins about the (1-6) axis the effect of $f_2(i)$ will be less; and with a (2-5) spin axis both constraints will be weakened.

Indeed, as soon as the die is unsymmetrical, all sorts of physical conditions that were irrelevant for a perfectly symmetrical one, become relevant. The frequencies will surely depend not only on its center of gravity but also on all the second moments of its mass distribution, the sharpness of its edges, the smoothness, elasticity, and coefficient of friction of the table, etc.

However, we conjecture that α, β depend very little on these factors within the small range of conditions usually employed (i.e., small angular momentum in tossing, etc.); and suspect that in that range the coefficient α is already well known to those who deal with loaded dice.

I really must thank Rowlinson for giving us (albeit inintentionally) such a magnificent test case by which the nature and power of the Principle of Maximum Entropy can be demonstrated, in a context entirely removed from the conceptual problems of quantum theory. And indeed, all the criticisms he made were richly deserved; for he was not, after all, criticizing the Principle of Maximum Entropy; only a gross misunderstanding of it. Rowlinson's criticisms were, however, taken up and extended by Lindhard (1974); in view of the long commentary above we may leave it as an exercise for the reader to deal with his arguments.

The Constraint Rule. There is a further point of logic about our use of maximum entropy that has troubled some who are able to see the distinction between probability and frequency. In imposing the mean-value constraint (B1) we are simply appropriating a sample average obtained from N measurements that yielded f_j on the j'th observation:

$$F = \overline{f} = \frac{1}{N}\sum_{j=1}^{N} f_j \qquad \text{(B70)}$$

and equating it to a probability average

$$\langle f \rangle = \sum_{i=1}^{n} p_i \, f(x_i) \quad . \tag{B71}$$

Is there not an element of arbitrariness about this? A cynic might say that after all these exhortations about the distinction between probability and frequency, we proceed to confuse them after all, by using the word "average" in two quite different senses.

Our rule can be justified in more than one way; in Section D below we argue in terms of what it means to say that certain information is "contained" in a probability distribution. Let us ask now whether the constraint rule (B1) is consistent with, or derivable from, the usual principles of Bayesian inference.

If we decide to use maximum entropy based on expectations of certain specified functions $\{f_1(x) \dots f_m(x)\}$, then we know in advance that our final distribution will have the mathematical form

$$p(x_i|H) = \frac{1}{Z(\lambda_1 \dots \lambda_m)} \exp\left[-\lambda_1 \, f_1(x_i) \, \dots - \lambda_m \, f_m(x_i)\right] \tag{B72}$$

and nothing prevents us from thinking of this as defining a class of sampling distributions parameterized by the Lagrange multipliers λ_k, the parameter space consisting of all values of $\{\lambda_1 \dots \lambda_m\}$ which lead to normalizable distributions (B72). Choosing a specific distribution from this class is then equivalent to making an estimate of the parameters λ_k. But parameter estimation is a standard problem of statistical inference.

The class C of hypothesis being considered is thus specified; any particular choice of the $\{\lambda_1 \dots \lambda_m\}$ may be regarded as defining a particular hypothesis $H \varepsilon C$. However, the class C does not determine any particular choice of the functions $\{f_1(x), \dots, f_m(x)\}$. For, if A is any nonsingular $(m \times m)$ matrix, we can carry out a linear transformation

$$\sum_{k=1}^{m} \lambda_k \, f_k(x) = \sum_{j=1}^{m} \lambda_j^* \, f_j^*(x) \tag{B73}$$

where

$$\lambda_j^* \equiv \sum_k \lambda_k \, A_{kj} \tag{B74a}$$

$$f_j^*(x) \equiv \sum_k (A^{-1})_{jk} \, f_k(x) \tag{B74b}$$

and the class of distributions (B72) can be written equally well as

$$P(x_i|H) = \frac{1}{Z(\lambda_1^* \ldots \lambda_m^*)} \exp\left\{-\lambda_1^* f_1^*(x_i) - \ldots - \lambda_m^* f_m^*(x_i)\right\}. \qquad \text{(B75)}$$

As the $\{\lambda_1^* \ldots \lambda_m^*\}$ vary over their range, we generate exactly the same family of probability distributions as (B72). The class C is therefore characteristic, not of any particular choice of the $\{f_1(x) \ldots f_m(x)\}$, but of the linear manifold M(C) spanned by them.

If the $f_k(x)$ are linearly independent, the manifold M(C) has dimensionality m. Otherwise, M(C) is of some lower dimensionality $m' < m$; the set of functions $\{f_1(x) \ldots f_m(x)\}$ is then redundant, in the sense that at least one of them could be removed without changing the class C. While the presence of redundant functions $f_k(x)$ proves to be harmless in that it does not affect the actual results of entropy maximization (Jaynes, 1968), it is a nuisance for present purposes [Eq. (B81) below]. In the following we assume that any redundant functions have been removed, so that $m' = m$.

Suppose now that x_i is the result of some random experiment that has been repeated r times, and we have obtained the data

$$D \equiv \{x_1 \text{ true } r_1 \text{ times}, x_2 \text{ true } r_2 \text{ times}, \ldots, x_n \text{ true } r_n \text{ times}\}. \qquad \text{(B76)}$$

Of course, $\Sigma r_i = r$. Out of all hypotheses $H \varepsilon C$, which is most strongly supported by the data D according to the Bayesian, or likelihood, criterion? To answer this, choose any particular hypothesis $H_o \equiv \{\lambda_1^{(0)} \ldots \lambda_m^{(0)}\}$ as the "null hypothesis" and test it against any other hypothesis $H \equiv \{\lambda_1 \ldots \lambda_m\}$ in C by Bayes' theorem. The log-likelihood ratio in favor of H over H_o is

$$L \equiv \log \frac{P(D|H)}{P(D|H_o)} = \sum_{i=1}^{n} r_i \log\left[p_i/p_i^{(0)}\right]$$

$$= r\left[\log(Z_o/Z) + \sum_{k=1}^{m} \left(\lambda_k^{(0)} - \lambda_k\right)\overline{f}_k\right] \qquad \text{(B77)}$$

where

$$\overline{f}_k \equiv \frac{1}{r} \sum_{i=1}^{n} r_i f_k(x_i) \qquad \text{(B78)}$$

is the measured average of $f_k(x)$, as found in the experiment. Out of all hypotheses in class C the one most strongly supported by the data D is the one for which the first variation vanishes:

$$\delta L = -r \sum_{k=1}^{m} \left[\frac{\partial}{\partial \lambda_k} \log Z + \overline{f}_k \right] \delta\lambda_k = 0 \quad . \tag{B79}$$

But from (B4), this yields just our constraint rule (B1):

$$\{\langle f_k \rangle = \overline{f}_k \ , \qquad 1 \le k \le m\} \quad . \tag{B80}$$

To show that this yields a true maximum, form the second variation and note that the covariance matrix

$$\frac{\partial^2 \log Z}{\partial \lambda_j \partial \lambda_k} = \langle f_j f_k \rangle - \langle f_j \rangle \langle f_k \rangle \tag{B81}$$

is positive definite almost everywhere on the parameter space if the $f_k(x)$ are linearly independent.

Evidently, this result is invariant under the aforementioned linear transformations (B74); i.e., we shall be led to the same final distribution satisfying (B80) however the $f_k(x)$ are defined. Therefore, we can state our conclusion as follows:

> Out of all hypotheses in class C, the data D support most strongly that one for which the expectation $\langle f(x) \rangle$ is equal to the measured average $\overline{f(x)}$ for every function f(x) in the linear manifold M(C). (B82)

This appears to the writer as a rather complete answer to some objections that have been raised to the constraint rule. We are not, after all, confusing two averages; it is a derivable consequence of probability theory that we <u>should</u> set them equal. Maximizing the entropy subject to the constraints (B80), is equivalent to (i.e., it leads to the same result as) maximizing the likelihood over the manifold of sampling distributions picked out by maximum entropy.

<u>Forney's Question.</u> An interesting question related to this was put to me by G. David Forney in 1963. The procedure (B1) uses only the <u>numerical value</u> of F, and it seems to make no difference whether this was a measured average over 20 observations, or 20,000. Yet there is surely a difference in our state of knowledge--our degree of confidence in the accuracy of F--that depends on N. The maximum-entropy method seems to ignore this. Shouldn't our final distribution depend on N as well as F?

It is better to answer a question 15 years late than not at all. We can do this on both the philosophical and the technical level. Philosophically, we are back to the question: "What is the specific problem being solved?"

In the problem I am considering F is simply a number given to us in the statement of the problem. *Within the context of that problem*, F is exact by definition and it makes no difference how it was obtained. It might, for example, be only the guess of an idiot, and not obtained from any measurement at all. Nevertheless, that is the number given to us, and our job is not to question it, but to do the best we can with it.

This may seem like an inflexible, cavalier attitude; I am convinced that nothing short of it can ever remove the ambiguity of *"What is the problem?"* that has plagued probability theory for two centuries.

Just as Rowlinson was impelled to invent an Urn Model that was not specified in the statement of the problem, you and I might, in some cases, feel the urge to put more structure into this problem than I have used. Indeed, we demand the right to do this. But then, let us recognize that we are considering a *different* problem than pure "classical" maximum entropy; and it becomes a technical question, not a philosophical one, whether with some new model structure we shall get different results. Clearly, the answer must be sometimes yes, sometimes no, depending on the specific model structure assumed. But it turns out that the answer is "no" far more often than one might have expected.

Perhaps the first thought that comes to one's mind is that any uncertainty as to the value of F ought to be allowed for by averaging the maximum-entropy distribution $p_i(F)$ over the possible values of F. But the maximum-entropy distribution is, by construction, already as "uncertain" as it can get for the stated mean value. Any averaging can only result in a distribution with still higher entropy, which will therefore necessarily violate the mean value number given to us. This hardly seems to take us in the direction wanted; i.e., we are already up against the wall from having maximized the entropy in the first place.

But such averaging was only an *ad hoc* suggestion; and in fact the Principle of Maximum Entropy already provides the proper means by which any testable information can be built into our probability assignments. If we wish only to incorporate information about the accuracy with which f is known, no new model structure is needed; the way to do this is to impose another constraint. In addition to $\langle f \rangle$ we may specify $\langle f^2 \rangle$; or indeed, any number of moments $\langle f^n \rangle$ or more general functions $\langle h(f) \rangle$. Each such constraint will be accompanied by its Lagrange multiplier λ, and the general maximum-entropy formalism already allows for this.

Of course, whenever information of this kind is available it should in principle be taken into account in this way. I would "hold it to be self-evident" that for any problem of inference, the ideal toward which we should aim is that all the relevant information we have ought to be incorporated explicitly into our equations; while at the same time, "objectivity" requires that we carefully avoid assuming any information that we do not possess. The Principle of Maximum Entropy, like Ockham, tells us to refrain from inventing Urn Models when we have no Urn.

But in practice, some kinds of information prove to be far more relevant than others, and this extra information about the accuracy of F usually affects our actual conclusions so little that it is hardly worth the effort. This is particularly true in statistical mechanics, due to the enormously high dimensionality of the phase space. Here the effect of specifying any reasonable accuracy in F is usually completely negligible. However, there are occasional exceptions; and whenever this extra information does make an appreciable difference it would, of course, be wrong to ignore it.

C. Speculations for the Future

The field of statistical Inference--in or out of Physics--is so wide that there is no hope of guessing every area in which new advances might be made. But we can indicate a few areas where progress may be predicted rather safely because it is already underway, with useful results being found at a rate proportional to the amount of effort invested.

Current progress is taking place at several different levels:

I Application of existing techniques to existing problems

II Extension of present theory to new problems.

III More powerful mathematical methods.

IV Further development of the basic theory of inference.

However, I shall concentrate on I and IV, because II is so enveloped in fog that nothing can be seen clearly, and III seems to be rather stagnant except for development of new specialized computer techniques, which I am not competent even to describe, much less predict.

There are important current areas that seem rather desperately in need of the same kind of house-cleaning that statistical mechanics has received. What they all have in common is: (a) long-standing problems, still unsolved after decades of mathematical efforts, (b) domination by a mental outlook that leads one to concentrate all attention on the analogs of ergodic theory. That is, in the belief that a probability is not respectable unless it is also a frequency, one attempts a direct calculation of frequencies, or tries to guess the right "statistical assumption" about frequencies, even though the available information does not consist of frequencies, but consists rather of partial knowledge of certain "macroscopic" parameters

$\{\alpha_i\}$; and the predictions desired are not frequencies, but estimates of certain other parameters $\{\theta_i\}$. It is not yet realized that, by looking at the problems this way one is not making efficient use of probability theory; by restricting its meaning one is denying himself nearly all its real power.

The real problem is not to determine frequencies, but to describe one's state of knowledge by a probability distribution. If one does this correctly, he will find that whatever frequency connections are relevant will appear automatically, not as "statistical assumptions" but as mathematical consequences of probability theory.

Examples are the theory of hydrodynamic turbulence, optical coherence, quantum field theory, and surprisingly, communication theory which after thirty years has hardly progressed beyond the stage of theorems which presuppose all the ten-gram frequencies known in advance.

In early 1978 I attended a Seminar talk by one of the current experts on turbulence theory. He noted that the basic theory is in a quandary because "Nobody knows what statistical assumptions to make." Yet the objectives of turbulence theory are such things as: given the density, compressibility, and viscosity of a fluid, predict the conditions for onset of turbulence, the pressure difference required to maintain turbulent flow, the rate of heat transfer in a turbulent fluid, the distortion and scattering of sound waves in a turbulent medium, the forces exerted on a body in the fluid, etc. Even if one's objective were only to predict some frequencies g_i related to turbulence, statements about the best estimate of g_i and the reliability of that estimate, can only be derived from probabilities that are not themselves frequencies.

We indicated a little of this above [Equations (B15)-(B23)]; now let us see in a more realistic case why the frequencies with which various things happen in a time-dependent process are not the same as their probabilities; but that, nevertheless, there are always definite connections between probability and frequency, derivable as consequences of probability theory.

Fluctuations. Consider some physical quantity f(t). What follows will generalize at once to field quantities f(x,t); but to make the present point it is sufficient to consider only time variations. Therefore, we may think of f(t) as the net force exerted on an area A by some pressure P(x,t):

$$f(t) = \int_A P(x,t)\, dA \tag{C1}$$

or the net force in the x-direction exerted by an electric field on a charge distributed with density $\rho(x)$: $f(t) = \int E_x(x,t)\rho(x)\,d^3x$

or as the total magnetic flux passing through an area A, or the number of molecules in an observed volume V; or the difference in magnetic and electrostatic energy stored in V:

$$f(t) = \frac{1}{8\pi}\int_V [H^2(x,t) - E^2(x,t)]d^3x \qquad \text{(C2)}$$

and so on! For any such physical meaning, the following considerations will apply.

Given any probability distribution (which we henceforth call, for brevity an ensemble) for f(t), the best prediction of f(t) that we can make from it--"best" in the sense of minimizing the expected square of the error--is the ensemble average

$$\langle f(t)\rangle = \langle f\rangle \qquad \text{(C3)}$$

which is independent of t if it is an equilibrium ensemble, as we henceforth assume. But this may or may not be a reliable prediction of f(t) at any particular time. The mean square expected deviation from the prediction (C3) is the variance

$$[\Delta f(t)]^2 = \langle f^2\rangle - \langle f\rangle^2 \qquad \text{(C4)}$$

again independent of t by our assumption. Only if $|\Delta f/\langle f\rangle| \ll 1$ is the ensemble making a sharp prediction of the measurable value of f.

Basically, the quantity Δf just defined represents only the uncertainty of the prediction; i.e., the degree of ignorance about f expressed by the ensemble. Yet Δf is held, almost universally in the literature of fluctuation theory, to represent also the measurable RMS fluctuations in f. Clearly, this is an additional assumption, which might or might not be true; for, obviously, the mere fact that I know f only to ±1% accuracy, is not enough to make it fluctuate by ±1%! Therefore, we note there is logically no room for any postulate that Δf is the measurable RMS fluctuation; whether this is or is not true is mathematically determined by the probability distribution. To understand this we need a more careful analysis of the relation between ⟨f⟩, Δf, and experimentally measurable quantities.

More generally, we can consider a large class of functionals of f(t) in some time interval $(0 < t < T)$; for example,

$$K[f(t)] \equiv T^{-n}\int_0^T dt_1 \ldots \int_0^T dt_n\; G[f(t_1)\ldots f(t_n)] \qquad \text{(C5)}$$

with $G(f_1 \ldots f_n)$ a real function. For any such functional, the ensemble will determine some probability distribution P(K)dK, and the best prediction we can make by the mean-square-error criterion is its expectation ⟨K⟩. What is the necessary and sufficient condition that, as $T \to \infty$, the ensemble predicts a sharp value for K? It is, as always, that

$$(\Delta K)^2 \equiv \langle K^2 \rangle - \langle K \rangle^2 \to 0 \quad . \tag{C6}$$

For any such functional, this condition may be written out explicitly; let us give two examples that will surprise some readers.

One of the sources of confusion in this field is that the word "average" is used in several different senses. We try to avoid this by using different notations for different kinds of average. For the single system that exists in the laboratory, the observable average is not the ensemble average $\langle f \rangle$, but a time average, which we denote by a bar (reserving the angular brackets to mean only ensemble averages):

$$\overline{f} \equiv \frac{1}{T}\int_0^T f(t)dt \tag{C7}$$

which corresponds to (C5) with $G \equiv f(t_1)$. The averaging time T is left arbitrary for the time being because the results (C8), (C11), (C18), to be derived next, being exact for any T, then provide a great deal of insight that would be lost if we pass to the limit too soon.

In the state of knowledge represented by the ensemble, the best prediction of $\overline{f}$ by the mean square error criterion, is

$$\langle \overline{f} \rangle = \left\langle \frac{1}{T}\int_0^T f(t)dt \right\rangle = \frac{1}{T}\int_0^T \langle f \rangle dt$$

or, for an equilibrium ensemble,

$$\langle \overline{f} \rangle = \langle f \rangle \quad , \tag{C8}$$

an example of a very general rule of probability theory; an ensemble average $\langle f \rangle$ is not the same as a measured value $f(t)$ or a measured average $\overline{f}$; but it is equal to the expectations of both of those quantities.

But (C8), like (C3), tells us nothing about whether the prediction is a reliable one; to answer this we must again consider the variance

$$(\Delta\overline{f})^2 \equiv \langle (\overline{f} - \langle \overline{f} \rangle)^2 \rangle$$
$$= \frac{1}{T^2}\int_0^T dt_1 \int_0^T dt_2 [\langle f(t_1)f(t_2)\rangle - \langle f(t_1)\rangle\langle f(t_2)\rangle] \quad . \tag{C9}$$

Only if $|\Delta\bar{f}/\langle\bar{f}\rangle| \ll 1$ is the ensemble making a sharp prediction of the measured average $\bar{f}$. Now, however, the time averaging can help us; for $\Delta\bar{f}$ may become very small compared to Δf, if we average over a long enough time.

Now in an equilibrium ensemble the integrand of (C9) is a function of $(t_2 - t_1)$ only, and defines the covariance function

$$\phi(\tau) \equiv \langle f(t)f(t+\tau)\rangle - \langle f(t)\rangle\langle f(t+\tau)\rangle$$
$$= \langle f(0)f(\tau)\rangle - \langle f\rangle^2 \qquad \text{(C10)}$$

from which (C9) reduces to a single integral:

$$(\Delta\bar{f})^2 = \frac{2}{T^2}\int_0^T (T-\tau)\phi(\tau)d\tau \quad . \qquad \text{(C11)}$$

A sufficient (stronger than necessary) condition for $\Delta\bar{f}$ to tend to zero is that the integrals

$$\int_0^\infty \phi(\tau)\tau d\tau \quad , \quad \int_0^\infty \phi(\tau)d\tau \qquad \text{(C12)}$$

converge; and then the characteristic correlation time

$$\tau_c \equiv \left[\int_0^\infty \phi(\tau)d\tau\right]^{-1}\left[\int_0^\infty \tau\,\phi(\tau)d\tau\right] \qquad \text{(C13)}$$

is finite, and we have asymptotically,

$$(\Delta\bar{f})^2 \sim \frac{2}{T}\int_0^\infty \phi(\tau)d\tau \quad . \qquad \text{(C14)}$$

$\Delta\bar{f}$ then tends to zero like $1/\sqrt{T}$, and the situation is very much as if successive samples of the function over non-overlapping intervals of length τ_c were independent. However, the slightest positive correlation, if it persists indefinitely, will prevent any sharp prediction of $\bar{f}$. For, if $\phi(\tau) \to \phi(\infty) > 0$, then from (C11) we have

$$(\Delta\bar{f})^2 \to \phi(\infty) \qquad \text{(C15)}$$

and the ensemble can never make a sharp prediction of the measured average; i.e., any postulate that the ensemble average equals the time average, violates the mathematical rules of probability theory. These results correspond to (B23).

Now everything we have said about measurable values of f can be repeated mutatis mutandis for the measurable fluctuations $\delta f(t)$; we need only take a step up the hierarchy of successively higher order correlations. For, over the observation time T, the measured mean-square fluctuation in f(t)--i.e., deviation from the measured mean--is

$$(\delta f)^2 \equiv \frac{1}{T}\int_0^T [f(t) - \bar{f}]^2 \, dt \tag{C16}$$

$$= \overline{f^2} - \bar{f}^2 \tag{C17}$$

which corresponds to the choice $G = f^2(t_1) - f(t_1)f(t_2)$ in (C5). The "best" prediction we can make of this from the ensemble, is its expectation, which reduces to

$$\langle(\delta f)^2\rangle = (\Delta f)^2 + (\Delta\bar{f})^2 \tag{C18}$$

as a short calculation using (C4), (C11) will verify. This is in itself a very interesting (and I am sure to many surprising) result. The predicted <u>measurable</u> fluctuation δf is not the same as the <u>ensemble</u> fluctuation Δf unless the ensemble is such, and the averaging time so long, that $\Delta\bar{f}$ is negligible compared to Δf.

But (C18) tells us nothing about whether the prediction $\langle(\delta f)^2\rangle$ is a reliable one; to answer this we must, once more, examine the variance

$$V = \langle(\delta f)^4\rangle - \langle(\delta f)^2\rangle^2 \quad . \tag{C19}$$

Unless (C19) is small compared to the square of (C18), the ensemble is not making any definite prediction of $(\delta f)^2$. After some computation we find that (C19) can be written in the form

$$V = \frac{1}{T^4}\int_0^T dt_1 \int_0^T dt_2 \int_0^T dt_3 \int_0^T dt_4 \; \psi(t_1,t_2,t_3,t_4) \tag{C20}$$

where ψ is a four-point correlation function:

$$\psi(t_1,t_2,t_3,t_4) = \langle f(t_1)f(t_2)f(t_3)f(t_4)\rangle - 2\langle f(t_1)f(t_2)f^2(t_3)\rangle + \langle f^2(t_1)f^2(t_2)\rangle - [(\Delta f)^2 + (\Delta\bar{f})^2]^2 \tag{C21}$$

which we have written in reduced form, taking advantage of the symmetry of the domain of integration in (C20).

As we see, the person who supposes that the RMS fluctuation Δf in the ensemble is also the experimentally measurable RMS fluctuation δf, is inadvertently supposing some rather non-trivial mathematical properties of that ensemble, which would seem to require some nontrivial justification! Yet to the best of my knowledge, no existing treatment of fluctuation theory even recognizes the distinction between δf and Δf.

In almost all discussions of random functions in the existing literature concerned with physical applications, it is taken for granted that (C6) holds for all functionals. One can hardly avoid this if one postulates, with Rowlinson, that "the probability of an event is the frequency with which it occurs in a given situation." But if it requires the computation (C20) to justify this for the mean-square fluctuation, what would it take to justify it in general? That is just the "ergodic" problem for this model.

Future progress in a number of areas will, I think, require that the relation between ensembles and physical systems be more carefully defined. The issue is not merely one of "philosphy of interpretation" that practical people may ignore; for not only the quantitative details, but even the qualitative kinds of physical predictions that a theory can make, depend on how these conceptual problems are resolved. For example, as was pointed out in my 1962 Brandeis Lectures, [loc. cit. Eqs. (83)-(93)], one cannot even state, in terms of the underlying ensemble, the criterion for a phase transition, or distinguish between laminar and turbulent flow, until the meaning of that ensemble is recognized.

A striking example of the need for clarifications in fluctuation theory is provided by quantum electrodynamics. Here one may calculate the expectation of an electric field at a point: $\langle E(x,t)\rangle = 0$, but the expectation of its square diverges: $\langle E^2(x,t)\rangle = \infty$. Thus $\Delta E = \infty$; in present quantum theory one interprets this as indicating that empty space is filled with "vacuum fluctuations," yielding an infinite "zero-point" energy density. But when we see the distinction between ΔE and δE, a different interpretation suggests itself. If $\Delta E = \infty$ that does not have to mean that any physical quantity is infinite; it means only that the present theory is totally unable to predict the field at a point, i.e., the only thing which is infinite is the uncertainty of the prediction.

It had been thought for 30 years that these vacuum fluctuations had to be real, because they were the physical cause of the Lamb shift; however it has been shown (Jaynes, 1978) that a classical calculation leads to just the same formula for this frequency shift without invoking any field fluctuations. Therefore, it appears that a reinterpretation of the "fluctuation laws" of quantum theory along these lines might clear up at least some of the paradoxes of present quantum theory.

The situation just noted is only one of a wide class of connections that might be called "generalized fluctuation-dissipation theorems," or "fluctuation-response theorems." These include all of the Kubo-type theorems relating transport coefficients to various "thermal fluctuations." I believe that relations of this type will become more general and more useful with a better understanding of fluctuation theory.

Biology. Perhaps the largest and most obvious beckoning new field for application of statistical thermodynamics is biology. At present, we do not have the input information needed for a useful theory, we do not know what simplifying assumptions are appropriate; and indeed we do not know what questions to ask. Nevertheless, molecular biology has advanced to the point where some preliminary useful results do not seem any further beyond us now than the achievement of an integrated circuit computer chip did thirty years ago.

In the case of the simplest organism for which a great deal of biochemical information exists, the bacterium *E. coli*, Watson (1965) estimated that "one-fifth to one-third of the chemical reactions in *E. coli* are known," and noted that additions to the list were coming at such a rate that by perhaps 1985 it might be possible to describe "essentially all the metabolic reactions involved in the life of an *E. coli* cell."

As a pure speculation, then, let us try to anticipate a problem that might just possibly be amenable to the biochemical knowledge and computer technology of the year 2000: Given the structure and chemical composition of *E. coli*, predict its experimentally reproducible properties, i.e., the range of environmental conditions (temperature, pH, concentrations of food and other chemicals) under which a cell can stay alive; the rate of growth as a function of these factors. Given a specific mutation (change in the DNA code), predict whether it can survive and what the reproducible properties of the new form will be.

Such a program would be a useful first step. It seems, in my very cloudy crystal ball, that (1) its realization might be a matter of decades rather than centuries, (2) success in one instance would bring about a rapid increase in our ability to deal with more complicated problems, because it would reveal what simplifying assumptions are permissible.

At present one could think of several thousand factors that might, as far as we know, be highly relevant for these predictions. If a single cell contains 20,000 ribosomes where protein synthesis is taking place, are they performing 20,000 different functions, each one essential to the life of the cell? This just seems unlikely. I would conjecture that of all the complicated detail that can be seen in a cell, the overwhelmingly greatest part is--like every detail of the hair on our heads, or our fingerprints--accidental to the history of that particular individual; and not at all essential for its biological function.

The problem seems terribly complicated at present, because in all this detail we do not know what is relevant, what is irrelevant. But success in one instance would show us how to judge this. It might turn out that prediction of biological activity requires information about only a dozen separate factors, instead of a million. If so, then one would have both the courage and the insight needed to attack more complicated problems.

This has been stated so as to bring out the close analogy with what has happened in the theory of irreversible processes. In the early 1950's the development of a general formalism for irreversible processes appeared to be a hopelessly complicated program, not to be thought of in the next thousand years, if ever. Thus, van Hove (1956) stated: "... in view of the unlimited diversity of possible nonequilibrium situations, the existence of such a set of equations seems rather doubtful." Yet, as noted in Section A above, the principle which has solved this problem already existed, unrecognized, at that time. And today it seems that our major problem is not the complications of detail, but the conceptual difficulty in understanding how such a complicated problem could have such a (formally) simple solution. The answer is that, while full dynamical information is extremely complicated, the _relevant_ information is not.

Perhaps there is a general principle--which we are conceptually unprepared to recognize today because it is too simple--that would tell us which features of an organism are its essential, relevant biological features; and which are not.

Of course, applications of statistical mechanics to biology may be imagined at many different levels, so widely separated that they have nothing to do with each other. Thus, while I have been speculating about complexities within a single cell, the contribution of E. H. Kerner to this Symposium goes after the opposite extreme, interaction of many organisms. At that level the relevant information is now so much simpler and more easily obtained that many interesting results are already available.

Basic Statistical Theory. From the standpoint of statistical theory in general, the principle of maximum entropy is only one detail, which arose in connection with the problem of generalizing Laplace's statistical practice from (A6), and we have examined it above only in the finite discrete case. As $n \to \infty$ a new feature is that for some kinds of testable information there is no upper bound to the entropy. For mean-value constraints, the partition function may diverge for all real λ, or the constraint equations (B4) may not have a solution. In this way, the theory signals back to us that we have not put enough information into the problem to determine any definite inferences. In the finite case, the mere enumeration of the possibilities $\{i = 1,2,\ldots n\}$ specifies enough to ensure that a solution exists. If $n \to \infty$, we have specified far less in the enumeration, and it is hardly surprising that this must be compensated by specifying more information in our constraints.

Rowlinson quotes Leslie Ellis (1842) to the effect that "Mere ignorance is no ground for any inference whatever. *Ex nihilo nihil*." I am bewildered as to how Rowlinson can construe this as an argument against maximum entropy, since as we see the maximum entropy principle immediately tells us the same thing. Indeed, it is the principle of maximum entropy--and not Leslie Ellis--that tells us precisely *how much* information must be specified before we have a normalizable distribution so that rational inferences *are* possible. Once this is recognized, I believe that the case $n \to \infty$ presents no difficulties of mathematics or of principle.

It is very different when we generalize to continuous distributions. We noted that Boltzmann was obliged to divide his phase space into discrete cells in order to get the entropy expression (A14) from the combinatorial factor (A11). Likewise, Shannons uniqueness proof establishing $-\Sigma p_i \log p_i$ as a consistent information measure, goes through only for a discrete distribution. We therefore approach the continuous case as the limit of a discrete one. This leads (Jaynes, 1963b, 1968) to the continuous entropy expression

$$S = -\int p(x) \log \frac{p(x)}{m(x)}\, dx \qquad \text{(C22)}$$

where the "measure function" $m(x)$ is proportional to the limiting density of discrete points (all this theory is readily restated in the notation of measure theory and Stieltjes integrals; but we have never yet found a problem that needs it). So, it is the entropy relative to some "measure" $m(x)$ that is to be maximized. Under a change of variables, the functions $p(x)$, $m(x)$ transform in the same way, so that the entropy so defined is invariant; and in consequence it turns out that the Lagrange multipliers and all our conclusions from entropy maximization are independent of our choice of variables. The maximum-entropy probability density for prescribed averages

$$\int f_k(x) p(x) dx = F_k \quad , \quad 1 \le k \le m \qquad \text{(C23)}$$

is

$$p(x) = \frac{m(x)}{Z(\lambda_1 \ldots \lambda_m)} \exp\left[-\sum_k \lambda_k f_k(x) \right] \tag{C24}$$

with the partition function

$$Z(\lambda_1 \ldots \lambda_m) \equiv \int dx\, m(x) \exp\left[-\sum_k \lambda_k f_k(x) \right]. \tag{C25}$$

An interesting fact, which may have some deep significance not yet seen, is that the class of maximum-entropy functions (C24) is, by the Pitman-Koopman theorem, identical with the class of functions admitting sufficient statistics; that is, if as in (B72)-(B81) we think of (C24) as defining a class of sampling distributions from which, given data D ≡ "N measurements of x yielded the results $\{x_1 \ldots x_N\}$," we are to estimate the $\{\lambda_1 \ldots \lambda_m\}$ by applying Bayes' theorem, we find that the posterior distribution of the λ's depends on the data only through the observed averages:

$$\bar{f}_k \equiv \frac{1}{N} \sum_{r=1}^{N} f_k(x_r) \tag{C26}$$

all other aspects of the data being irrelevant. This seems to strengthen the point of view noted before in (B72)-(B81). For many more details, see Huzurbazar (1976).

But now let us return to our usual viewpoint, that (C24) is not a sampling distribution but a prior distribution from which we are to make inferences about x, which incorporate any testable prior information. If the space S_x in which the continuous variable x is defined, is not the result of any obvious limiting process, there seems to be an ambiguity; for what now determines m(x)?

This problem was discussed in some detail before (Jaynes, 1968). If there are no constraints, maximization of (C22) leads to p(x) = Am(x) where A is a normalization constant; thus m(x) has the intuitive meaning that it is the distribution representing "complete ignorance" and we are back, essentially, to Bernoulli's problem from where it all started. In the continuous case, then, before we can even apply maximum entropy we must deal with the problem of complete ignorance.

Suppose a man is lost in a rowboat in the middle of the ocean. What does he mean by saying that he is "completely ignorant" of his position? He means that, if he were to row a mile in any direction he would still be lost; he would be just as ignorant as before. In other words, ignorance of one's location is a state of knowledge which is not changed by a small change in that location. Mathematically, "complete ignorance" is an invariance property.

The set of all possible changes of location forms a group of translations. More generally, in a space S_x of any structure, one can define precisely what he means by "complete ignorance" by specifying some group of transformations of S_x onto itself under which an element of probability m(x)dx is to be invariant. If the group is transitive on S_x (i.e., from any point x, any other point x' can be reached by some element of the group), this determines m(x), to within an irrelevant multiplicative constant, on S_x.

This criterion follows naturally from the basic desideratum of consistency: In two problems where we have the same state of knowledge, we should assign the same probabilities. Any transformation of the group defines a new problem in which a "completely ignorant" person would have the same state of prior knowledge. If we can recognize a group of transformations that clearly has this property of transforming the problem into one that is equivalent in this sense, then the ambiguity in m(x) has been removed. Quite a few useful "ignorance priors" have been found in this way; and in fact for most real problems that arise the solutions are now well--if not widely--known.

But while the notion of transformation groups greatly reduces the ambiguity in m(x), it does not entirely remove it in all cases. In some problems no appropriate group may be apparent; or there may be more than one, and we do not see how to choose between them. Therefore, still more basic principles are needed; and active research is now underway and is yielding promising results.

One of these new approaches, and the one on which there is most to report, is the method of marginalization (Jaynes, 1979). The basic facts pointing to it were given already by Jeffreys (1939; §3.8), but it was not realized until 1976 that this provides a new, constructive method for defining what is meant by "ignorance," with the advantage that everything follows from the basic rules (A8), (A9) of probability theory, with no need for any such desiderata as entropy or group invariance. We indicate the basic idea briefly, using a bare skeleton notation to convey only the structure of the argument.

Marginalization. We have a sampling distribution $p(x|\theta)$ for some observable quantity x, depending on a parameter θ, both multidimensional. From an observed value x we can make inferences about θ; with prior information I_1, prior probability distribution $p(\theta|I_1)$ Bayes' theorem (B24) yields the posterior distribution

$$p(\theta|xI_1) = Ap(\theta|I_1)p(x|\theta) \quad . \tag{C27}$$

In the following, A always stands for a normalization constant, not necessarily the same in all equations.

But now we learn that the parameter θ can be separated into two components: $\theta = (\zeta,\eta)$ and we are to make inferences only about ζ. Then we discover that the data x may also be separated: $x = (y,z)$ in such a way that the sampling distribution of z depends only on ζ:

$$p(z|\zeta\eta) = \int p(yz|\eta\zeta)dy = p(z|\zeta) \quad . \tag{C28}$$

Then, writing the prior distribution as $p(\theta|I_1) = \pi(\zeta)\pi(\eta)$, (C27) gives for the desired marginal posterior distribution

$$p(\zeta|y,z\, I_1) = A\ \pi(\zeta)\int p(y,z|\zeta\eta)\pi(\eta)d\eta \tag{C29}$$

which must in general depend on our prior information about η. This is the solution given by a "conscientious Bayesian" B_1.

At this point there arrives on the scene an "ignorant Bayesian" B_2, whose knowledge of the experiment consists only of the sampling distribution (C28); i.e., he is unaware of the existence of the components (y,η). When told to make inferences about ζ, he confidently applies Bayes' theorem to (C28), getting the result

$$p(\zeta|z\, I_2) = A\ \pi(\zeta)p(z|\zeta) \quad . \tag{C30}$$

This is what was called a "pseudoposterior distribution" by Geisser and Cornfield (1963).

B_1 and B_2 will in general come to different conclusions because B_1 is taking into account extra information about (y,η). But now suppose that for some particular prior $\pi(\eta)$, B_1 and B_2 happen to agree after all; what does that mean? Clearly, B_2 is not incorporating any information about η; he doesn't even know it exists. If, nevertheless, they come to the same conclusions, then it must be that B_1 was not incorporating any information about η either. In other words, a prior $\pi(\eta)$ that leaves B_1 and B_2 in agreement must be, within the context of this model, a <u>completely uninformative</u> prior; it contains no information <u>relevant</u> to questions about ζ.

Now the condition for equality of (C29), (C30) is just a Fredholm integral equation:

$$\int p(y,z|\zeta,\eta)\pi(\eta)d\eta = \lambda(y,z)p(z|\zeta) \tag{C31}$$

where $\lambda(y,z)$ is a function to be determined from (C31). Therefore, the rules of probability theory already contain the criterion for defining what is meant by "completely uninformative."

Mathematical analysis of (C31) proves to be quite involved; and we do not yet know a necessary and sufficient condition on $p(y,z|\zeta\eta)$ for it to have solutions, or unique solutions, although a number of isolated results are now in (Jaynes, 1979). We indicate one of them.

Suppose y and η are positive real, and η is a scale parameter for y; i.e., we have the functional form

$$p(y,z|\zeta,\eta) = \eta^{-1} f(z,\zeta;y/\eta) \tag{C32}$$

for the sampling density function. Then, (C31) reduces to

$$y^{-1} \int f(z,\zeta;\alpha)\left[\frac{y}{\alpha}\,\pi\left(\frac{y}{\alpha}\right)\right] d\alpha = \lambda(y,z)\int f(z,\zeta;\alpha)\,d\alpha \tag{C33}$$

where we have used (C28). It is apparent from this that the Jeffreys prior

$$\pi(\eta) = \eta^{-1} \tag{C34}$$

is always a solution, leading to $\lambda(y,z) = y^{-1}$. Thus (C34) is "completely uninformative" for all models in which η appears as a scale parameter; and it is easily shown (Jaynes, 1979) that one can invent specific models for which it is unique.

We have therefore, the result that the Jeffreys prior is uniquely determined as the only prior for a scale parameter that is "completely uninformative" without qualifications. We can hardly avoid the inference that it represents, uniquely, the condition of "complete ignorance" for a scale parameter.

This example shows how marginalization is able to give results consistent with those found before, but in a way that springs directly out of the principles of probability theory without any additional appeal to intuition (as is involved in choosing a transformation group). At the moment, this approach seems very promising as a means of rigorizing and extending the basic theory. However, there are enough complicated technical details not noted here, so that it will require quite a bit more research before we can assess its full scope and power. In fact, the chase is at present quite exciting, because it is still mathematically an open question whether the integral equations may in some cases become overdetermined, so that no uninformative prior exists. If so, this would call for some deep clarification, and perhaps revision, of present basic statistical theory.

D. An Application: Irreversible Statistical Mechanics.

The calculation of an irreversible process usually involves three distinct stages; (1) Setting up an "ensemble," i.e., choosing a density matrix $\rho(0)$, or an N-particle distribution function, which is to describe our initial knowledge about the system of interest; (2) Solving the dynamical problem; i.e., applying the microscopic equations of motion to obtain the time-evolution of the system $\rho(t)$; (3) Extracting the final physical predictions from the time-developed ensemble $\rho(t)$.

Stage (3) has never presented any procedural difficulty; to predict the quantity F from the ensemble ρ, one follows the practice of equilibrium theory, and computes the expectation value $\langle F \rangle = \mathrm{Tr}(\rho F)$. While the ultimate justification of this rule has been much discussed (ergodic theory), no alternative procedure has been widely used.

In this connection, we note the following. Suppose we are to choose a number f, representing our estimate of the physical quantity F, based on the ensemble ρ. A reasonable criterion for the "best" estimate is that the expected square of the error, $\langle (F-f)^2 \rangle$ shall be made a minimum. The solution of this simple variational problem is: $f = \langle F \rangle$. Thus, if we regard statistical mechanics, not in the "classical" sense of a means for calculating time averages in terms of ensemble averages, but rather as an example of statistical estimation theory based on the mean square error criterion, the usual procedure is uniquely determined as the optimal one, independently of ergodic theory. A justification not depending on ergodic theory is in any event necessary as soon as we try to predict the time variation of some quantity F(t); for the physical phenomenon of interest then consists just of the fact that the ensemble average $\langle F(t) \rangle$ is not equal to a time average.

The dynamical problem of stage (2) is the most difficult to carry out, but it is also the one in which most recent progress has been booked (Green's function methods). While the present work is not primarily concerned with these techniques, they are available, and needed, in carrying out the calculations indicated here for all but the simplest problems.

It is curious that stage (1), which must logically precede all the others, has received such scant attention since the pioneering work of Gibbs, in which the problem of ensemble construction was first recognized. Most recent discussions of irreversible processes concentrate all attention on stage (2); many fail to note even the existence of stage (1). One consequence of this is that the resulting theories apply unambiguously only to the case of "response functions," in which the nonequilibrium state is one resulting from a dynamical perturbation (i.e., an explicitly given term in the

Hamiltonian), starting from thermal equilibrium at some time in the past; the initial density matrix is then given by conventional equilibrium theory, and so the problem of ensemble construction is evaded.

If, however, the nonequilibrium state is defined (as it usually is from the experimental standpoint) in terms of temperature or concentration gradients, rate of heat flow, shearing stress, sound wave amplitudes, etc., such a procedure does not apply, and one has resorted to various ad hoc devices. An extreme example is provided by some problems in astrophysics, in which it is clear that the system of interest has never, in the age of the universe, been in a state approximating thermal equilibrium. Such cases have been well recognized as presenting special difficulties of principle.

We show here that recognition of the existence of the stage (1) problem, and that its general solution is available, can remove such ambiguities and reduce the labor of stage (2). In the case of the nonequilibrium steady state, stage (2) can be dispensed with entirely if stage (1) has been properly treated.

Background. To achieve a certain unity within the present volume, we shall take the review article of Mori, Oppenheim, and Ross (1962)--hereafter denoted MOR--as indicating the level to which nonequilibrium theory had been brought before the introduction of Information Theory notions. This work is virtually unique in that the Stage 1 problem, and even the term "ensemble construction" appear explicitly. The earlier work of Kirkwood, Green, Callen, Kubo and others, directly related to ours, is noted in MOR, Sec. 6.

To fix ideas, consider the calculation of transport properties in systems close to equilibrium (although our final results will be far more general). In the treatments discussed by MOR, dissipative-irreversible effects did not appear in the ensemble initially set up. For example, a system of N particles of mass m, distributed with macroscopic density $\rho(x)$, local temperature $T(x)$, is often described in classical theory by an N-particle distribution function, or Liouville function, of the form:

$$W_N(x_1 p_1 \dots x_N p_N) = \prod_{i=1}^{N} \frac{\rho(x_i)}{Nm} [2\pi mkT(x_i)]^{3/2} \exp\left\{-\frac{p_i^2}{2mkT(x_i)}\right\} \quad \text{(D1)}$$

where x_i, p_i denote the (vector) position and momentum of the ith particle. But, although this distribution represents nonvanishing density and temperature gradients $\nabla\rho$, ∇T, the diffusion current or heat flow computed from (D1) is zero.

Likewise, in quantum theory MOR described such a physical situation by the "local equilibrium," or "frozen-state" density matrix:

$$\rho_t = \frac{1}{Z} \exp\left\{-\int d^3x\ \beta(x)[H(x) - \mu(x)n(x)]\right\} \qquad \text{(D2)}$$

where H(x), n(x) are the Hamiltonian density and number density operators. Again, although (D2) describes gradients of temperature, concentration, and chemical potential, the fluxes computed from (D2) are zero.

Mathematically, it was found that dissipative effects appear in the equations only after one has carried out the following operations: (a) approximate forward integration of the equations of motion for a short "induction time," and (b) time smoothing or other coarse-graining of distribution functions or Heisenberg operators.

Physically, it has always been somewhat of a mystery why either of these operations is needed; for one can argue that, in most experimentally realizable cases, irreversible flows (A) are already "in progress" at the time the experiment is started, and (B) take place slowly, so that the low-order distribution functions and expectation values of measurable quantities must be already slowly-varying functions of time and position; and thus not affected by coarse-graining. In cases where this is not true, coarse-graining would result in loss of the physical effects of interest.

The real nature of the forward integration and coarse-graining operations is therefore obscure; in a correctly formulated theory neither should be required. We are led to suspect the choice of initial ensemble; i.e., that ensembles such as (D1) and (D2) do not fully describe the conditions under which irreversible phenomena are observed, and therefore do not represent the correct solution of the stage (1) problem. [We note that (D1) and (D2) were not "derived" from anything more fundamental; they were written down intuitively, by analogy with the grand canonical ensemble of equilibrium theory.] The forward integration and coarse-graining operations would, on this view, be regarded as corrective measures which in some way compensate for the error in the initial ensemble.

This conclusion is in agreement with that of MOR. These authors never claimed that ρ_t in (D2) was the correct density matrix, but supposed that it differed by only a small amount from another matrix $\rho(t)$, which they designate as the "actual distribution." They further supposed that after a short induction time, ρ_t relaxes into $\rho(t)$, which would explain the need for forward integration.

Such relaxation undoubtedly takes place in the low-order distribution functions derived from ρ, as was first suggested by Bogoliubov for the analogous classical problem. However, this is not possible for the full "global" density matrix; if ρ_t and $\rho(t)$ differ at $t = 0$ and undergo the same unitary transformation in their time development, they cannot be equal at any other time. Furthermore, $\rho(t)$ was never uniquely defined; given two different candidates $\rho_1(t)$, $\rho_2(t)$ for this role, MOR give no criterion by which one could decide which is indeed the "actual" distribution.

For reasons already explained in earlier Sections and in Jaynes (1967), we believe that such criteria do not exist; i.e., that the notion of an "actual distribution" is illusory, since different density matrices connote only different states of knowledge. In the following Section we approach the problem in a different way, which yields a definite procedure for constructing a density matrix which is to replace ρ_t, and will play approximately the same role in our theory as the $\rho(t)$ of MOR.

The Gibbs Algorithm. If the above reasoning is correct, a reexamination of the procedures by which ensembles are set up in statistical mechanics is indicated. If we can find an algorithm for constructing density matrices which fully describe nonequilibrium conditions, we should find that transport and other dissipative effects are obtainable by direct quadratures over the initial ensemble.

This algorithm, we suggest, was given already by Gibbs (1902). The great power and scope of the methods he introduced have not been generally appreciated to this day; until recently it was scarcely possible to understand the rationale of his method for constructing ensembles. This was (*loc. cit.*, p. 143) to assign that probability distribution which, while agreeing with what is known, "gives the least value of the average index of probability of phase," or as we would describe it today, *maximizes the entropy*. This process led Gibbs to his canonical ensemble for describing closed systems in thermal equilibrium, the grand canonical ensemble for open systems, and (*loc. cit.*, p. 38) an ensemble to represent a system rotating at angular velocity $\vec{\omega}$ in which the probability density is proportional to

$$\exp[-\beta(H - \vec{\omega}\cdot\vec{M})] \tag{D3}$$

where H, M are the phase functions representing Hamiltonian and total angular momentum.

Ten years later, the Ehrenfests (1912) dismissed these ensembles as mere "analytical tricks," devoid of any real sigificance, and asserted the physical superiority of Boltzmann's methods, thereby initiating a school of thought which dominated statistical mechanics for decades. It is one of the major tragedies of science that Gibbs did not live long enough to answer these objections, as he could have so easily.

The mathematical superiority of the canonical and grand canonical ensembles for calculating equilibrium properties has since become firmly established. Furthermore, although Gibbs gave no applications of the rotational ensemble (D3), it was shown by Heims and Jaynes (1962) that this ensemble provides a straightforward method of calculating the gyromagnetic effects of Barnett and Einstein-de Haas. At the present time, therefore, the Gibbs methods--like the Laplace methods and the Jeffreys methods--stand in a position of proven success in applications, independently of all the conceptual problems regarding their justification, which are still being debated.

The development of Information Theory made it possible to see the method of Gibbs as a general procedure for inductive reasoning, independent of ergodic theory or any other physical hypotheses, and whose range of validity is therefore not restricted to equilibrium problems; or indeed to physics. In the following we show that the Principle of Maximum Entropy is sufficient to construct ensembles representing a wide variety of nonequilibrium conditions, and that these new ensembles yield transport coefficients by direct quadratures. Indeed, we shall claim--for reasons already explained in Jaynes (1957b), that this is the <u>only</u> principle needed to construct ensembles which predict any experimentally reproducible effect, reversible or irreversible.

The general rule for constructing ensembles is as follows. The available information about the state of a system consists of results of various macroscopic measurements. Let the quantities measured be represented by the operators $F_1, F_2, \ldots F_m$. The results of the measurements are, of course, simply a set of <u>numbers</u>: $\{f_1, \ldots, f_m\}$. These numbers make no reference to any probability distribution. The ensemble is then a <u>mental construct</u> which we invent in order to describe the range of <u>possible</u> microscopic states compatible with those numbers, in the following sense.

If we say that a density matrix ρ "contains" or "agrees with" certain information, we mean by this that, if we communicate the density matrix to another person he must be able, by applying the usual procedure of stage (3) above, to recover this information from it. In this sense, evidently, the density matrix agrees with the given information if and only if it is adjusted to yield expectation values equal to the measured numbers:

$$f_k = \mathrm{Tr}(\rho F_k) = \langle F_k \rangle \quad , \qquad k = 1, \ldots, m \tag{D4}$$

and in order to ensure that the density matrix describes the full range of possible microscopic states compatible with (D4), and not just some arbitrary subset of them (in other words, that it describes only the information given, and contains no hidden arbitrary assumptions about the microscopic state), we demand that, while satisfying the constraints (D4), it shall maximize the quantity

$$S_I = -\mathrm{Tr}(\rho \log \rho) \quad . \tag{D5}$$

A great deal of confusion has resulted from the fact that, for decades, the single word "entropy" has been used interchangeably to stand for either the quantity (D5) or the quantity measured experimentally (in the case of closed systems) by the integral of dQ/T over a reversible path. We shall try to maintain a clear distinction here by following the usage introduced in my 1962 Brandeis lectures (Jaynes, 1963b); referring to S_I as the "information entropy" and denoting the experimentally measured entropy by S_E. These quantities are different in general; in the equilibrium case (the only one for which S_E is defined in conventional thermodynamics) the relation between them was shown (loc. cit.) to be: for all density matrices ρ which agree with the macroscopic information that defines the thermodynamic state; i.e., which satisfy (D4),

$$kS_I \leq S_E \tag{D6}$$

where k is Boltzmann's constant, with equality in (D6) if and only if S_I is computed from the canonical density matrix

$$\rho = \frac{1}{Z(\lambda_1 \ldots \lambda_m)} \exp[\lambda_1 F_1 + \ldots + \lambda_m F_m] \tag{D7}$$

where the λ_k are unspecified real constants. In the nonequilibrium theory we find it easier to change our sign convention, so that all λ's here are the negative of the usual ones; otherwise, from this point on it would be invariably $(-\lambda)$ rather than λ that we need. For normalization ($\mathrm{Tr}\,\rho = 1$) we have

$$Z(\lambda_1 \ldots \lambda_m) = \mathrm{Tr}\exp[\lambda_1 F_1 + \ldots + \lambda_m F_m] \tag{D8}$$

which quantity will be called the partition function. It remains only to choose the λ_k [which appear as Lagrange multipliers in the derivation of (D7) from a variational principle] so that (D4) is satisfied. This is the case of

$$f_k = \langle F_k \rangle = \frac{\partial}{\partial \lambda_k} \log Z \quad , \qquad k = 1,2,\ldots,m \quad . \tag{D9}$$

If enough constraints are specified to determine a normalizable density matrix, it will be found that these relations are just sufficient to determine the unknowns λ_k in terms of the given data $\{f_1 \ldots f_m\}$; indeed, we can then solve (D9) explicitly for the λ_k as follows. The maximum attainable value of S_I is, from (D7), (D8),

$$(S_I)_{max} = \log Z - \sum_{k=1}^{m} \lambda_k \langle F_k \rangle \quad . \tag{D10}$$

If this quantity is expressed as a function of the given data, $S(f_1 \ldots f_m)$, it is easily shown from the above relations that

$$\lambda_k = - \frac{\partial S}{\partial f_k} \quad . \tag{D11}$$

It has been shown (Jaynes, 1963b, 1965) that the second law of thermodynamics, and a generalization thereof that tells which nonequilibrium states are accessible reproducibly from others, follow as simple consequences of the inequality (D6) and the dynamical invariance of S_I.

We note an important property of the maximum entropy ensemble, which is helpful in gaining an intuitive understanding of this theory. Given any density matrix ρ and any ε in $0 < \varepsilon < 1$, one can define a "high-probability linear manifold" (HPM) of finite dimensionality $W(\varepsilon)$, spanned by all eigenvectors of ρ which have probability greater than a certain amount $\delta(\varepsilon)$, and such that the eigenvectors of ρ spanning the complementary manifold have total probability less than ε. Viewed in another way, the HPM consists of all state vectors ψ to which ρ assigns an "array probability" as defined in Jaynes (1957b), Sec. 7, greater than $\delta(\varepsilon)$. Specifying the density matrix ρ thus amounts to asserting that, with probability $(1-\varepsilon)$, the state vector of the system lies somewhere in this HPM. As ε varies, any density matrix ρ thus defines a nested sequence of HPM's.

For a macroscopic system, the information entropy S_I may be related to the dimensionality $W(\varepsilon)$ of the HPM in the following sense: if N is the number of particles in the system, then as $N \to \infty$ with the intensive parameters held constant, $N^{-1} S_I$ and $N^{-1} \log W(\varepsilon)$ approach the same limit independently of ε. This is a form of the asymptotic equipartition theorem of Information Theory, and generalizes Boltzmann's $S = k \log W$. The process of entropy maximization therefore amounts, for all practical purposes, to the same thing as finding the density matrix which, while agreeing with the available information, defines the <u>largest</u> possible HPM; this is the basis of the remark following (D4). An analogous result holds in classical theory (Jaynes, 1965), in which $W(\varepsilon)$ becomes the phase volume of the "high-probability region" of phase space, as defined by N-particle distribution function.

The above procedure is sufficient to construct the density matrix representing equilibrium conditions, provided the quantities F_k are chosen to be constants of the motion. The extension to nonequilibrium cases, and to equilibrium problems in which we wish to incorporate information about quantities which are not intrinsic constants of the motion (such as stress or magnetization) requires mathematical generalization which we give in two steps.

It is a common experience that the course of a physical process does not in general depend only on the present values of the observed macroscopic quantities; it depends also on the past history of the system. The phenomena of magnetic hysteresis and spin echoes are particularly striking examples of this. Correspondingly, we must expect that, if the F_k are not constants of the motion, an ensemble constructed as above using only the present values of the $\langle F_k \rangle$ will not in general suffice to predict either equilibrium or nonequilibrium behavior. As we will see presently, it is just this fact which causes the error in the "local equilibrium" density matrix (D2).

In order to describe time variations, we extend the F_k to the Heisenberg operators

$$F_k(t) = U^{-1}(t)F_k(0)U(t) \tag{D12}$$

in which the time-development matrix U(t) is the solution of the Schrödinger equation

$$i\hbar\dot{U}(t) = H(t)U(t) \tag{D13}$$

with $U(0) = 1$, and H(t) is the Hamiltonian. If we are given data fixing the $\langle F_k(t_i) \rangle$ at various times t_i, then each of these must be considered a separate piece of information, to be given its Lagrange multiplier λ_{ki} and included in the sum of (D7). In the limit where we imagine information given over a continuous time interval, $-\tau < t < 0$, the summation over the time index i becomes an integration and the canonical density matrix (D7) becomes

$$\rho = \frac{1}{Z} \exp\left\{\sum_{k=1}^{m} \int_{-\tau}^{0} \lambda_k(t)F_k(t)\, dt\right\} \tag{D14}$$

where the partition function has been generalized to a partition functional

$$Z[\lambda_1(t)\ldots\lambda_m(t)] \equiv \mathrm{Tr}\, \exp\left\{\sum_{k=1}^{m} \int_{-\tau}^{0} \lambda_k(t)F_k(t)\, dt\right\} \tag{D15}$$

and the unknown Lagrange multiplier functions $\lambda_k(t)$ are determined from the condition that the density matrix agree with the given data $\langle F_k(t)\rangle$ over the "information-gathering" time interval:

$$\langle F_k(t)\rangle = \mathrm{Tr}[\rho F_k(t)] = f_k(t) \quad , \quad -\tau \le t \le 0 \quad . \tag{D16}$$

By the perturbation methods developed below, we find that (D16) reduces to the natural generalization of (D9):

$$\langle F_k(t)\rangle = \frac{\delta}{\delta\lambda_k(t)} \log Z \quad , \quad -\tau \le t \le 0 \tag{D17}$$

where δ denotes the functional derivative.

Finally, if the operators F_k depend on position as well as time, as in (D2), Eq. (D12) is changed to

$$F_k(x,t) = U^{-1}(t) F_k(x,0) U(t) \tag{D18}$$

and the values of these quantities at each point of space and time now constitute the independent pieces of information, which are coupled into the density matrix via the Lagrange multiplier function $\lambda_k(x,t)$. If we are given macroscopic information about $F_k(x,t)$ throughout a space-time region R_k (which can be a different region for different quantities F_k), the ensemble which incorporates all this information, while locating the largest possible HPM of microscopic states, is

$$\rho = \frac{1}{Z} \exp\left\{\sum_k \int_{R_k} dt\, d^3x\, \lambda_k(x,t) F_k(x,t)\right\} \tag{D19}$$

with the partition functional

$$Z = \mathrm{Tr} \exp\left\{\sum_k \int_{R_k} dt\, d^3x\, \lambda_k(x,t) F_k(x,t)\right\} \tag{D20}$$

and the $\lambda_k(x,t)$ determined from

$$\langle F_k(x,t)\rangle = \frac{\delta}{\delta\lambda_k(x,t)} \log Z \quad , \quad (x,t) \text{ in } R_k \quad . \tag{D21}$$

Prediction of any quantity $J(x,t)$ is then accomplished by calculating

$$\langle J(z,t)\rangle = \mathrm{Tr}[\rho J(x,t)] \tag{D22}$$

The form of equations (D19)-(D22) makes it appear that stages (1) and (2), discussed in the Introduction, are now fused into a single stage. However, this is only a consequence of our using the Heisenberg representation. According to the usual conventions, the Schrödinger and Heisenberg representations coincide at time $t = 0$; thus we may regard the steps (D19)-(D21) equally well as determining the density matrix $\rho(0)$ in the Schrödinger representation; i.e., as solving the stage (1) problem. If, having found this initial ensemble, we switch to the Schrodinger representation, Eq. (D22) is then replaced by

$$\langle J(x)\rangle_t = \mathrm{Tr}[J(x)\rho(t)] \tag{D23}$$

in which the problem of stage (2) now appears explicitly as that of finding the time-evolution of $\rho(t)$. The form (D23) will be more convenient if several different quantities $J_1, J_2, \ldots$ are to be predicted.

Discussion. In equations (D19)-(D23) we have the generalized Gibbs algorithm for calculating irreversible processes. They represent the three stages: (1) finding the ensemble which has maximum entropy subject to the given information about certain quantities $\{F_k(x,t)\}$; (2) Utilizing the full dynamics by working out the time evolution from the microscopic equations of motion; (3) making those predictions of certain other quantities of interest $\{J_i(x,t)\}$ which take all the above into account, and minimize the expected square of the error. We do not claim that the resulting predictions must be correct; only that they are the best (by the mean-square error criterion) that could have been made from the information given; to do any better we would require more initial information.

Of course this algorithm will break down, as it should, and refuse to give us any solution if we ask a foolish, unanswerable question; for example, if we fail to specify enough information to determine any normalizable density matrix, if we specify logically contradictory constraints, or if we specify space-time variations incompatible with the Hamiltonian of the system.

The reader may find it instructive to work this out in detail for a very simple system involving only a (2×2) matrix; a single particle of spin 1/2, gyromagnetic ratio γ, placed in a constant magnetic field $\vec{B}$ in the z-direction, Hamiltonian $H = -(1/2)\hbar\,\gamma(\vec{\sigma}\cdot\vec{B})$. Then the only dynamically possible behavior is uniform precession about $\vec{B}$ at the Larmor frequency $\omega_0 = \gamma B$. If we specify any time variation for $\langle\sigma_x\rangle$ other than sinusoidal at this frequency, the above equations will break down; while if we specify $\langle\sigma_x(t)\rangle = a\cos\omega_0 t + b\sin\omega_0 t$, we find a unique solution whenever $(a^2 + b^2) \leq 1$.

Mathematically, whether the ensemble ρ is or is not making a sharp prediction of some quantity J is determined by whether the variance $\langle J^2 \rangle - \langle J \rangle^2$ is sufficiently small. In general, information about a quantity F would not suffice to predict some other quantity J with deductive certainty (unless J is a function of F). But in inductive reasoning, Information Theory tells us the precise extent to which information about F is relevant to predictions of J. In practice, due to the enormously high dimensionality of the spaces involved, the variance $\langle J^2 \rangle - \langle J \rangle^2$ usually turns out to be very small compared to any reasonable mean-square experimental error; and therefore the predictions are, for all practical purposes, deterministic.

Experimentally, we impose various constraints (volume, pressure, magnetic field, gravitational or centrifugal forces, sound level, light intensity, chemical environment, etc.) on a system and observe how it behaves. But only when we reach the degree of control where reproducible response is observed, do we record our data and send it off for publication. Because of this sociological convention, it is not the business of statistical mechanics to predict everything that can be observed in nature; only what can be observed reproducibly. But the experimentally imposed macroscopic constraints surely do not determine any unique microscopic state; they ensure only that the state vector is somewhere in the HPM. If effect A is, nevertheless, reproducible, then it must be that A is characteristic of each of the overwhelming majority of possible states in the HPM; and so averaging over those states will not change the prediction.

To put it another way, the macroscopic experimental conditions still leave billions of microscopic details undetermined. If, nevertheless, some result is reproducible, then those details must have been irrelevant to the phenomenon; and so with proper understanding we ought to be able to eliminate them mathematically. This is just what Information Theory does for us; it removes irrelevant details by averaging over them, while retaining what is relevant to the particular question being asked [i.e., the particular quantity J(x,t) that we want to predict].

It is clear, then, why the maximum entropy prescription works in such generality. If the constraints used in the calculation are the same as those actually operative in the experiment, then the maximum-entropy density matrix will locate the same HPM as did the experimental conditions; and will therefore make sharp predictions of any reproducible effect, provided that our assumed microscopic physics (enumeration of possible states, equations of motion) is correct.

For these reasons--as stressed in Jaynes (1957b),--if the class of phenomena predictable from the maximum entropy principle is found to differ in any way from the class of reproducible phenomena, that would constitute evidence for new microscopic laws of physics, not presently known. Indeed (Jaynes, 1968) this is just what did happen early in this Century; the failure of Gibbs' classical statistical mechanics to predict the correct heat capacities and vapor pressures provided the first clues pointing to the quantum theory. Any successes make this theory useful in an "engineering" sense; but for a research physicist its failures would be far more valuable than its successes.

We emphasize that the basic physical and conceptual formulation of the theory is complete at this point; what follows represents only the working out of various mathematical consequences of this algorithm.

Perturbation Theory. For systems close to thermal equilibrium, the following general theorems are useful. We denote an "unperturbed" density matrix ρ_o, by

$$\rho_o = \frac{e^A}{Z_o} \quad , \qquad Z_o \equiv \mathrm{Tr}\left(e^A\right) \quad , \tag{D24}$$

a "perturbed one by

$$\rho = \frac{e^{A+\varepsilon B}}{Z} \, , \qquad Z \equiv \mathrm{Tr}\left(e^{A+\varepsilon B}\right) \tag{D25}$$

where A, B are Hermitian. The expectation values of any operator C over these ensembles are respectively

$$\langle C\rangle_o = \mathrm{Tr}(\rho_o C) \quad , \qquad \langle C\rangle = \mathrm{Tr}(\rho C) \quad . \tag{D26}$$

The cumulant expansion of <C> to all orders in ε is derived in Heims and Jaynes (1962), Appendix B. The n'th order term may be written as a covariance in the unperturbed ensemble:

$$\langle C\rangle - \langle C\rangle_o = \sum_{n=1}^{\infty} \varepsilon^n \left[\langle Q_n C\rangle_o - \langle Q_n\rangle_o \langle C\rangle_o\right] \quad . \tag{D27}$$

Here Q_n is defined by $Q_1 \equiv S_1$, and

$$Q_n \equiv S_n - \sum_{k=1}^{n-1} \langle Q_k\rangle_o \, S_{n-k} \quad , \qquad n > 1 \tag{D28}$$

in which S_n are the operators appearing in the well-known expansion

$$e^{A+\varepsilon B} = e^A \left[1 + \sum_{n=1}^{\infty} \varepsilon^n \, S_n\right] \quad . \tag{D29}$$

More explicitly,

$$S_n = \int_0^1 dx_1 \int_0^{x_1} dx_2 \cdots \int_0^{x_{n-1}} dx_n \; B(x_1) \cdots B(x_n) \tag{D30}$$

where

$$B(x) \equiv e^{-xA} \, B \, e^{xA} \quad . \tag{D31}$$

The first-order term is thus

$$\langle C \rangle - \langle C \rangle_o = \varepsilon \int_0^1 dx \left[\langle e^{-xA} B \, e^{xA} C \rangle_o - \langle B \rangle_o \langle C \rangle_o \right] \tag{D32}$$

and it will appear below that all relations of linear transport theory are special cases of (D32).

For a more condensed notation, define the average of any operator B over the sequence of similarity transformations as

$$\overline{B} \equiv \int_0^1 dx \; e^{-xA} B \, e^{xA} \tag{D33}$$

which we will call the Kubo transform of B. Then (D32) becomes

$$\langle C \rangle - \langle C \rangle_o = \varepsilon \, K_{CB} \tag{D34}$$

in which, for various choices of C, B, the quantities

$$K_{CB} \equiv \langle \overline{B} C \rangle_o - \langle \overline{B} \rangle_o \langle C \rangle_o \tag{D35}$$

are the basic covariance functions of the linear theory.

We list a few useful properties of these quantities; in all cases, the result is proved easily by writing out the expressions in the representation where A is diagonal. Let F, G be any two operators; then

$$\langle \overline{F} \rangle_o = \langle F \rangle_o \tag{D36}$$

$$K_{FG} = K_{GF} \quad . \tag{D37}$$

If F, G are Hermitian, then

$$K_{FG} \text{ is real} \quad , \qquad K_{FF} \geq 0 \quad . \tag{D38}$$

If ρ_o is a projection operator representing a pure state, then $K_{FG} \equiv 0$. If ρ_o is not a pure state density matrix, then with Hermitian F, G,

$$K_{FF}K_{GG} - K_{FG}^2 \geq 0 \tag{D39}$$

with equality if and only if $F = qG$, where q is a real number. If G is of the form

$$G(u) = e^{-uA}\, G(0)\, e^{uA} \tag{D40}$$

then

$$\frac{d}{du} K_{FG} = \langle [F,G] \rangle_o \quad . \tag{D41}$$

This identity, with u interpreted as a time, provides a general connection between statistical and dynamical problems.

Near-Equilibrium Ensembles. A closed system in thermal equilibrium is described, as usual, by the density matrix

$$\rho_o = \frac{e^{-\beta H}}{Z_o(\beta)} \tag{D42}$$

which maximizes S_I for prescribed $\langle H \rangle$, and is a very special case of (D19). The thermal equilibrium prediction for any quantity F is, as usual,

$$\langle F \rangle_o = \mathrm{Tr}(\rho_o F) \quad . \tag{D43}$$

But suppose we are now given the value of $\langle F(t) \rangle$ throughout the "information-gathering" interval $-\tau \leq t \leq 0$. The ensemble which includes this new information is of the form (D19), which maximizes S_I for prescribed $\langle H \rangle$ and $\langle F(t) \rangle$. It corresponds to the partition functional

$$Z[\beta,\lambda(t)] = \mathrm{Tr}\, \exp\left[-\beta H + \int_{-\tau}^{0} \lambda(t)\, F(t)\, dt\right] \quad . \tag{D44}$$

If, during the information-gathering interval, this new information was simply $\langle F(t) \rangle = \langle F \rangle_o$, it is easily shown from (D17) that we have identically

$$\int_{-\tau}^{0} \lambda(t)\, F(t)\, dt = 0 \quad . \tag{D45}$$

In words: if the new information is redundant (in the sense that it is only what we would have predicted from the old information), then it will drop out of the equations and the ensemble is unchanged. This is a general property of the formalism here presented. In applications it means that there is never any need, when setting up an ensemble, to ascertain whether the different pieces of information used are independent; any redundant parts will drop out automatically.

If, therefore, we treat the integral in (D44) as a small perturbation, we are expanding in powers of the departure from equilibrium. For validity of the perturbation scheme it is not necessary that $\lambda(t)F(t)$ be everywhere small; it is sufficient if the integral is small. First-order effects in the departure from equilibrium, such as linear diffusion or heat flow, are then predicted using the general formula (D32), with the choices $A = -\beta H$, and

$$\varepsilon B = \int_{-\tau}^{0} \lambda(t)\, F(t)\, dt \quad . \tag{D46}$$

With constant H, the Heisenberg operator F(t) reduces to

$$F(t) = \exp(iHt/\hbar)\, F(0)\, \exp(-iHt/\hbar) \tag{D47}$$

and its Kubo transform (D33) becomes

$$\overline{F}(t) = \frac{1}{\beta}\int_{0}^{\beta} du\, F(t - i\hbar u) \quad , \tag{D48}$$

the characteristic quantity of the Kubo (1957, 1958) theory.

In the notation of (D34), the first-order expectation value of any quantity C(t) will then be given by

$$\langle C(t)\rangle - \langle C\rangle_0 = \int_{-\tau}^{0} K_{CF}(t,t')\, \lambda(t')\, dt' \tag{D49}$$

where K_{CF} is now indicated as a function of the parameters t,t' contained in the operators:

$$K_{CF}(t,t') \equiv \langle \overline{F}(t')C(t)\rangle_0 - \langle F\rangle_0 \langle C\rangle_0 \tag{D50}$$

Remembering that the parameters t,t' are part of the operators C, F, the general reciprocity law (D37) now becomes

$$K_{CF}(t,t') = K_{FC}(t',t) \quad . \tag{D51}$$

When H is constant, it follows also from (D47) that

$$K_{CF}(t,t') = K_{CF}(t - t') \tag{D52}$$

and (D41) becomes

$$i\hbar \frac{\partial}{\partial t} K_{CF}(t,t') = \langle [C(t), F(t')]\rangle_0 \; \beta^{-1} \tag{D53}$$

Integral Equations for the Lagrange Multipliers. We wish to find the $\lambda_k(x,t)$ to first order in the given departures from equilibrium, $\langle F_k(x,t)\rangle - \langle F_k(x,t)\rangle_0$. This could be done by direct application of the formalism; by finding the perturbation expansion of log Z to second order in the λ's and taking the functional derivative explicitly according to (D21). It will be sufficient to do this for the simpler case described by Equations (D44)-(D53); but on carrying through this calculation we discover that the result is already contained in our perturbation-theory formula (D49). This is valid for any operator C(t); and therefore in particular for the choice C(t) = F(t). Then (D49) becomes

$$\int_{-\tau}^{0} K_{FF}(t,t')\lambda(t')dt' = \langle F(t)\rangle - \langle F\rangle_0 \tag{D54}$$

If t is in the "information-gathering interval" $(-\tau \le t \le 0)$ this is identical with what we get on writing out (D17) explicitly with log Z expanded to second order. In lowest order, then, taking the functional derivative of log Z has the effect of constructing a linear Fredholm integral equation for $\lambda(t)$, in which the "driving force" is the given departure from equilibrium.

However, from that direct manner of derivation it would appear that (D54) applies only when $(-\tau \le t \le 0)$; while the derivation of (D49) makes it clear that (D54) has a definite meaning for all t. When t is in $-\tau \le t \le 0$, it represents the integral equation from which $\lambda(t)$ is to be determined; when $t > 0$, it represents the predicted future of F(t); and when $t < -\tau$, it represents the retrodicted past of F(t).

If the information about $\langle F(t)\rangle$ is given in the entire past, $\tau = \infty$, (D54) becomes a Wiener-Hopf equation. Techniques for the solution, involving matching functions on strips in the complex fourier transform space are well known; we remark only that the solution $\lambda(t)$ will in general contain a δ-function singularity at $t = 0$; it is essential to retain this in order to get correct physical predictions. In other words, the upper limit of integration in (D54) must be taken as (0+). The case of finite τ, where we generally have δ-functions at both end-points, is discussed by Middleton (1960).

For example, with a Lorentzian correlation function

$$K_{FF}(t-t') = \frac{a}{2}\exp[-a|t-t'|] \tag{D55}$$

and $\tau = \infty$, the solution is found to be

$$\lambda(t) = \left(1 - \frac{1}{a^2}\frac{d^2}{dt^2}\right)f(t) + \frac{1}{a^2} f(0)[a\delta(t) - \delta'(t)] \tag{D56}$$

where

$$f(t) \equiv \begin{cases} \langle F(t)\rangle - \langle F\rangle_o & t \leq 0 \\ 0 & t > 0 \end{cases} \tag{D57}$$

Then we find

$$\int_{-\infty}^{o+} K_{FF}(t-t')\lambda(t')dt' = \begin{cases} f(t) \ , & t < 0 \\ f(0)e^{-at} \ , & t > 0 \end{cases} \tag{D58}$$

in which it is necessary to note the δ-functions in f"(t) at the upper limit. The nature and need for these δ-functions becomes clear if we approach the solution as the limit of the solutions for a sequence $\{f_n\}$ of "good" driving functions each of which satisfies the same boundary conditions as K_{FF} at the upper limit:

$$\left[f_n'(t')K_{FF}(t-t') - f_n(t')\frac{\partial}{\partial t'}K_{FF}(t-t')\right]_{t'=0} = 0 \ , \quad t < 0 \ . \tag{D59}$$

The result (D58) thus predicts the usual exponential approach back to equilibrium, with a relaxation time $\tau = a^{-1}$. The particular correlation function (D55) is "Markoffian" in that the predicted future decay depends only on the specified departure from equilibrium at $t = 0$; and not on information about its past history. With other forms of correlation function we get a more complicated prediction, with in general more than one relaxation time.

Relation to the Wiener Prediction Theory. This problem is so similar conceptually to Wiener's (1949) problem of optimal prediction of the future of a random function whose past is given that one would guess them to be mathematically related. However, this is not obvious from the above, because the Wiener theory was stated in entirely different terms. In particular, it contained no quantity such as λ(t) which enables us to express both the given past and predicted future of F(t) in a single equation (D54). To establish the connection between these two theories, and to exhibit an alternative form of our theory, we may eliminate λ(t) by the following purely formal manipulations.

If the resolvent $K_{FF}^{-1}(t,t')$ of the integral equation (D54) can be found so that

$$\int_{-\tau}^{0} K_{FF}(t,t'')K_{FF}^{-1}(t'',t')dt'' = \delta(t-t'), \quad -\tau \leq t,t' \leq 0 \tag{D60}$$

$$\int_{-\tau}^{0} K_{FF}^{-1}(t,t'')K_{FF}(t'',t')dt'' = \delta(t-t') \ , \quad -\tau \leq t,t' \leq 0 \tag{D61}$$

then

$$\lambda(t) = \int_{-\tau}^{0} K_{FF}^{-1}(t,t')[\langle F(t')\rangle - \langle F\rangle_o]dt' \quad , \quad -\tau \le t \le 0 \tag{D62}$$

and the predicted value (D49) of any quantity C(t) can be expressed directly as a linear combination of the given departures of F from equilibrium:

$$\langle C(t)\rangle - \langle C\rangle_o = \int_{-\tau}^{0} R_{CF}(t,t')[\langle F(t')\rangle - \langle F\rangle_o]dt' \tag{D63}$$

in which

$$R_{CF}(t,t') \equiv \int_{-\tau}^{0} K_{CF}(t,t'')K_{FF}^{-1}(t'',t')dt'' \tag{D64}$$

will be called the relevance function.

In consequence of (D61), the relevance function is itself the solution of an integral equation:

$$K_{CF}(t) = \int_{-\tau}^{0} R_{CF}(t,t')K_{FF}(t')dt' \quad , \quad -\infty < t < \infty \tag{D65}$$

so that, in some cases, the final prediction formula (D63) can be obtained directly from (D65) without the intermediate step of calculating $\lambda(t)$.

In the Wiener theory we have a random function f(t) whose past is known. For any "lead time" $h > 0$, we are to try to predict the value of f(t+h) by a linear operation on the past of f(t), i.e., the prediction is

$$\hat{f}(t+h) = \int_{0}^{\infty} f(t-t')W(t')dt' \tag{D66}$$

and the problem is to find that W(t) which minimizes the mean square error of the prediction:

$$I[W] = \lim_{T\to\infty} \frac{1}{2T}\int_{-T}^{T} |f(t+h) - \hat{f}(t+h)|^2 \, dt \quad . \tag{D67}$$

We readily find that the optimal W satisfies the Wiener-Hopf integral equation

$$\phi(t+h) = \int_{0}^{\infty} \phi(t-t')W(t')dt' \quad , \quad t \ge 0 \tag{D68}$$

where

$$\phi(t) \equiv \lim_{T\to\infty} \frac{1}{2T}\int_{-T}^{T} f(t+t')f(t')dt' \tag{D69}$$

is the autocorrelation function of f(t), assumed known.

Evidently, the response function W(t) corresponds to our relevance function $R_{FF}(t,t')$; and to establish the formal identity of the two theories, we need only show that R also satisfies the integral equation (D68). But, with the choice $C(t) = F(t)$, this is included in (D65) making the appropriate changes in notation; our "quantum covariance function" $K_{FF}(t)$ corresponding to Wiener's autocorrelation function $\phi(t)$. In the early stages of this work, the discovery of the formal identity between Wiener's prediction theory and this special case of the maximum-entropy prediction theory was an important reassurance.

The relevance function $R_{CF}(t,t')$ summarizes the precise extent to which information about F at time t' is relevant to prediction of C at time t. It is entirely different from the physical impulse-response function $\phi_{CF}(t-t')$ discussed, for example, by Kubo (1958), Eq. (2.18). The latter represents the dynamical response $\langle C(t)\rangle - \langle C\rangle_0$ at a time $t > t'$, to an impulsive force term in the Hamiltonian applied at $t = t'$: $H(t) = H_0 + F\,\delta(t-t')$, while in (D63) the "input" $\langle F(t')\rangle - \langle F\rangle_0$ consists only of _information_ concerning what the system, with a fixed Hamiltonian but in a nonequilibrium state, was doing in the interval $-\tau \le t' \le 0$. This distinction is perhaps brought out most clearly by emphasizing again that (D63) is valid for an arbitrary time t, which may be before, within, or after this information-gathering interval. Thus, while our conception of causality is based on the postulate that a force applied at time t' can exert physical influences only at _later_ times, there is no such limitation in (D63). It therefore represents an explicit statement of the fact that, while _physical influences_ propagate only forward in time, _logical inferences_ propagate equally well in either direction; i.e., new information about the present affects our knowledge of the past as well as the future. Although relations such as (D63) have been rather rare in physics, the situation is, of course, commonplace in other fields; sciences such as geology depend on logical connections of this type.

Space-Time Variations. Suppose the particle density $n(x,t)$ departs slightly from its equilibrium value in a space-time region R. Defining $\delta n(x,t) \equiv n(x,t) - \langle n(x)\rangle_0$, the ensemble containing this information corresponds to the partition functional

$$Z[\beta,\lambda(x,t)] = \mathrm{Tr}\, e\left[-\beta H + \int_R \lambda(x,t)\,\delta n(x,t)\,d^3x\,dt\right] \qquad \text{(D70)}$$

and (D34) becomes

$$\langle \delta n(x,t)\rangle = \int_R K_{nn}(x-x';t-t')\lambda(x',t')\,d^3x'\,dt' \quad . \qquad \text{(D71)}$$

When (x,t) are in R this represents the integral equation determining $\lambda(x',t')$; when (x,t) are outside R it gives the predicted nonequilibrium behavior of $\langle n(x,t)\rangle$ based on this information, and the predicted departure from equilibrium of any other quantity J(x,t) is

$$\langle J(x,t)\rangle - \langle J\rangle_o = \int_R K_{Jn}(x-x';t-t')\lambda(x',t')d^3x'dt' \quad . \tag{D72}$$

To emphasize the generality of (D72), note that it contains no limitation on time scale or space scale. Thus it encompasses both diffusion and ultrasonic propagation.

In (D71) we see the deviation $\langle\delta n\rangle$ expressed as a linear superposition of basic relaxation functions $K_{nn}(x,t) = \langle\delta\bar{n}(0,0)\delta n(x,t)\rangle_o$ with $\lambda(x',t')$ as the "source" function. The class of different nonequilibrium ensembles based on information about $\langle\delta n\rangle$ is in 1:1 correspondence with different functions $\lambda(x,t)$. In view of the linearity, we may superpose elementary solutions in any way, and while to solve a problem with specific given information would require that we solve an integral equation, we can extract the general laws of nonequilibrium behavior from (D71), (D72) without this, by considering $\lambda(x,t)$ as the independent variable.

For example, let J be the α-component of particle current, and for brevity write the covariance function in (D72) as

$$\langle\delta\bar{n}(x',t')J_\alpha(x,t)\rangle_o = K_\alpha(x-x';t-t') \quad . \tag{D73}$$

Now choose R as all space and all time $t < 0$, and take

$$\lambda(x,t) = \mu(x)\dot{q}(t) \quad , \quad t < 0 \tag{D74}$$

where $\mu(x)$, $q(t)$ are "arbitrary" functions (but of course, sufficiently well-behaved so that what we do with them makes sense mathematically). In this ensemble, the current is

$$\langle J_\alpha(x,t)\rangle = \int d^3x' \ \mu(x')\int_{-\infty}^{0} dt' \ \dot{q}(t')K_\alpha(x-x',t-t') \quad . \tag{D75}$$

Integrate by parts on t and use the identity $\dot{n} + \nabla\cdot J = 0$: the RHS of (D75) becomes

$$\int d^3x' \ \mu(x')\left[q(0)K_\alpha(x-x',0) + \frac{\partial}{\partial x'_\beta}\int_{-\infty}^{0} dt' \ q(t)K_{\alpha\beta}(x-x',t-t')\right] \tag{D76}$$

where $K_{\alpha\beta}$ is the current-current covariance:

$$K_{\alpha\beta}(x-x',t-t') \equiv \langle\bar{J}_\beta(x',t')J_\alpha(x,t)\rangle_o \quad . \tag{D77}$$

But from symmetry $K_\alpha(x-x',0) \equiv 0$. Another integration by parts then yields

$$<J_\alpha(x,t)> = -\int_{-\infty}^{0} dt' q(t') \int d^3x' \; K_{\alpha\beta}(x-x',t-t') \frac{\partial \mu}{\partial x'_\beta} \tag{D78}$$

and thus far no approximations have been made.

Now let us pass to the "long wavelength" limit by supposing $\mu(x)$ so slowly varying that $\partial\mu/\partial x'\beta$ is essentially a constant over distances in which $K_{\alpha\beta}$ is appreciable:

$$<J_\alpha(x,t)> \cong -\frac{\partial \mu}{\partial x\beta}\int_{-\infty}^{0} dt' \; q(t') \int d^3x' \; K_{\alpha\beta}(x-x',t-t') \tag{D79}$$

and in the same approximation (D71) becomes

$$<\delta n(x,t)> \simeq q(0)\mu(x) \int d^3x' \; K_{nn}(x',0) \; . \tag{D80}$$

Therefore, the theory predicts the relation

$$<J_\alpha> = -D_{\alpha\beta} \frac{\partial}{\partial x_\beta} <\delta n> \tag{D81}$$

with

$$D_{\alpha\beta} \equiv \frac{\int_{-\infty}^{0} dt' \; q(t') \int d^3x' \; K_{\alpha\beta}(x-x',t-t')}{q(0) \int d^3x' \; K_{nn}(x-x',0)} \; . \tag{D82}$$

If the ensemble is also quasi-stationary, $q(t)$ very slowly varying, only the value of $q(t)$ near the upper limit matters, and the choice $q(t) = \exp(\varepsilon t)$ is as good as any. This leads to just the Kubo expression for the diffusion coefficient.

If instead of taking the long-wavelength limit we choose a plane wave: $\mu(x) = \exp(ik \cdot x)$, (D75), (D71) become

$$<J_\alpha(x,t)> = e^{ik\cdot x} \int_{-\infty}^{0} dt' \; \dot{q}(t') K_\alpha(k;t-t') \tag{D83}$$

$$<\delta n(x,t)> = e^{ik\cdot x} \int_{-\infty}^{0} dt' \; \dot{q}(t') K_{nn}(k,t-t') \tag{D84}$$

where $K_\alpha(k,t)$, $K_{nn}(k,t)$ are the space fourier transforms. These represent the decay of sound waves as linear superpositions of many characteristic decays $\sim K_\alpha(k,t)$, $K_{nn}(k,t)$ with various "starting times" t'. If we take time fourier transforms, (D84) becomes

$$<\delta n(x,t)> = e^{ik\cdot x} \int \frac{d\omega}{2\pi} K_{nn}(k,\omega) Q(\omega) e^{-i\omega t} \tag{D85}$$

which shows how the exact details of the decay depend on the method of preparation (past history) as summarized by $Q(\omega)$. Now, however, we find that $K_{nn}(k,\omega)$ usually has a sharp peak at some frequency $\omega = \omega_o(k)$, which arises mathematically from a pole near the real axis in the complex ω-plane. Thus if $\omega_1 = \omega_o - i\alpha$ and $K_{nn}(k,\omega)$ has the form

$$K_{nn}(k,\omega) = \frac{-iK_1}{\omega-\omega_1} + \hat{K}(k,\omega) \tag{D86}$$

where $\hat{K}(k,\omega)$ is analytic in a neighborhood of ω_1, this pole will give a contribution to the integral (D86) of

$$K_1\, Q(\omega_1)\, e^{i(k\cdot x-\omega_o t)}\, e^{-\alpha t} \quad , \qquad t > 0 \; . \tag{D87}$$

Terms which arise from parts of $K_{nn}(k,\omega)Q(\omega)$ that are not sharply peaked as a function of ω, decay rapidly and represent short transient effects that depend on the exact method of preparation. If α is small, the contribution (D87) will quickly dominate them, leading to a long-term attenuation and propagation velocity essentially independent of the method of preparation.

Thus, the laws of ultrasonic dispersion and attenuation are contained in the location and width of the sharp peaks in $K_{nn}(k,\omega)$.

Other Forms of the Theory. Thus far we have considered the application of maximum entropy in its most general form: given some arbitrary initial information, to answer an arbitrary question about reproducible effects. Of course, we may ask any question we please; but maximum entropy can make sharp predictions only of reproducible things (that is in itself a useful property; for maximum entropy can tell us which things are and are not reproducible, by the sharpness of its predictions). Maximum entropy separates out what is relevant for predicting reproducible phenomena, and discards what is irrelevant (we saw this even in the example of Wolf's die where, surely, the only reproducible events in his sample space of $6^{20,000}$ points were the six face frequencies or functions of them; just the things that maximum entropy predicted).

Likewise, in the stage 2 techniques of prediction from the maximum entropy distribution, if we are not interested in every question about reproducible effects, but only some "relevant subset" of them, we may seek a further elimination of details that are irrelevant to those particular questions. But this kind of problem has come up before in mathematical physics; and Dicke (1946) introduced an elegant projection operator technique for calculating desired external elements of a scattering matrix while discarding irrelevant internal details. Our present problem, although entirely different in physical and mathematical details, is practically identical formally; and so this same technique must be applicable.

Zwanzig (1962) introduced projection operators for dealing with two interacting systems, only one of which is of interest, the other serving only as its "heat bath." Robertson (1964) recognized that this will work equally well for any kind of separation, not necessarily spatial; i.e., if we want to predict only the behavior of a few physical quantities $\{F_1 \ldots F_m\}$ we can introduce a projection operator P which throws away everything that is irrelevant for predicting those particular things; allowing, in effect, everything else to serve as a kind of "heat bath" for them.

In the statistical theory this dichotomy may be viewed in another way: instead of "relevant" and "irrelevant" read "systematic" and "random." Then, referring to Robertson's presentation in this volume, it is seen that Eq. (9.3), which could be taken as the definition of his projection operator P(t), is formally just the same as the solution of the problem of "reduction of equations of condition" given by Laplace for the optimal estimate of systematic effects. A modern version can be found in statistics textbooks, under the heading: "multiple regression." Likewise, his formulas (9.8), (9.9) for the "subtracted correlation functions" have a close formal correspondence to Dicke's final formulas.

Of course, this introduction of projection operators is not absolutely required by basic principles; it is in the realm of art, and any work of art may be executed in more than one way. All kinds of changes in detail may still be thought of; but needless to say, most of them have been thought of already, investigated, and quietly dropped. Seeing how far Robertson has now carried this approach, and how many nice results he has uncovered, it is pretty clear that anyone who wants to do it differently has his work cut out for him.

Finally, I should prepare the reader for his contribution. When Baldwin was a student of mine in the early 1960's, I learned that he has the same trait that Lagrange and Fermi showed in their early works: he takes delight in inventing tricky variational arguments, which seem at first glance totally wrong. After long, deep thought it always developed that what he did was correct after all. A beautiful example is his derivation of (4.3), where most readers would leave the track without this hint from someone with experience in reading his works: you are not allowed to take the trace and thus prove that $a = 1$, invalidating (4.4), because this is a formal argument in which the symbol $\langle F \rangle$ stands for $\mathrm{Tr}(F\sigma)$ even when σ is not normalized to $\mathrm{Tr}(\sigma) = 1$. For a similar reason, you are not allowed to protest that if $F_o \equiv 1$, then $\delta \langle F_o \rangle \equiv 0$. One of my mathematics professors once threw our class into an uproar by the same trick; evaluating an integral, correctly, by differentiating with respect to π. For those with a taste for subtle trickery, variational mathematics is the most fun of all.

References

T. Bayes (1763), Phil. Trans. Roy. Soc. 330-418. Reprint, with biographical note by G. A. Barnard, in Biometrika 45, 293 (1958) and in Studies in the History of Statistics and Probability, E. S. Pearson and M. G. Kendall, editors, C. R. Griffin and Co. Ltd., London (1970). Also reprinted in Two Papers by Bayes with Commentaries, W. E. Deming, editor, Hafner Publishing Co., New York (1963).

E. T. Bell (1937), Men of Mathematics, Simon and Schuster, N.Y.

L. Boltzmann (1871), Wien. Ber. 63, 397, 679, 712.

G. Boole (1854), The Laws of Thought, Reprinted by Dover Publications, Inc., New York.

R. T. Cox (1946), Am. J. Phys. 17, 1. See also The Algebra of Probable Inference, Johns Hopkins University Press (1961); reviewed by E. T. Jaynes in Am. J. Phys. 31, 66 (1963).

E. Czuber (1908), Wahrscheinlichkeitsrechnung und Ihre Anwendung auf Fehlerausgleichung, Teubner, Berlin. Two Volumes.

P. and T. Ehrenfest (1912), Encykl. Math. Wiss. English translation by M. J. Moravcsik, The Conceptual Foundations of the Statistical Approach in Mechanics. Cornell University Press, Ithaca, N. Y. (1959).

P. L. Ellis (1842), "On the Foundations of the Theory of Probability", Camb. Phil. Soc. vol. viii. The quotation given by Rowlinson actually appears in a later work: Phil. Mag. vol. xxxvii (1850). Both are reprinted in his Mathematical and other Writings (1863).

K. Friedman and A. Shimony (1971), J. Stat. Phys. 3, 381. See also M. Tribus and H. Motroni, ibid. 4, 227; A. Hobson, ibid. 6, 189; D. Gage and D. Hestenes, ibid. 7, 89; A. Shimony, ibid. 9, 187; K. Friedman, ibid. 9, 265.

J. W. Gibbs (1902), Elementary Principles in Statistical Mechanics, Reprinted in Collected works and commentary, Yale University Press (1936), and by Dover Publications, Inc. (1960).

I. J. Good (1950), Probability and the Weighing of Evidence, C. Griffin and Co., London.

D. Heath and Wm. Sudderth (1976), "de Finetti's Theorem on Exchangeable Variables," The American Statistician, 30, 188.

d. ter Haar (1954), Elements of Statistical Mechanics, Rinehart and Co., New York.

S. P. Heims and E. T. Jaynes (1962), Rev. Mod. Phys. 34, 143.

V. S. Huzurbazar (1976), Sufficient Statistics (Volume 19 in Statistics Textbooks and Monographs Series) Marcel Dekker, Inc., New York.

E. T. Jaynes (1957a,b), Phys. Rev. 106, 620; 108, 171.

E. T. Jaynes (1963a), "New Engineering Applications of Information Theory," in Proceedings of the First Symposium on Engineering Applications of Random Function Theory and Probability, J. L. Bogdanoff and F. Kozin, editors; J. Wiley and Sons, New York, pp. 163-203.

E. T. Jaynes (1963b), "Information Theory and Statistical Mechanics," in Statistical Physics (1962 Brandeis Lectures) K. W. Ford, ed. W. A. Benjamin, Inc., New York, pp. 181-218.

E. T. Jaynes (1965), "Gibbs vs. Boltzmann Entropies," Am. J. Phys. 33, 391.

E. T. Jaynes (1967), "Foundations of Probability Theory and Statistical Mechanics," in Delaware Seminar in the Foundations of Physics, M. Bunge, ed., Springer-Verlag, Berlin; pp. 77-101.

E. T. Jaynes, (1968), "Prior Probabilities," IEEE Trans. on Systems Science and Cybernetics, SSC-4, pp. 227-241. Reprinted in Concepts and Applications of Modern Decision Models, V. M. Rao Tummala and R. C. Henshaw, eds., (Michigan State University Business Studies Series, 1976).

E. T. Jaynes (1971), "Violation of Boltzmann's H-theorem in Real Gases," Phys. Rev. A 4, 747.

E. T. Jaynes (1973), "The Well-Posed Problem," Found. Phys. 3, 477.

E. T. Jaynes (1976), "Confidence Intervals vs. Bayesian Interv Intervals," in Foundations of Probability Theory, Statistical Inference, and Statistical Theories of Science, W. L. Harper and C. A. Hooker, eds., D. Reidel Publishing Co., Dordrecht-Holland, pp. 175-257.

E. T. Jaynes (1978), "Electrodynamics Today," in Proceedings of the Fourth Rochester Conference on Coherence and Quantum Optics, L. Mandel and E. Wolf, eds., Plenum Press, New York.

E. T. Jaynes (1979), "Marginalization and Prior Probabilities," in Studies in Bayesian Econometrics and Statistics, A. Zellner, ed., North-Holland Publishing Co., Amsterdam (in press).

H. Jeffries (1939), Theory of Probability, Oxford University Press.

J. M. Keynes (1921), A Treatise on Probability, MacMillan and Co., London.

R. Kubo (1957), J. Phys. Soc. (Japan) 12, 570.

R. Kubo (1958), "Some Aspects of the Statistical-Mechanical Theory of Irreversible Processes," in Lectures in Theoretical Physics, Vol. 1, W. E. Brittin and L. G. Dunham, eds, Interscience Publishers, Inc., New York; pp. 120-203.

S. Kullback (1959), Information Theory and Statistics, J. Wiley and Sons, Inc.

J. Lindhard (1974), "On the Theory of Measurement and Its Consequences in Statistical Dynamics," Kgl. Danske. Vid. Selskab Mat.-Fys. Medd. 39, 1.

J. C. Maxwell (1859), "Illustrations of the Dynamical Theory of Gases," Collected Works, W. D. Niven, ed., London (1890), Vol. I, pp. 377-409.

J. C. Maxwell (1876), "On Boltzmann's Theorem on the Average Distribution of Energy in a System of Material Points," ibid. Vol. II, pp. 713-741.

D. Middleton (1960), Introduction to Statistical Communication Theory, McGraw-Hill Book Company, Inc., New York, pp. 1082-1102.

H. Mori, I. Oppenheim, and J. Ross (1962), "Some Topics in Quantum Statistics," in *Studies in Statistical Mechanics*, Volume 1, J. de Boer and G. E. Uhlenbeck, eds., North-Holland Publishing Company, Amsterdam; pp. 213-298.

G. Polya (1954), *Mathematics and Plausible Reasoning*. Princeton University Press, two volumes.

B. Robertson (1964), Stanford University, Ph.D. Thesis.

J. S. Rowlinson (1970), "Probability, Information, and Entropy," Nature, 225, 1196.

L. J. Savage (1954), *The Foundations of Statistics*, John Wiley and Sons, Inc., Revised Edition by Dover Publications, Inc., 1972.

E. Schrodinger (1948), *Statistical Thermodynamics*, Cambridge University Press.

C. Shannon (1948), Bell System Tech. J. 27, 379, 623. Reprinted in C. E. Shannon and W. Weaver, *The Mathematical Theory of Communication*, University of Illinois Press, Urbana (1949).

C. Truesdell (1960), "Ergodic Theory in Classical Statistical Mechanics," in *Ergodic Theories*, Proceedings of the International School of Physics "Enrico Fermi," P. Caldirola, ed., Academic Press, New York, pp. 21-56.

J. Venn (1866), *The Logic of Chance*

R. von Mises (1928), *Probability, Statistics and Truth*, J. Springer. Revised English Edition, H. Geiringer, ed., George Allen and Unwin Ltd., London (1957).

A. Wald (1950), *Statistical Decision Functions*, John Wiley and Sons, Inc., New York.

J. D. Watson (1965), *Molecular Biology of the Gene*, W. A. Benjamin, Inc., New York. See particularly pp. 73-100.

N. Wiener (1949), *Extrapolation, Interpolation, and Smoothing of Stationary Time Series*, Technology Press, M.I.T.

S. S. Wilks (1961), Forward to Geralomo Cardano, *The Book on Games of Chance*, translated by S. H. Gould. Holt, Rinehart, and Winston, New York.

R. Zwanzig (1960), J. Chem. Phys. 33, 1338; Phys. Rev. 124. 983 (1961).

Additional Reference

Dicke, R. H., "Measurement of Thermal Radiation at Microwave Frequencies," The Review of Scientific Instruments, Ed: Gaylord P. Harnwell, a publication of the American Institute of Physics, New York, Vol. 17, No. 7, pp. 268-275, July 1946.

OF INFERENCE AND INQUIRY, An Essay In Inductive Logic

Richard T. Cox

"What song the Sirens sang, or what name Achilles assumed when he hid himself among women, though puzzling questions, are not beyond all conjecture."

Sir Thomas Browne, Hydriotaphia : Urn-Burial

1. Conjunction, Disjunction And Implication

It is my hope and aim in this essay to employ the familiar means of Boolean algebra in the discovery of some elements of an inductive logic, a logic of inference and inquiry. I do not offer the essay as a complete treatise on the subject, but I hope it may provide insights which others will find helpful in such an undertaking [1].*

Let us begin by considering the algebra. Small letters as a, b, c, ..., will represent assertions. The signs, =, $\wedge$, $\vee$, will denote equality, conjunction and disjunction, respectively, according to the definitions [2]:

> Two assertions are equal if they tell the same thing. The conjunction of two assertions is the assertion which tells what they tell jointly; their disjunction is the assertion which tells what they tell in common.

It may be objected that things are told by assertors rather than by assertions, but expressions such as we have used here are familiar and accepted in ordinary discourse. For example, we should be more apt to say that the Westminster Catechism rather than the Westminster Assembly of Divines asked what is the chief end of man, or that the Declaration of Independence rather than the Continental Congress said that all men are created free and equal. And as far as the algebra is concerned, an assertion is not changed by being paraphrased with no change in its meaning. Thus the assertions, "He is my sister's son," and "I am his mother's brother," are the same assertion. We may represent them, if we wish, by the same letter or, if we prefer, we may represent them by different letters, a and b, and express their identity by the equation, a = b.

By these definitions, we can prove, as rules of the algebra, the equations:

$a \wedge a = a$ (1.1L) $a \vee a = a$ (1.1R)

$a \wedge b = b \wedge a$ (1.2L) $a \vee b = b \vee a$ (1.2R)

*A bracketed number, as appears here, refers to a note in the set of notes collected at the end of this essay.

$(a \wedge b) \wedge c = a \wedge (b \wedge c) = a \wedge b \wedge c$ (1.3L) $(a \vee b) \vee c = a \vee (b \vee c) = a \vee b \vee c$ (1.3R)

$(a \wedge b) \vee c = (a \vee c) \wedge (b \vee c)$ (1.4L) $(a \vee b) \wedge c = (a \wedge c) \vee (b \wedge c)$ (1.4R)

$(a \wedge b) \vee b = b$ (1.5L) $(a \vee b) \wedge b = b$ (1.5R)

The first pair of rules, (1.1L) and (1.1R), are so nearly obvious by the definitions as scarcely to require proof. What an assertion tells jointly with itself, what it tells in common with itself, what it tells by itself are all the same. The signs $\wedge$ and $\vee$ can be read, with some qualifications, as "and" and "or". To say "He is my sister's son and I am his mother's brother" tells no more than either assertion tells alone, and to say "He is my sister's son or I am his mother's brother" tells no less.

Obvious though they are, these two rules are worthy of remark as the most striking peculiarity of Boolean algebra in contrast to ordinary algebra. In some respects ordinary addition and multiplication resemble Boolean disjunction and conjunction, but they are subject to the very different rules, $a + a = 2a$ and $a \times a = a^2$. The reason for the difference is plain enough. Ordinary algebra deals with other things than assertions and of necessity deals with them differently. "Nothing so like as eggs," but their likeness does not imply that they are all the same egg [3]: two eggs, however alike, are still one egg and another egg. Or, if we speak of multiplying something by itself, we do not mean "by itself" in the idiomatic sense in which a thing by itself is a solitary thing, as when we speak of a tree standing by itself in a clearing. We mean rather that we multiply it by something to which it is equal as a quantity, but from which it is distinct as an entity. For example, the area of a rectangle with sides a and b is $a \times b$, and that of an ellipse with semi-axes a and b is $\pi \times a \times b$. If the rectangle is a square, its sides are equal and its area is a^2, and if the ellipse is a circle, its semi-axes are equal and its area is $\pi \times a^2$; but a^2, like $a \times b$, is the product of one side or one semi-axis by another, not by itself.

The equations (1.2L) and (1.2R), which express the commutative rule in Boolean algebra, also follow directly from the definitions of conjunction and disjunction in that reversing the order of a and b does not alter what they tell jointly or what they tell in common. But it is well to note that the rendering of $\wedge$ as "and," though convenient for brevity's sake, is not always consistent with either the definition of conjunction or the commutative rule. There is no difficulty if a represents "Adam delved" and b represents "Eve span". For we make no distinction of meaning between "Adam delved and Eve span" and "Eve span and Adam delved". We may write $a \wedge b$ or $b \wedge a$ for either one, because each tells what a and b tell jointly and tells no more. But Eve's apology, "The serpent beguiled me, and I did eat," does not convey the same sense as "I did eat, and the serpent beguiled me;" for we are accustomed, in the narration of successive events, to being told the earlier before the later, especially when they are cause and effect. By reason of this unstated implication, "a and b" is not the same assertion in this instance as "b and a," and neither one is the simple conjunction represented indifferently by $a \wedge b$ or $b \wedge a$. Thus, not every pair of assertions joined by "and" satisfies the logical definition of conjunction. Nor is a conjunction expressible only by such a pair. "Sarai had no child" tells all that is told jointly by the two assertions "Sarai had no son," "Sarai had no daughter," and is their conjunction.

The rendering of V, the sign of disjunction, as "or" is also subject to exceptions. An example of a disjunction expressed otherwise than as a pair of assertions joined by "or" is the assertion "Jacob had parental help in defrauding Esau." The help Jacob received was parental whether it came from both parents or only one, and if from one, whether from his mother or his father. Consequently, this assertion, although it does not contain the word "or," is nevertheless the disjunction of two assertions, "Jacob had his mother's help in defrauding Esau," "Jacob had his father's help in defrauding Esau;" because it tells what the two assertions tell in common, as the definition of disjunction requires. It can indeed be paraphrased in the assertion "Jacob had the help of his mother or his father in defrauding Esau," but only on the understanding that "or" is used in its inclusive sense, sometimes expressed by "and/or" to distinguish it from the exclusive sense, expressible by "either...or...but not both." [4]

A conjunction of assertions, being itself an assertion, can be conjoined to yet another assertion, and so on, to any number we please. Indeed the two lines,

> "Three hundred sixty days a year
> Drunk I lie, like mud every day," [5]

addressed to his wife by the Chinese poet Li Po, can reasonably be construed as a conjunction of 360 assertions, one for each day on the calendar of his time and country. Any conjunction of more than two assertions can be made in steps which show it as a conjunction of simpler conjunctions. Thus the conjunction of a, b, and c can be taken as the conjunction of $(a \wedge b)$ with c and written $(a \wedge b) \wedge c$, or as the conjunction of a with $(b \wedge c)$ and written $a \wedge (b \wedge c)$. But written either way, the conjunction of three assertions tells what all three tell jointly. Either way, therefore, it is the same assertion, which can be written more simply without parentheses, as a∧b∧c. It is the same with the disjunction of three assertions; it tells what all three tell in common and needs parentheses no more than their conjunction. Hence the pair of rules, (1.3L) and 1.3R).

The case is quite different with mixed conjunction and disjunction, shown in the rules (1.4L) and (1.4R). On the left side of (1.4L), (a∧b) tells what a and b tell jointly, and (a∧b)∨c tells what they jointly tell in common with c. What a tells in common with c is what a∨c tells, what b tells in common with c is what (b∨c) tells, and what they jointly tell in common with c is what (a∨c)∧(b∨c) tells. Thus (1.4L). On the right side of (1.4R), (a∧c) tells what a and c tell jointly, (b∧c) tells what b and c tell jointly, and (a∧c)∨(b∧c) tells what these two conjunctions tell in common. This is everything that c tells, since c is common to both conjunctions, and whatever a and b tell in common; therefore it is, as the rule states, whatever (a∨b)∧c tells.

As an example of the need, in mixed conjunction and disjunction, for parentheses to distinguish a∨(b∧c) from (a∨b)∧c, and both of these from the ambiguous form a∨b∧c, consider the remark, "The bus will be on time, or late and crowded." It can be represented by a∨(b∧c), where a, b, and c represent the three assertions, "The bus will be on time," "The bus will be late," "The bus will be crowded." It cannot be represented by (a∨b)∧c, because a and b (if it is granted that there is a bus coming sometime) tell nothing in common, and their disjunction, a∨b, is the truism, "The bus will be on time or late." Consequently (a∨b)∧c tells only what c tells alone, that the bus (whether on time or late) will be crowded.

In either (1.4L) or (1.4R), if we replace c by b and recall that b∨b, b∧b and b are all the same assertion, we find that (a∧b)∨b = (a∨b)∧b. Indeed each side of this equation is simply b by the definition of conjunction and disjunction. For on the left, (a∧b) can not tell less than b, because it tells all that a and b tell jointly; but its disjunction with b can not tell more than b, because a disjunction of assertions tells only what they tell in common. Thus (a∧b)∨b tells whatever b tells, no less and no more. Similarly on the right, (a∨b)∧b tells no more and no less than b: no more, because (a∨b) can not tell more; no less, because its conjunction with b can not tell less. Hence the pair of rules (1.5L) and (1.5R).

The rules we have just been considering are general rules, valid for all assertions. We find their first use in the definition and description of implication, a special relation which holds between some assertions.

In the familiar meaning of implication, the assertion "He is my sister's son" implies the assertion "He is my nephew," because every sister's son is a nephew by the mere meaning of the words, although not every nephew is a sister's son. This is an obvious example, but one typical of implication in general. If we represent "He is my sister's son" by x and "He is my nephew" by y we can represent "He is my sister's son and he is my nephew" by $x \wedge y$. But with every sister's son a nephew, y is clearly redundant in the conjunction. Indeed its redundancy is the very condition of its implication, which we may therefore express by the equation $x = x \wedge y$.

If we disjoin y on both sides of this equation, we have on the right $(x \wedge y) \vee y$, which is equal simply to y by (1.5L). So we have $x \vee y = y$, whence, conjoining x on both sides, we can recover the equation $x = x \wedge y$ by (1.2R), (1.5R) and (1.2L). Thus the two equations,

$x = x \wedge y$ (1.6L) and $y = x \vee y$, (1.6R)

are the same equation, and we may take either one as the condition of implication, with x as implicant and y as implicate.

We may now define implication:

> If one of two assertions is equal to their conjunction, the other is equal to their disjunction, and conversely. The relation between the assertions is implication: the assertion equal to their conjunction is the implicant; the assertion equal to their disjunction is the implicate. [6]

Or we can put the definition in more ordinary language by saying that implication is a relation in which the implying assertion tells by itself all that it tells jointly with the implied assertion, and the implied assertion tells by itself only what it tells in common with the implying assertion.

Every rule of conjunction or disjunction provides an instance or proves a theorem of implication. Thus the rules (1.1L) and (1.1R) are instances of implication in the special case in which the same assertion is both implicant and implicate. Because the rules are general, holding for any assertion, they prove that every assertion implies itself, or as it may be said:

Implication is reflexive.

The rule (1.2L) proves that:

Implication is never mutual between different assertions.

For by (1.6L), if a implies b, $a = a \wedge b$; and if b implies a, $b = b \wedge a$. The commutative rule makes these conditions incompatible unless a and b are the same assertion. This theorem can be proved as well by the rule (1.2R) and the condition of implication as expressed by (1.6R).

The rule (1.3L), or its counterpart (1.3R), proves that:

Implication is transitive,

in the sense that, in any series of assertions, if each implies the one next following, then each implies every one following and is implied by every one preceding.

For proof, consider three assertions, a, b, c, and let a imply b and b imply c. By (1.6L), the conditions of these implications are $a = a \wedge b$ and $b = b \wedge c$. Substituting in (1.3L), we find that $a \wedge c = a \wedge b = a$, and thus a implies c. The theorem follows by repeated application of this result.

The rules (1.4L) and (1.4R) serve to prove the theorems:

> An assertion which implies both of two others implies their conjunction;

and:

> An assertion which is implied by both of two others is implied by their disjunction.

For proof of the first of these, let c be an assertion implying two assertions a and b. The conditions of these implications can be expressed as $a \vee c = a$ and $b \vee c = b$, whence $(a \vee c) \wedge (b \vee c) = a \wedge b$. By substitution in the rule (1.4L), we have $(a \wedge b) \vee c = a \wedge b$, and see that c implies $(a \wedge b)$.

To prove the second theorem of this pair, let c be implied by both a and b. Expressing the conditions of these implications as $a = a \wedge c$ and $b = b \wedge c$, we find, by substitution in the rule (1.4R), that $(a \vee b) \wedge c = a \vee b$, and see that c is implied by $(a \vee b)$.

Finally, we come to the last pair of rules and note that (1.5L) is the condition by which b is implied by $(a \wedge b)$, and that (1.5R) is the condition by which b implies $(a \vee b)$, whence the theorem:

> Every assertion is implied by its conjunction, and implies its disjunction with any other assertion.

To complete the chapter, we have now to notice how conjunction, disjunction and implication are involved in the truth and falsity of assertions.

In the familiar canons of Aristotelian logic:

> Some assertions are true; some assertions are false.

> Every assertion is either true or false; no assertion is both true and false.
>
> A conjunction of assertions is true if all of them are true and is false if any of them is false.
>
> A disjunction of assertions is false if all of them are false and is true if any of them is true.

Hence it follows that:

> Every assertion implying a false assertion is false; every assertion implied by a true assertion is true.

For in any implication of one assertion by another, the implicant, being equal to their conjunction, must be false if the implicate is false; and the implicate, being equal to their disjunction, must be true if the implicant is true. Thus, in the obvious example with which we introduced implication, "He is my sister's son" is false if "He is my nephew" is false, and "He is my nephew" is true if "He is my sister's son" is true. On the other hand, a false implicant can have a true implicate: there are nephews who are not sister's sons.

The canon of Aristotelian logic which assigns every assertion to one and only one of the two classes, true and false, is sometimes denied. At first sight this denial might seem to be sustained by the ambiguities which abound in our ordinary discourse. There are many assertions which admit more than one interpretation; and if one interpretation is true and another false, it might be supposed that the assertion is both true and false or that it is neither true nor false. But an assertion which admits two interpretations tells whatever they tell in common and is therefore, by definition, their disjunction. And if one of the interpretations is, in its turn, ambiguous, it also is a disjunction. Thus ultimately any assertion, however ambiguous, can be construed as a disjunction of unambiguous assertions. If any of them is true, the original assertion is true; if all of them are false, the original assertion is false.

2. Contradiction, Exclusion And Exhaustion

A useful theorem simply proved is that:

> Two assertions are the same if they have the same conjunction and the same disjunction with a given assertion.

Thus let a be the given assertion, and let x and y be two other assertions. The theorem states that x = y if

$$a \wedge x = a \wedge y \qquad \text{and} \qquad a \vee x = a \vee y.$$

For proof, we disjoin x on both sides of the first equation and have on the left $(a \wedge x) \vee x$, which is simply x, by (1.5L), and on the right $(a \wedge y) \vee x$, which is equal to $(a \vee x) \wedge (y \vee x)$ by (1.4L). Thus

$$x = (a \vee x) \wedge (y \vee x).$$

Similarly, by the disjunction of y instead of x, we obtain the equation

$$(a \vee y) \wedge (x \vee y) = y,$$

which indeed can be inferred immediately from the preceding equation by the symmetry of the hypothesis with respect to x and y.

Now comparing the right side of the first of these equations with the left side of the second, we note that $a \vee x = a \vee y$ by hypothesis, and $y \vee x = x \vee y$ by (1.2R). Therefore x = y and the theorem is proved.

As an instance of this theorem, let us consider pairs of assertions which we may call mutually complementary, defining complementary assertions as a pair whose conjunction is equal to the conjunction of all assertions and whose disjunction is equal to the disjunction of all assertions. By this definition and the theorem just proved, a given assertion has only one assertion complementary to it and is itself complementary only to that assertion: every assertion is the complement of its complement. Thus if we let ~a represent the complement of a, we have the rule,

$$\sim\sim a = a. \tag{2.1}$$

Other rules governing the use of the sign ~ are

$$(a \wedge \sim a) \wedge b = a \wedge \sim a \quad (2.2L) \qquad (a \vee \sim a) \vee b = a \vee \sim a \quad (2.2R)$$

$$(a \wedge \sim a) \vee b = b \quad (2.3L) \qquad (a \vee \sim a) \wedge b = b \quad (2.3R)$$

$$\sim(a \wedge b) = \sim a \vee \sim b \quad (2.4L) \qquad \sim(a \vee b) = \sim a \wedge \sim b \quad (2.4R)$$

The rules (2.2L) and (2.3L) are the condition by which the conjunction, $(a \wedge \sim a)$, of complementary assertions implies every assertion, as it must, being equal to the conjunction of all assertions. And (2.2R) and (2.3R) are the condition by which the disjunction, $(a \vee \sim a)$, of complementary assertions, being equal to the disjunction of all assertions, is implied by every assertion.

The proof of the last two rules is more involved. For (2.4L), it consists in showing that the conjunction of (a∧b) and (~a∨~b) is equal to the conjunction of all assertions, and that their disjunction is equal to the disjunction of all assertions; for only so can (a∧b) and (~a∨~b) be proved mutually complementary. It is easily found by the rules of conjunction and disjunction that

$$(a\wedge b)\wedge(\sim a\vee\sim b) = (a\wedge b\wedge\sim a)\vee(a\wedge b\wedge\sim b).$$

The two parentheses on the right can be written as (a∧~a)∧b and (b∧~b)∧a, and so, by (2.2L), as (a∧~a) and (b∧~b). Thus we have

$$(a\wedge b)\wedge(\sim a\vee\sim b) = (a\wedge\sim a)\vee(b\wedge\sim b).$$

Because each of the conjunctions on the right in this equation is equal to the conjunction of all assertions, so also is their disjunction, and so also therefore is (a∧b)∧(~a∨~b). This is half of the proof of (2.4L).

The other half consists in showing that the disjunction of (a∧b) and (~a∨~b) is equal to the disjunction of all assertions. Returning to the rules of conjunction and disjunction, we find that

$$(a\wedge b)\vee(\sim a\vee\sim b) = (a\vee\sim a\vee\sim b)\wedge(b\vee\sim a\vee\sim b).$$

Here the two parentheses on the right can be written as (a∨~a)∨~b and (b∨~b)∨~a, and so, by (2.2R), as (a∨~a) and (b∨~b). Thus

$$(a\wedge b)\vee(\sim a\vee\sim b) = (a\vee\sim a)\wedge(b\vee\sim b);$$

and because (a∨~a), (b∨~b) and (a∨~a)∧(b∨~b) are all equal to the disjunction of all assertions, so is (a∧b)∨(~a∨~b). So (2.4L) is proved.

The same reasoning, except that conjunction and disjunction are everywhere exchanged, proves (2.4R).

These two rules can be expressed in words by saying that the complement of a conjunction of assertions is equal to the disjunction of their complements, and the complement of a disjunction of assertions is equal to the conjunction of their complements.

If now we review the seventeen rules we have found in this chapter and the one before, we discover that all of them, with the single exception of (2.1), ~~a = a, come in pairs, and that every pair is symmetrical in respect to the signs of conjunction and disjunction. If the signs are exchanged in any rule, it is thereby transformed into the other rule of the same pair. Therefore, if the exchange is made in both rules of a pair, each is transformed into the other, and the pair remains the same. If the exchange is made in the whole set, the set as a whole is unchanged.

The single rule, $\sim\sim a = a$, because it contains neither sign, makes no exception to this general symmetry. Indeed it is this rule, itself unaltered by the exchange of signs, which allows the exchange to be made in all the other rules. To see this, let us return to the last two rules we found: (2.4L), $\sim(a \wedge b) = \sim a \vee \sim b$, and (2.4R) $\sim(a \vee b) = \sim a \wedge \sim b$. Because they hold for every meaning of a and b, they remain valid if we replace a by $\sim a$ and b by $\sim b$, so obtaining the equations,

$$\sim(\sim a \wedge \sim b) = \sim\sim a \vee \sim\sim b \quad \text{and} \quad \sim(\sim a \vee \sim b) = \sim\sim a \wedge \sim\sim b$$

which, by (2.1), we can write as

$$\sim(\sim a \wedge \sim b) = a \vee b \quad \text{and} \quad \sim(\sim a \vee \sim b) = a \wedge b.$$

Because an assertion has only one complement, it follows that, if two assertions are equal, their complements are equal. Taking complements on both sides of these equations and applying the rule (2.1) on each left side, we have

$$\sim a \wedge \sim b = \sim(a \vee b) \quad \text{and} \quad \sim a \vee \sim b = \sim(a \wedge b),$$

and we see that these are the same two rules with which we started, but transposed. We have derived each of them from the other by using only the rule $\sim\sim a = a$. From this point we can go on to derive each rule from the other in all the pairs. So we discover that from any of the numerous sets of rules composed by choosing one from each pair, we can derive the rest by means of the singular rule.

Let us notice here that the conjunction of all assertions is a false assertion, and that it is false not just as a matter of observed fact (as it is false, for example, that chalk is harder than diamond) but by plain logical necessity. For some assertions are false, and a conjunction of assertions is true only if they are all true. The disjunction of all assertions, on the other hand, and equally by logical necessity, is true; for some assertions are true, and a disjunction of assertions is false only if they are all false. Hence it follows that if an assertion is true, its complement must be false; and if it is false, its complement must be true. For if an assertion and its complement were both true, their conjunction would be true, but it is false; and if they were both false, their disjunction would be false, but it is true.

Thus mutually complementary assertions, as we have defined them, are mutually contradictory, as contradiction is understood in scholastic logic:

> "If one proposition is the contradictory of another, then both propositions cannot be true nor can they both be false: If one is true, then the other is false; and if one is false, the other is true. For example, if the proposition expressed by 'All men are animals' is true, then the proposition expressed by 'Some men are not animals' cannot be true. On the other hand, if the proposition expressed by 'Some men are not animals' is true, then the proposition expressed by 'All men are animals' is false. This is so clear that further explanation would only obscure the point."

This passage from the Port Royal Logic [7] is indeed admirably clear, and it leaves no doubt that the assertions we have called mutually complementary are mutually contradictory. But further explanation is needed if we are to show that mutually contradictory assertions are, in every instance, mutually complementary.

Taking the instance from the passage just quoted, let a be the assertion "All men are animals," and b the assertion "Some men are not animals." Let c be the conjunction and d the disjunction of a and b, and for a random assertion, let e be "The moon will be full on Friday." Then $c \vee e$ is "All men are animals and some men are not animals, or the moon will be full on Fridays." To make this assertion is to say nothing else than that the moon will be full on Friday, because c, the alternative to e in the disjunction, is logically absurd. Then we may reasonably say that $c \vee e = e$, and note that this equation is the condition of implication, with c as implicant and e as implicate. If a and b were any other pair of contradictories, or e were another assertion, the case would be the same. So we may reasonably conclude that the conjunction of any two contradictories implies every assertion and is thus equal to the conjunction of all assertions.

Consider next the assertion $d \wedge e$, "All men are animals or some men are not animals, and the moon will be full on Friday." Because d is a logical truism, the sense of $d \wedge e$, like that of $c \vee e$, is simply that the moon will be full on Friday. So $d \wedge e = e$, and we note that this equation also is a condition of implication, in this instance with d as implicate and e as implicant. So we may further conclude that the disjunction of any two contradictories is implied by every assertion and is therefore equal to the disjunction of all assertions.

The whole matter can be put in a nutshell. We have seen that mutually complementary assertions can neither both be true nor both be false; therefore they are mutually contradictory. But an assertion has one and only one contradictory. Therefore its contradictory must be its complement.

Now that we have identified as contradiction the relation between mutually complementary assertions, it becomes appropriate to read the symbol ~a as "not a." It also becomes convenient to interpret, in terms of contradiction, the rules which govern the use of the sign ~. Thus:

> Every assertion has one, and only one, assertion contradictory to it; in any pair of contradictories, one must be true and one must be false.

> The conjunction of a pair of contradictories implies every assertion and is the logical absurdity. The disjunction is implied by every assertion and is the logical truism.

> The contradictory of a conjunction of assertions is the disjunction of their contradictories. The contradictory of a disjunction of assertions is the conjunction of their contradictories.

There now remain to be considered the several possibilities of implication between an assertion in one pair of contradictories and an assertion in another pair.

Let the two pairs of contradictories be x, ~x and y, ~y. First let us suppose that x implies y, and let us look for the relation between ~x and ~y consequent on this implication. The equations of the assumed implication are

$$x = x \wedge y \qquad \text{and} \qquad y = y \vee x$$

Taking contradictories of both sides of both equations, we find that

$$\sim x = \sim x \vee \sim y \qquad \text{and} \qquad \sim y = \sim y \wedge \sim x,$$

and we recognize these equations as the condition by which ~x is implied by ~y. So we have the theorem:

> If one assertion implies another, its contradictory is implied by the other's contradictory.

Next let us make a different assumption by supposing that x implies ~y, so that

$$x = x \wedge {\sim}y \qquad \text{and} \qquad {\sim}y = {\sim}y \vee x.$$

Now taking contradictories of both sides of both equations, we find that

$${\sim}x = {\sim}x \vee y \qquad \text{and} \qquad y = y \wedge {\sim}x.$$

In these equations we recognize the condition by which y implies ~x. And if we now conjoin the sides of the equation on the left above with the corresponding sides of the equation on the right below, we obtain the equation,

$$x \wedge y = (x \wedge {\sim}x) \wedge (y \wedge {\sim}y).$$

Because the right side is the logical absurdity, it follows that x and y can not both be true. This relation between them is called exclusion, according to the definition:

> Two assertions are mutually exclusive if they can not both be true.

So now we have the theorem:

> If one assertion implies the contradictory of another, each implies the contradictory of the other, and the two assertions are mutually exclusive.

Again changing the assumption, let us suppose now that x, instead of implying ~y, is implied by it. The equations of this implication are

$$x = x \vee {\sim}y \qquad \text{and} \qquad {\sim}y = {\sim}y \wedge x.$$

We find, by taking contradictories, that

$${\sim}x = {\sim}x \wedge y \qquad \text{and} \qquad y = y \vee {\sim}x,$$

whence it follows that y is implied by ~x. Also, by disjoining corresponding sides of the upper left and lower right equations, we obtain the equation

$$x \vee y = (x \vee {\sim}x) \vee (y \vee {\sim}y),$$

in which the right side is the logical truism. In this relation (which is called exhaustion) x and y can not both be false. So we have the definition:

Two assertions are an exhaustive pair if they can not both be false;

and the theorem:

If one assertion is implied by the contradictory of another, each is implied by the contradictory of the other, and the two assertions are an exhaustive pair.

Comparing the definitions of the three relations, contradiction, exclusion and exhaustion, we see that contradiction is the combination of the other two. But either subordinate relation can hold without the other. Consider the two conjunctions, $(a \wedge b)$ and $(\sim a \wedge b)$, where a and b are any assertions. Their conjunction is easily found equal to $(a \wedge \sim a)$, their disjunction to b. By the first result, $(a \wedge b)$ and $(\sim a \wedge b)$ are mutually exclusive; by the second, they are not an exhaustive pair (unless b is the logical truism). Or consider the disjunctions, $(a \vee b)$ and $(\sim a \vee b)$. Their disjunction can be proved equal to $(a \vee \sim a)$, their conjunction to b. They are an exhaustive pair, but not mutually exclusive (unless b is the logical absurdity).

"All men are animals and the moon will be full on Friday" and "Some men are not animals and the moon will be full on Friday" can not both be true, but they are both false if the moon is not full on Friday. And "All men are animals or the moon will be full on Friday" and "Some men are not animals or the moon will be full on Friday" can not both be false, but they are both true if the moon is full on Friday.

3. Probability

Implication in our first chapter and exclusion in our second were treated as two distinct relations. In this chapter we adopt another point of view and consider them as the two extremes of a single relation, which is capable of every degree from one extreme to the other, and is measurable, with high precision or low, according to certain rules which constitute the calculus of probabilities.

From this point of view, probability is a relation between assertions, a premise and an inference; greatest when the premise implies the inference, least when premise and inference are mutually exclusive. The estimation of probabilities attains its highest precision in games of chance, where the premise comprises the rules of the game, together with a description of the cards, dice or other implements of play. These are the contrived instances which brought the calculus of probabilities into being. The precision is generally lower, although the reckoning may require more shrewdness, when the premise is an inconclusive reminiscence of a varied experience.

Here is Shylock speaking of Antonio in A Merchant of Venice:

> "... my meaning in saying he is a good man is to have you understand me that he is sufficient. Yet his means are in supposition: he hath an argosy bound to Tripolis, another to the Indies; I understand, moreover, upon the Rialto he hath a third at Mexico, a fourth for England - and other ventures he hath, squandered abroad. But ships are but boards, sailors but men: there be land-rats and water-rats, water-thieves and land-thieves, I mean pirates; and then there is the peril of waters, winds and rocks. The man is, notwithstanding, sufficient: - three thousand ducats: I think I may take his bond."

Although the rules of probability are subject to choice, the choice is not entirely free. For one thing, the rules must conform to some indispensable demands of common sense, which we may take as axioms of probability [8]. For another, they must be consistent with Boolean algebra.

First let us consider an axiom. The nursery rhyme, "How many miles to Babylon?" [9] provides an example apt to our purpose. The premise is that you are seventy miles from Babylon and your heels are nimble and light. The inference is that you can get to Babylon and back again before candle-light. By any reckoning, the probability of this inference on the given premise is determined by two subordinate probabilities. The first of these is the probability, on the given premise, of the inference that you can get to Babylon with time to spare. The second is the probability of your getting back before candle-light, as estimated on the given premise in conjunction with the assertion that you will already have run one way.

To generalize this example, let us express it in symbols, representing the probability of an inference b on a premise a by $b|a$. In the example, let a be the given premise, b the assertion that you can get to Babylon in good time, and c the assertion that you can get from Babylon back to your starting-point before candle-light.

We take it now, not only as a condition for this instance but as an axiom for all instances, that

$$b \wedge c|a = \underline{F}(b|a,\ c|a \wedge b). \qquad (3.1)$$

The function $\underline{F}$ is unspecified except that it is an ordinary algebraic function of two variables. Thus our axiom is only that:

> The probability on a given premise of an inference which is the conjunction of two assertions is determined by the probability of the first assertion on the given premise and the probability of the second assertion on a premise which is the conjunction of the given premise and the first assertion.

Let us conjoin another assertion d in the inference and see what restriction is imposed on the function $\underline{F}$ by the rule (1.3L), which can be written as

$$(a \wedge b) \wedge c = a \wedge (b \wedge c) = a \wedge b \wedge c$$

or as

$$(b \wedge c) \wedge d = b \wedge (c \wedge d) = b \wedge c \wedge d.$$

For the sake of a simpler notation, let

$$b|a = \underline{x}, \qquad c|a \wedge b = \underline{y}, \qquad d|a \wedge b \wedge c = \underline{z}.$$

Then, by (3.1):

$$b \wedge c \wedge d|a = (b \wedge c) \wedge d|a = \underline{F}(b \wedge c|a, \underline{z})$$

and

$$b \wedge c|a = \underline{F}(\underline{x}, \underline{y}),$$

whence

$$b \wedge c \wedge d|a = \underline{F}[\underline{F}(\underline{x}, \underline{y}), \underline{z}].$$

Similarly,

$$b \wedge c \wedge d|a = b \wedge (c \wedge d)|a = \underline{F}(\underline{x}, c \wedge d|a \wedge b)$$

and

$$c \wedge d|a \wedge b = \underline{F}(\underline{y}, \underline{z}),$$

whence

$$b \wedge c \wedge d|a = \underline{F}[\underline{x}, \underline{F}(\underline{y}, \underline{z})].$$

Equating the two expressions for $b \wedge c \wedge d|a$, we have the functional equation [10]

$$\underline{F}[\underline{F}(\underline{x}, \underline{y}), \underline{z}] = \underline{F}[\underline{x}, \underline{F}(\underline{y}, \underline{z})], \tag{3.2}$$

which $\underline{F}$ must satisfy for the sake of consistency with Boolean algebra.

A particular solution is to let $\underline{F}$ be the product of its two variables. Then $\underline{F}(\underline{x}, \underline{y}) = \underline{x}\ \underline{y}$, $\underline{F}(\underline{y}, \underline{z}) = \underline{y}\ \underline{z}$, and each side of (3.2) is equal simply to $\underline{x}\ \underline{y}\ \underline{z}$, and therefore equal to the other side. With this choice, (3.1) becomes

$$b\wedge c|a = (b|a)(c|a\wedge b), \qquad (3.3)$$

which is simply the expression in our notation of the familiar rule for the probability of the conjunction of two assertions.

In our choice of the particular solution,

$$\underline{F}(\underline{x}, \underline{y}) = \underline{x}\ \underline{y},$$

instead of the general solution of (3.2), there was no real loss of generality as far as the calculus of probabilities is concerned. The general solution is found to be [14]

$$\underline{F}\ (\underline{x}, \underline{y}) = \underline{G}^{-1}[\underline{G}(\underline{x})\underline{G}(\underline{y})],$$

where $\underline{G}$ is an arbitrary function of a single variable.

Substituting this expression in (3.1) and taking the function $\underline{G}$ on both sides, we obtain, in place of (3.3),

$$\underline{G}(b\wedge c|a) = \underline{G}(b|a)\ \underline{G}(c|a\wedge b).$$

Every equation of the calculus which we might derive, in the pages to follow, by the use of this equation would be the same as we shall obtain by the use of (3.3), except for the difference that the symbol $\underline{G}(b|a)$ would everywhere replace the symbol $b|a$. With no more use for the replaced symbol, $b|a$, we should most likely leave it nameless and give the name 'probability' to the symbol $\underline{G}(b|a)$. Thus the only consequence of using the general rather than the particular solution of the functional equation (3.2) would be an unnecessary complexity in our notation.

Because every assertion has one and only one assertion contradictory to it, we may reasonably take it as a second axiom of probability that:

> The probability of an inference on a given premise determines the probability of its contradictory on the same premise.

Thus

$$\sim b|a = \underline{f}(b|a),$$

where $\underline{f}$ is a function as yet unspecified but which must be chosen

consistently with the rules of Boolean algebra.
In this equation, if we replace b by ~b, we have

$$\sim\sim b|a = \underline{f}(\sim b|a) = \underline{f}[\underline{f}(b|a)];$$

whence, because ~~b = b (every assertion being the contradictory of its contradictory) we see that

$$\underline{f}[\underline{f}(b|a)] = b|a. \qquad (3.4)$$

Therefore $\underline{f}$ must satisfy the functional equation

$$\underline{f}[\underline{f}(\underline{x})] = \underline{x}.$$

A particular solution is evidently

$$\underline{f}(\underline{x}) = 1 - \underline{x}.$$

If we make this choice for simplicity's sake (and no solution is simpler) we have the familiar rule of the calculus of probabilities,

$$(b|a) + (\sim b|a) = 1. \qquad (3.5)$$

More generally, (3.4) is satisfied if

$$\underline{g}(b|a) + \underline{g}(\sim b|a) = 1,$$

where $\underline{g}$ is an arbitrary function. But if we require consistency not only with Boolean algebra but also with (3.3), our choice of $\underline{g}$ is much more restricted [15]: we find that $\underline{g}(\underline{x}) = \underline{x}^r$, where $\underline{r}$ is an arbitrary number. Under this constraint, therefore, the widest allowable generalization of the usual rule, (3.5), is

$$(b|a)^r + (\sim b|a)^r = 1.$$

Now, raising to the power $\underline{r}$ both sides of (3.3), we have

$$(b \wedge c|a)^r = (b|a)^r (c|a \wedge b)^r.$$

If we were to take these two equations as the rules of probability, we could derive the whole calculus from them by Boolean algebra with no change except that the exponent $\underline{r}$ (the value of which we would never have to determine) would be attached to every symbol (b|a). Nothing would be achieved except a more difficult notation. Therefore it is both permissible and simpler to take $\underline{r}$ equal to 1, thereby adopting (3.3),

$$(b \wedge c|a) = (b|a)(c|a \wedge b).$$

and (3.5),

$$(b|a) + (\sim b|a) = 1,$$

as the rules of probability.

If we let c be equal to b in (3.3) and recall that $b \wedge b = b$, we have, for any premise a and inference b,

$$(b|a) = (b|a)\ (b|a \wedge b);$$

whence

$$b|a \wedge b = 1.$$

If now a implies b, then $a \wedge b = a$. In this case, therefore, $b|a = 1$. So we have the theorem:

> Every probability in which the premise implies the inference is equal to 1.

It follows, by the definition of exclusion, that:

> Every probability in which the premise and the inference are mutually exclusive is equal to 0.

For if a and b are mutually exclusive, each implies the contradictory of the other. In this case, by the previous theorem, $\sim b\,|\,a = 1$, and hence, by (3.5), $b|a = 0$. These then are the extremes of probability.

It is worth remarking, as corollaries to these two theorems, that the truism, being implied by every assertion, has the probability 1 as an inference from every premise; and the absurdity, being mutually exclusive with every assertion, has the probability 0 on every premise.

At this point it is worth while to notice the consequences of taking the absurdity itself as premise. As the conjunction of all assertions, it implies them all. Thus every inference has the probability 1 on the absurdity as premise. But if $b|a \wedge \sim a = 1$ and $\sim b|a \wedge \sim a = 1$, their sum is 2, whereas it is 1 by (3.5). This inconsistency need not disturb us unduly: in assuming an absurd premise, we assumed the risk of an absurd conclusion. It would be a strange logic indeed which did not allow us to recognize an absurd premise by the absurdity of its logical consequences.

The two axioms by which we obtained (3.3) for the probability of a conjunctive inference and (3.5) for the probabilities of mutually contradictory inferences suffice also for dealing with the disjunctive inference. We do not require a third: disjunction is sufficiently described in terms of conjunction and contradiction by either of the rules (2.4L) or (2.4R) to make a third axiom unnecessary.

Let us proceed to the consideration of disjunction by means of a lemma easily derived from (3.3) and (3.5):

$$(b\wedge c|a) + (b\wedge{\sim}c|a) = b|a. \tag{3.6}$$

This can also be written as

$$(b\wedge c|a) + ({\sim}b\wedge c|a) = c|a$$

by the exchange of b and c, or as

$$({\sim}b\wedge c|a) + ({\sim}b\wedge{\sim}c|a) = {\sim}b|a$$

by the substitution of ~b for b; whence we find by subtracting the sides of the second equation from those of the first, that

$$(b\wedge c|a) - ({\sim}b\wedge{\sim}c|a) = (c|a) - ({\sim}b|a).$$

Recalling now that $\sim b\wedge\sim c=\sim(b\vee c)$, we find by substituting, as we may by (3.5), $1 - (b\vee c|a)$ for $\sim b\wedge\sim c|a$, and $1 - (b|a)$ for $\sim b|a$, that

$$(b\wedge c|a) + (b\vee c|a) = (b|a) + (c|a).$$

This equation, written in the form

$$(b\vee c|a) = (b|a) + (c|a) - (b\wedge c|a), \tag{3.7}$$

is a special case of an equation for the probability of the disjunction of all the assertions of a set, b_i, ... b_m, where <u>m</u> is any integer greater than 1. The general equation is

$$b_1\vee\ldots\vee b_m|a = \sum_{i=1}^{m}(b_i|a) - \sum_{i=1}^{m-1}\sum_{j=i+1}^{m}(b_i\wedge b_j|a)$$
$$+ \sum_{i=1}^{m-2}\sum_{j=i+1}^{m-1}\sum_{k=j+1}^{m}(b_i\wedge b_j\wedge b_k|a) - \ldots$$
$$\pm (b_1\wedge \ldots \wedge b_m|a). \tag{3.8}$$

The limits of the summations in this equation are such that none of the assertions, b_1, ... b_m, is conjoined with itself in any inference, and also that no two inferences in any summation are conjunctions of the same assertions in different order. In the three-fold summation, for example, there is no such term as $b_1\wedge b_1\wedge b_2|a$, and the only conjunction of b_1, b_2, and b_3 is in the term $b_1\wedge b_2\wedge b_3|a$, because the limits exclude probabilities such as $b_2\wedge b_1\wedge b_3|a$, obtained from this one by permuting the assertions. Of the m - fold summation, therefore, nothing is left except the single term $b_1 \wedge \ldots \wedge b_m | a$, because there is only one allowed order of all m assertions. The sign is plus if m is odd, minus if m is even.

The proof of this equation is by mathematical induction and consists in proving that it holds for the disjunction of m + 1 assertions if it holds for the disjunction of m.

As the first step in the proof, we replace b in (3.7) by $b_1 \vee \ldots \vee b_m$ and so obtain the equation,

$$b_1 \vee \ldots \vee b_m \vee c | a = (b_1 \vee \ldots \vee b_m | a) + (c|a) - [(b_1 \vee \ldots \vee b_m) \wedge c | a].$$

By (1.4R),

$$(b_1 \vee \ldots \vee b_m) \wedge c = (b_1 \wedge c) \vee \ldots \vee (b_m \wedge c),$$

whence, by substitution in the equation just obtained, we have

$$b_1 \vee \ldots \vee b_m \vee c | a = (b_1 \vee \ldots \vee b_m | a) + (c|a) - [(b_1 \wedge c) \vee \ldots \vee (b_m \wedge c) | a]. \tag{3.9}$$

Of the three probabilities on the right, both the first and the third are those of disjunctions of m assertions, for which we now assume, for the sake of the mathematical induction, that (3.8) is valid. For $(b_1 \vee \ldots \vee b_m | a)$ we can immediately substitute the expression given by (3.8). For $(b_1 \wedge c) \vee \ldots \vee (b_m \wedge c) | a$, (3.8) provides a somewhat more complex expression, which we may simplify if we recall that c, no matter how many times conjoined with itself, remains c, by (1.1L). In the simpler form,

$$(b_1 \wedge c) \vee \ldots \vee (b_m \wedge c) | a = \sum_{i=1}^{m} (b_i \wedge c | a) - \sum_{i=1}^{m-1} \sum_{j=i+1}^{m} (b_i \wedge b_j \wedge c | a) + \ldots \pm (b_1 \wedge \ldots \wedge b_m \wedge c | a).$$

After these substitutions (and some rearrangement of terms) (3.9) becomes

$$\begin{aligned} b_1 \vee \ldots \vee b_m \vee c | a = {} & [\sum_{i=1}^{m} (b_i | a) + (c|a)] \\ & - [\sum_{i=1}^{m-1} \sum_{j=i+1}^{m} (b_i \wedge b_j | a) + \sum_{i=1}^{m} (b_i \wedge c | a)] \\ & + \ldots \pm (b_1 \wedge \ldots \wedge b_m \wedge c | a). \end{aligned}$$

In this equation, if we let c be b_{m+1}, we may replace the first bracket on the right by $\sum_{i=1}^{m+1} (b_i | a)$, the second by $\sum_{i=1}^{m} \sum_{j=i+1}^{m+1} (b_i \wedge b_j | a)$, and so on. So we have the result,

$$b_1 \mathbf{V} \ldots \mathbf{V} b_{m+1} | a = \sum_{i=1}^{m+1} (b_i | a) - \sum_{i=1}^{m} \sum_{j=i+1}^{m+1} (b_i \mathbf{\Lambda} b_j | a) + \ldots \pm (b_1 \mathbf{\Lambda} \ldots \mathbf{\Lambda} b_{m+1} | a).$$

This completes the demonstration that (3.8) holds for a set of $\underline{m}$ + 1 assertions if it holds for a set of $\underline{m}$. To perfect the mathematical induction, we have only to replace b by b_1 and c by b_2 in (3.7). The resulting equation,

$$b_1 \mathbf{V} b_2 | a = (b_1 | a) + (b_2 | a) - (b_1 \mathbf{\Lambda} b_2 | a),$$

shows that (3.8) holds when $\underline{m}$ = 2, and so completes the proof that it holds for all integers greater than 1.

The rather elaborate way in which the limits of summation were indicated in (3.8) was convenient for the proof, but a simpler notation may serve for the ordinary uses of the equation. In the simpler form,

$$\begin{aligned} b_1 \mathbf{V} \ldots \mathbf{V} b_m | a = \Sigma_i (b_i | a) &- \Sigma_{i,j>i} (b_i \mathbf{\Lambda} b_j | a) \\ &+ \Sigma_i \Sigma_{j>i} \Sigma_{k>j} (b_i \mathbf{\Lambda} b_j \mathbf{\Lambda} b_k | a) - \ldots \\ &\pm (b_1 \mathbf{\Lambda} \ldots \mathbf{\Lambda} b_m | a). \end{aligned} \tag{3.10}$$

Earlier we agreed to call two assertions mutually exclusive if they can not both be true, and an exhaustive pair if they can not both be false. Let us now modify these definitions in two respects: instead of limiting them to a pair of assertions, let us extend them to a set of any number; and instead of making exclusion and exhaustion absolute relations, defined by immediate reference to truth and falsity, let us make them conditional on the truth of a premise, with reference to which they are defined. Thus:

> The assertions of a set are mutually exclusive on a given premise if the premise excludes every conjunction of different assertions of the set;

and:

> A set of assertions is exhaustive on a given premise if the premise implies the disjunction of the set.

By the first of these definitions, if $b_1, \ldots b_m$ are assertions so related among themselves and to a premise a that

$$b_i \mathbf{\Lambda} b_j | a = 0 \tag{3.11}$$

for all unequal values of $\underline{i}$ and $\underline{j}$, the assertions are mutually exclusive on the premise. It then follows by (3.10) that

$$b_1 \mathbf{V} \ldots \mathbf{V} b_m | a = \Sigma_i (b_i | a). \tag{3.12}$$

Thus the probability of a disjunction of assertions is equal to the sum of their probabilities when they are mutually exclusive. By the second definition, if a is a premise and $b_1, \ldots b_m$ are assertions such that

$$b_1 \vee \ldots \vee b_m | a = 1, \tag{3.13}$$

the assertions are an exhaustive set on the premise.

When $b_1, \ldots b_m$ and a are such as to satisfy both conditions, then $b_1, \ldots b_m$ are an exhaustive set of assertions mutually exclusive on a, and

$$\Sigma_i (b_i | a) = 1. \tag{3.14}$$

The former definitions of exclusion and exhaustion are not in conflict with these: on the contrary, they are the simplest examples. For two assertions are enough to make a set. When they can not both be true, it is because their conjunction is the absurdity, which is excluded by every premise. When they can not both be false, it is because their disjunction is the truism, which is implied by every premise. Thus two assertions which are mutually exclusive by our former definition are, by our present definition, mutually exclusive on every premise. And an exhaustive pair by our former definition is now a set of two assertions exhaustive on every premise.

Of a set of assertions mutually exclusive on a given premise, no more than one can be true if the premise is true. For if any two were true, their conjunction would be true. But the premise excludes their conjunction, and a true assertion can not exclude a true assertion. The assertions of a set which is exhaustive on a given premise can not all be false if the premise is true. For if they were all false, their disjunction would be false. But the premise implies their disjunction, and a true assertion can not imply a false assertion.

Suppose then that all three conditions are fully met: the premise is true, the assertions are mutually exclusive on the premise, and on the premise they are an exhaustive set. Then one assertion must be true and all the others false.

For a classical instance, we may cite the dog of Chrysippus in the words of Sir Walter Raleigh [11]:

> "This creature, saith Chrysippus (of the dog), is not void of Logick: for when in following any beast he cometh to three several ways, he smelleth to the one, and then to the second; and if he find that the beast which he pursueth be not fled one of these two ways, he presently, without smelling any further to it, taketh the third way; which

saith the same Philosopher, is as if he reasoned thus: the Beast must be gone either this, or this, or the other way; but neither this nor this; Ergo, the third: so away he runneth."

4. Entropy

If we divide the equation for the probability of a conjunction,

$$b\wedge c|a = (b|a\wedge c)(c|a) = (c|a\wedge b)\ (b|a),$$

by $(b|a)(c|a)$, we see that the two ratios,

$$(b|a\wedge c)/(b|a) \text{ and } (c|a\wedge b)/(c|a),$$

are equal, and each is equal to

$$\frac{b\wedge c|a}{(b|a)(c|a)} .$$

Here we have three expressions of a quotient which measures the mutual influence of the assertions b and c, each upon the probability of the other. When the quotient is greater than 1, so that $b|a\wedge c>b|a$ and $c|a\wedge b>c|a$, it shows the probability of each assertion increased by conjoining the other in the premise with a. Less than 1, it shows a diminished probability. Equal to 1, it shows the probability unchanged.

A more convenient measure of this influence is the logarithm of the quotient: it is positive when the influence is favorable, negative when it is unfavorable, and zero when there is no influence. So, taking logarithms, we have

$$\ln(b\wedge c|a) - \ln(b|a) - \ln(c|a)$$

$$= \ln(b|a\wedge c) - \ln(b|a) = \ln(c|a\wedge b) - \ln(c|a). \quad (4.1)$$

We may proceed rather simply from this equation to the consideration of entropy, of which the familiar definition is $-\Sigma_i\ \underline{p}_i \ln \underline{p}_i$ where $\underline{p}_i$ is the probability on a given premise of the $\underline{i}$ th one of an exhaustive set of mutually exclusive assertions, and the summation is over all values of $\underline{i}$.

We may think of the assertions as proposed answers to a question [12], and take the entropy as measuring the information expected, on a given initial premise, to be forthcoming from the eventual confirmation of one of them. For example, let D be the question, "What day of the week is this?" and let $d_1, \ldots d_7$ be the seven assertions beginning with "This is Sunday" and running through the other six days of the week. Taking η as the symbol of entropy and representing the initial premise by a, we write the equation,

$$-\Sigma_i \underline{p}_i \ln \underline{p}_i = \eta(D|a) = -\Sigma_i (d_i|a)\ln(d_i|a),$$

and we read η (D|a) as "the entropy of the question D on the premise a." On any premise apt to be entertained, in this example, by Rip van Winkle, just awake from his twenty years of sleep, the seven probabilities would be equal, giving the entropy of the question its maximum value, ln 7. A better informed and more discriminating premise would diminish the entropy by leaving less information contingent on the event. On a premise which actually told the day of the week, the question would be an idle one and its entropy would be zero.

Returning now to (4.1), let us replace b and c by b_i and c_j, where b_1, ... b_m and c_1, ... c_n are two sets of assertions exhaustive on the premise a, and the assertions of each set are mutually exclusive within the set. With these replacements and a shift in the order of terms, (4.1) becomes

$$-\ln(b_i|a) - \ln(c_j|a) + \ln(b_i \wedge c_j|a)$$

$$= -\ln(b_i|a) + \ln(b_i|a \wedge c_j) = -\ln(c_j \wedge a) + \ln(c_j|a \wedge b_i). \quad (4.2)$$

As each of these three expressions measures the mutual influence of the assertions b_i and c_j, each upon the probability of the other, so each provides, by a suitably weighted average over all the values of $\underline{i}$ and $\underline{j}$, an appropriate measure of the mutual influence of the two sets of assertions, b_i, ... b_m and c_1, ...c_n. Moreover each of the three expressions of this average proves to be a function of the entropies of the questions B and C, to which the sets of assertions are answers.

In the weighted average, let us assign to the mutual influence of the assertions b_i and c_j a weight equal to the probability of their conjunction, $b_i \wedge c_j|a$, for which we may substitute, as proves convenient, either $(c_j|a \wedge b_i)$ $(b_i|a)$ or $(b_i|a \wedge c_j)$ $(c_j|a)$. So from (4.2) we obtain, after some minor manipulations, the equation:

$$-\Sigma_i\Sigma_j(c_j|a \wedge b_i)(b_i|a)\ln(b_i|a) - \Sigma_j\Sigma_i(b_i|a \wedge c_j)(c_j|a)\ln(c_j|a)$$

$$+\Sigma_{i,j}(b_i \wedge c_j|a)\ln(b_i \wedge c_j|a)$$

$$= -\Sigma_i\Sigma_j(c_j|a \wedge b_i)(b_i|a)\ln(b_i|a) + \Sigma_j[(c_j|a)\Sigma_i(b_i|a \wedge c_j)\ln(b_i|a \wedge c_j)]$$

$$= -\Sigma_j\Sigma_i(b_i|a \wedge c_j)(c_j|a)\ln(c_j|a) + \Sigma_i[(b_i|a)\Sigma_j(c_j|a \wedge b_i)\ln(c_j|a \wedge b_i)]. \quad (4.3)$$

This can be made much simpler. First let us notice that the assertions $c_1, \ldots c_n$, being an exhaustive set on the premise a, are a like set on the premise $a\wedge b_i$ for every value of i. Hence $\Sigma_j(c_j|a\wedge b_i) = 1$, and so

$$-\Sigma_i\Sigma_j(c_j|a\wedge b_i)(b_i|a)\ln(b_i|a) = -\Sigma_i(b_i|a)\ln(b_i|a) = \eta(B|a).$$

Also the assertions $b_1, \ldots b_m$ are an exhaustive set on the premise $a\wedge c_j$ for every value of j, so that $\Sigma_i(b_i|a\wedge c_j) = 1$ and

$$-\Sigma_j\Sigma_i(b_i|a\wedge c_j)(c_j|a)\ln(c_j|a) = \eta(C|a).$$

Next let us notice that the conjunctions $b_i\wedge c_j$, when all the values of i from 1 to m and of j from 1 to n are included, form a set of answers to the question $B\wedge C$, if the conjunction of two questions is defined as the question which asks what they ask jointly. Hence it follows that

$$\Sigma_{i,j}(b_i\wedge c_j|a)\ln(b_i\wedge c_j|a) = -\eta(B\wedge C|a).$$

There remain only the summations on the right in the third and fourth lines of (4.3). Examining these, we observe that they can be written, respectively, as $-\Sigma_j(c_j|a)\eta(B|a\wedge c_j)$ and $-\Sigma_i(b_i|a)\eta(C|a\wedge b_i)$.

So we have finally three expressions for the mutual influence of the questions B and C:

$$\eta(B|a) + \eta(C|a) - \eta(B\wedge C|a)$$

$$= \eta(B|a) - \Sigma_j(c_j|a)\eta(B|a\wedge c_j) = \eta(C|a) - \Sigma_i(b_i|a)\,\eta(C|a\wedge b_i). \qquad (4.4)$$

Having found a joint entropy, $\eta(B\wedge C|a)$, for the questions B and C, we may reasonably expect to find that they have also a common entropy, $\eta(B\vee C|a)$; and as the joint entropy is a function of the assertions $b_i\wedge c_j$, which answer both questions, so we should expect the common entropy to be the same function of the assertions which answer either question and so form the set $b_1, \ldots b_m, c_1, \ldots c_n$.

Here, however, we face a dilemma. The familiar definition of entropy requires the assertions to be both mutually exclusive and an exhaustive set. The requirement of mutual exclusion, if applied to the set $b_i, \ldots b_m, c_1, \ldots c_n$, would exclude not only the conjunctions $b_i\wedge b_j$ and $c_i\wedge c_j$; it would exclude also all the conjunctions $b_i\wedge c_j$. But these are the very assertions which answer $B\wedge C$, and a set of assertions can not be exhaustive on a premise which excludes them all. Thus, if the assertions which answer $B \vee C$ are mutually exclusive, those which answer $B \wedge C$ can not be an exhaustive set.

To escape this dilemma, we dispense with the requirement of mutual exclusion and adopt a more general definition of entropy, suggested by a comparison of the equations (3.12) and (3.10), which express the probability of the disjunction of assertions: in (3.12) when they are mutually exclusive, in (3.10) when they are not necessarily so. This comparison suggests, as the general definition of entropy, the equation: [16]

$$\eta(D|a) = - \Sigma_i (d_i|a)\ln(d_i|a) + \Sigma_{i,j>i}(d_i \wedge d_j|a)\ln(d_i \wedge d_j|a)$$
$$- \Sigma_{i,j>i,k>j}(d_i \wedge d_j \wedge d_k|a)\ln(d_i \wedge d_j \wedge d_k|a) + \ldots$$
$$\pm (d_1 \wedge \ldots \wedge d_w|a)\ln(d_1 \wedge \ldots \wedge d_w|a), \qquad (4.5)$$

where $\pm$ is to be read as + if w is even, - if w is odd.

In the special case in which $d_1, \ldots d_w$ are all mutually exclusive on the premise a, the probabilities of all their conjunctions are zero. Consequently the summations after the first all vanish and the equation becomes simply $\eta(D|a) = - \Sigma_i(d_i|a)\ln(d_i|a)$, in agreement with the familiar $-\Sigma_i p_i \ln p_i$. So the general definition does not change the meaning of entropy in any instance in which the usual definition gives it a meaning, but extends the meaning to include instances to which the usual definition does not apply.

To find an expression for $\eta(B \vee C|a)$ from the generalized definition, we let $w = m + n$ and replace $d_1, \ldots d_w$ by $b_1, \ldots b_m$, $c_1, \ldots c_n$. For simplicity's sake, we retain for the present the assumption of mutual exclusion within each of the sets $b_1, \ldots b_m$ and $c_1, \ldots c_n$, though not for the set which includes the assertions of both sets. On this assumption, we find for the summations in (4.5) that:

$$- \Sigma_i (d_i|a) \ln (d_i|a) = - \Sigma_i(b_i|a)\ln(b_i|a) - \Sigma_j(c_j|a)\ln(c_j|a)$$
$$= \eta(B|a) + \eta(C|a),$$

$$\Sigma_{i,j>i}(d_i \wedge d_j|a)\ln(d_i \wedge d_j|a) = \Sigma_{i,j}(b_i \wedge c_j|a)\ln(b_i \wedge c_j|a) = -\eta(B \wedge C|a),$$

and all the other summations vanish, because every probability in which an assertion of either set is conjoined with another of its own set is zero by the simplifying assumption. So we have the result that

$$\eta(B \vee C|a) = \eta(B|a) + \eta(C|a) - \eta(B \wedge C|a). \qquad (4.6)$$

The right side of this equation is the same expression as the first member of (4.4). Thus the weighted average which we took there as the measure of the mutual influence of two questions appears here as their common entropy.

The simplifying assumption used in deriving (4.6) was convenient but unnecessary. By a longer calculation, the same result could have been obtained without the assumption. The calculation can be well enough understood without going through all its details. It is evident that the replacement in (4.5) of the set $d_1, \ldots d_w$ by the set $b_1, \ldots b_m, c_1, \ldots c_n$ will produce terms on the right of three kinds: those which are functions only of $b_1, \ldots b_m$; those which are functions only of $c_1, \ldots c_n$; and those in which assertions of both sets are conjoined. It becomes evident on reflection that the first kind and the second kind are equal respectively to $\eta(B|a)$ and $\eta(C|a)$, as these are given by the generalized definition of entropy. The terms of the third kind are found equal to $-\eta(B \wedge C|a)$ by means of the equations $b_i \wedge b_j \wedge c_k = (b_i \wedge c_k) \wedge (b_j \wedge c_k)$, $b_i \wedge c_j \wedge c_k = (b_i \wedge c_j) \wedge (b_i \wedge c_k)$, and others like these but involving more numerous assertions. So (4.6) holds in the most general case as in the more restricted case assumed in its derivation.

A lemma helpful to the understanding of entropy in its extended meaning is that:

> If one answer to a question implies another, the implicant answer makes no contribution to the entropy of the question.

Since we are free to number the assertions as we please, there will be no loss of generality in the proof if we call the implicant d_1 and the implicate d_2. Let us recall that the condition of this implication has the two equivalent forms, $d_2 = d_1 \vee d_2$ and $d_1 = d_1 \wedge d_2$.

From the first of these it follows that the set $d_2, \ldots d_w$ is exhaustive on every premise on which the set $d_1, d_2. \ldots d_w$ is exhaustive. For the former set is exhaustive if the premise implies $d_2 \vee \ldots \vee d_w$, the latter if the premise implies $d_1 \vee d_2 \vee \ldots \vee d_w$. Because $d_2 = d_1 \vee d_2$, neither condition can hold without the other.

For the rest of the proof, we have to see how the condition of implication in the alternative form, $d_1 = d_1 \wedge d_2$, affects the expression for the entropy given in (4.5). So we note that the term $-(d_1|a)\ln(d_1|a)$, occurring in the single summation, $-\Sigma_i(d_i|a)\ln(d_i|a)$, is cancelled by the term $(d_1 \wedge d_2|a)\ln(d_1 \wedge d_2|a)$ in the double summation, $\Sigma_{i,j>i}(d_i \wedge d_j|a)\ln(d_i \wedge d_j|a)$. Similarly, every term $(d_1 \wedge d_j|a)\ln(d_1 \wedge d_j|a)$, with $j>2$, in the double summation is cancelled by a term $-(d_1 \wedge d_2 \wedge d_j|a)\ln(d_1 \wedge d_2 \wedge d_j|a)$ in the triple summation; and so on, until d_1, the implicant assertion, entirely disappears from the expression for the entropy.

In order to employ this lemma to best advantage, we define a set of assertions called a system, and other sets called the irreducible set and defining sets of the system:

> A system of assertions is a set which includes every assertion implying any assertion of the set.
> The irreducible set of a system of assertions is the set of all the assertions each of which is an assertion of the system and none of which implies any assertion of the system except itself.
> A defining set of a system is any set which is included in the system and includes its irreducible set.

Because implication is a transitive relation, it follows that every assertion of a system implies at least one assertion of the irreducible set of the system. If the assertion in question is itself one of the irreducible set, it implies no other one of the system, but it still implies itself. By these definitions, therefore, any defining set of a system can be enlarged, by the inclusion of its implicants, to form that system and no other; or it can be diminished, by the exclusion of every assertion which implies another assertion of the set, to form the irreducible set of that system and of no other system. Thus a given system has an unlimited number of defining sets, among which the system itself is the most inclusive and its irreducible set is the least inclusive.

Now let us notice that any assertion which implies an answer to a question is itself an answer to that question. Thus the question, "What day of the week is this?" can be answered by the assertion, "It is Saturday, February 1," or by any other implicant of one of the seven assertions which name days of the week. The seven may be called the irreducible answers to the question; and together with their implicants, in whatever number, all or some or none, they define the question.

More generally:

> The set of all the answers to a question is a system of assertions: the irreducible answers to the question form the irreducible set of the system, and every defining set of the system defines the question.

The lemma is our warrant of consistency:

> All the defining sets of a given question are

> exhaustive on the same premises, and the entropy of the question is the same, by whatever one of its defining sets it is reckoned.

The definitions of equality, conjunction and disjunction of assertions given in the first chapter can be altered so as to apply to questions by changing "assertion" to "question" and "tell" to "ask," thus:

> Two questions are equal if they ask the same thing.
> The conjunction of two questions is the question which asks what they ask jointly; their disjunction is the question which asks what they ask in common.

We find, as before, but expressible now in more clearly defined terms, that:

> If assertions $b_1, \ldots b_m$ define a question B, and assertions $c_1, \ldots c_n$ define a question C, then

$$\begin{matrix} b_1 \wedge c_1, & \ldots & b_m \wedge c_1 \\ \ldots & \ldots & \ldots \\ b_1 \wedge c_n, & \ldots & b_m \wedge c_n \end{matrix}$$

define $B \wedge C$, and

$$b_1, \ldots b_m, c_1, \ldots c_n$$

define $B \vee C$.

We may anticipate, but we are obliged to prove, that this prescription for the defining sets of $B \wedge C$ and $B \vee C$ will lead to the same rules of Boolean algebra for the conjunction and disjunction of questions as were given in the first chapter for the conjunction and disjunction of assertions.

Let us begin with the rule $B \wedge B = B$. The set of conjunctions $b_i \wedge b_j$, which defines $B \wedge B$, includes those in which $\underline{j} = \underline{i}$. But $b_i \wedge b_i = b_i$, and thus the defining set of $B \wedge B$ includes the assertions $b_1, \ldots b_m$, which define B. The other conjunctions $b_i \wedge b_j$ are implicants of these, and they can therefore be discarded from the defining set of $B \wedge B$ or included in the defining set B without changing either question. Thus $B \wedge B$ and B, being definable by the same assertions, are the same question.

Passing on to the rule $B \wedge C = C \wedge B$, we have $B \wedge C$ defined by the set of all the conjunctions $b_i \wedge c_j$, and $C \wedge B$ by the set of all the conjunctions $c_j \wedge b_i$. These conjunctions being equal in the algebra of assertions, so also are the questions they define. Similar reasoning leads to the rule $(B \wedge C) \wedge D = B \wedge (C \wedge D) = B \wedge C \wedge D$.

Corresponding to these three are the rules of disjunction, $B \vee B = B$, $B \vee C = C \vee B$, $(B \vee C) \vee D = B \vee (C \vee D) = B \vee C \vee D$. The agreement here is evident almost on inspection of the defining sets. A defining set of assertions is not changed, insofar as the defined question is concerned, by repetition of the assertions or by any change in their order in the set. Thus $B \vee B$, defined by the assertions $b_1, \ldots b_m, b_1, \ldots b_m$, is the same question as B, defined by $b_1, \ldots b_m$; and $C \vee B$, defined by $c_1, \ldots c_n, b_1, \ldots b_m$, is the same question as $B \vee C$, defined by $b_1, \ldots b_m, c_1, \ldots c_n$. Also the assertions $b_1, \ldots b_m, c_1, \ldots c_n, d_1, \ldots d_w$ define the question which is represented indifferently as $(B \vee C) \vee D$, $B \vee (C \vee D)$, or $B \vee C \vee D$.

Agreement with the rules for mixed conjunction and disjunction,

$$(B \wedge C) \vee D = (B \vee D) \wedge (C \vee D), \qquad (B \vee C) \wedge D = (B \wedge D) \vee (C \wedge D),$$
$$(B \wedge C) \vee C = C, \qquad (B \vee C) \wedge C = C,$$

is less directly evident.

Let us begin by considering a set of assertions defining the question $(B \vee D) \wedge (C \vee D)$. We have $B \vee D$ defined by $b_1, \ldots b_m, d_1, \ldots d_w$ and $C \vee D$ by $c_1, \ldots c_n, d_1, \ldots d_w$. Their conjunction is therefore defined by four sets of conjunctions of assertions: those of $b_1, \ldots b_m$ with $c_1, \ldots c_n$; those of $b_1, \ldots b_m$ with $d_1, \ldots d_w$; those of $d_1, \ldots d_w$ with $c_1, \ldots c_n$; and those of $d_1, \ldots d_w$ each with itself and with one another. But a diminished set will suffice for the definition. For the conjunctions of $d_1, \ldots d_w$ each with itself are simply $d_1, \ldots d_w$; and each of the other conjunctions, except those of $b_1, \ldots b_m$ with $c_1, \ldots c_n$, implies one of the assertions $d_1, \ldots d_w$ and can therefore be omitted from the defining set without changing the defined question. So $(B \vee D) \wedge (C \vee D)$ is sufficiently defined by the set which includes the conjunctions $b_i \wedge c_j$ and the assertions $d_1, \ldots d_w$. This set defines also the question $(B \wedge C) \vee D$, and (since all questions defined by the same set are equal) the prescription for the defining sets leads to the rule of Boolean algebra, $(B \wedge C) \vee D = (B \vee D) \wedge (C \vee D)$.

Considering next the rule $(B \vee C) \wedge D = (B \wedge D) \vee (C \wedge D)$, we have $B \vee C$ defined by $b_1, \ldots b_m, c_1, \ldots c_n$ and D by $d_1, \ldots d_w$. Hence we find $(B \vee C) \wedge D$ defined by assertions which comprise two sets: one the set of all the conjunctions $b_i \wedge d_j$, the other the set of all the conjunctions $c_i \wedge d_j$. The first set defines $B \wedge D$, the second defines $C \wedge D$. Taken together, as one set, they define $(B \wedge D) \vee (C \wedge D)$. Thus we have proof of agreement with the rule $(B \vee C) \wedge D = (B \wedge D) \vee (C \wedge D)$.

Last we consider the rules $(B \wedge C) \vee C = C$ and $(B \vee C) \wedge C = C$. The rules already dealt with permit a proof that $(B \wedge C) \vee C = (B \vee C) \wedge C$. It is identical with the proof found for assertions in the first chapter and it need not be repeated here. And because each of the conjunctions, $b_i \wedge c_j$, of the set defining $B \wedge C$ implies one of the set $c_1, \ldots c_n$, it follows that $(B \wedge C) \vee C = C$.

This completes the demonstration that the defining sets prescribed for the conjunction and disjunction of questions are in accord with the rules of Boolean algebra.

For a very simple example, which will serve also as an instance of the implication of questions, let us suppose that a card is drawn from a standard deck. The four assertions, "The card is a heart," "... a diamond," "... a spade," "... a club," define the question, "What is the suit of the card?" Let the question be represented by B, the assertions by b_1, b_2, b_3, b_4. The two assertions, "The card is red," "The card is black," define the question, "What is the color of the card?" Let this question be C, these assertions, c_1, c_2. Then $B \wedge C$ asks both suit and color and is defined by the conjunctions $b_i \wedge c_j$. But all hearts and diamonds in a standard deck are red, and all spades and clubs are black. Therefore the conjunctions $b_1 \wedge c_2$, $b_2 \wedge c_2$, $b_3 \wedge c_1$, $b_4 \wedge c_1$ are impossible by hypothesis, so that $B \wedge C$ is defined by the remaining conjunctions, $b_1 \wedge c_1$, $b_2 \wedge c_1$, $b_3 \wedge c_2$, $b_4 \wedge c_2$. Now the suit implies the color, so that $b_1 \wedge c_1 = b_1$, $b_2 \wedge c_1 = b_2$, $b_3 \wedge c_2 = b_3$, $b_4 \wedge c_2 = b_4$. Thus the assertions, b_1, b_2, b_3, b_4, which define B, are a defining set of $B \wedge C$ also, and $B = B \wedge C$.

The disjunction, $B \vee C$, is defined by b_1, b_2, b_3, b_4, c_1, c_2. But by the lemma proved earlier, the implicant assertions can all be discarded, leaving $B \vee C$ defined by c_1, c_2. Thus $C = B \vee C$.

Here is nothing surprising. There is redundance in the conjunction, "What is the suit of the card and what is its color?" and also in the disjunction, "What is the suit of the card or what is its color?" Neither question is changed if the second part is omitted from the first question or the first part from the second question. The implication of questions can be defined in the same words as that of assertions except that "asks" must replace "tells." It is a relation in which the implicant asks by itself all that it asks jointly with the implicate, and the implicate asks by itself only what it asks in common with the implicant.

Let us remark here that some questions are hypothetical and some hypotheses are false, whether by design or from ignorance. "What song the Sirens sang," wrote Sir Thomas Browne, "or what name Achilles assumed when he hid himself among women, though puzzling questions, are not beyond all conjecture." But if the Sirens and Achilles never lived, or living if the Sirens were songless and Achilles never hid among women, or if he hid nameless among them, the questions have no true answers.

Let us therefore define two classes of questions, to which we may give the names "real" and "vain." [13]

A question is real if it can be answered by a true assertion, and is vain if it can be answered by no true assertion.

It follows from this definition that:

An assertion is false if it can answer a vain question, and is true if it can answer no vain question.

In analogy to the familiar canons of Aristotelian logic concerning true and false assertions, we have for real and vain questions:

Some questions are real; some questions are vain.

Every question is either real or vain; no question is both real and vain.

A conjunction of questions is real if all of them are real and is vain if any of them is vain.

A disjunction of questions is vain if all of them are vain and is real if any of them is real.

Hence it follows that:

Every question implying a vain question is vain; every question implied by a real question is real.

These principles are supported by more than mere analogy. Given the Aristotelian canons and the definition of real and vain questions, these are their necessary consequences by our conclusions about the conjunction, disjunction and implication of questions and their defining sets of assertions.

We are now led to the consideration of complementary questions, with the help of a lemma:

Two questions are the same if they have the same conjunction and the same disjunction with a given question.

For proof of the lemma, we need only turn back to the beginning of the second chapter, where its analog is proved for assertions, and substitute "question" for "assertion" in the proof we found there. Thence we define mutually complementary questions as a pair whose conjunction is equal to the conjunction of all questions, and whose disjunction is equal to the disjunction of all questions. Just as complementary assertions can neither both be true nor both be false, so complementary questions can neither both be real nor both be vain.

Familiar usage provides no verbal expression serving as complement to a given question as an assertion is served by, and serves, its contradictory. Perhaps such an expression could be invented, but none is required for the helpful use of the sign ~ in the expression of entropies. The lemma has given us a working definition of complementary questions, which is consistent with the rules of Boolean algebra, the same ones we found applicable to assertions in the second chapter.

For some consequences of the definition and the rules, let us return to (4.6),

$$\eta(B \vee C|a) = \eta(B|a) + \eta(C|a) - \eta(B \wedge C|a),$$

and let $B = X \vee Y$ and $C = X \vee {\sim}Y$. Then

$$B \vee C = (X \vee Y) \vee (X \vee {\sim}Y) = Y \vee {\sim}Y,$$

and

$$B \wedge C = (X \vee Y) \wedge (X \vee {\sim}Y) = X \vee (Y \wedge {\sim}Y) = X.$$

With these substitutions, (4.6) becomes

$$\eta(Y \vee {\sim}Y|a) = \eta(X \vee Y|a) + \eta(X \vee {\sim}Y|a) - \eta(X|a).$$

As the disjunction of all questions, $Y \vee {\sim}Y$ is answered by every assertion, including the truism. It is therefore definable by the truism, whence its entropy, on whatever premise, is zero.

So we find that $\eta(X|a) = \eta(X \vee Y|a) + \eta(X \vee {\sim}Y|a)$. This holds for any questions X and Y. Consequently we may return to our former symbols, B and C, and write:

$$\eta(B|a) = \eta(B \vee C|a) + \eta(B \vee {\sim}C|a). \qquad (4.7)$$

If we now return to (4.4) and recall that each of its three members, though taken first as a measure of the mutual influence of B and C, was later identified with their common entropy, we have, from the second member, the equation,

$$\eta(B \vee C|a) = \eta(B|a) - \Sigma_j (c_j|a)\eta(B|a \wedge c_j).$$

Between this equation and (4.7) we can eliminate both $\eta(B \vee C|a)$ and $\eta(B|a)$ to find that

$$\eta(B \vee {\sim}C|a) = \Sigma_j (c_j|a)\, \eta(B|a \wedge c_j).$$

However, this is not the general expression for $\eta(B\vee{\sim}C|a)$, because it was assumed in deriving (4.4) that C is defined by a set of mutually exclusive assertions.

The general expression may be surmised by its analogy to others obtained in this chapter and the one before. It is:

$$\eta(B\vee{\sim}C|a) = \Sigma_j(c_j|a)\eta(B|a\wedge c_j) - \Sigma_{j,k>j}(c_j\wedge c_k|a)\eta(B|a\wedge C_j\wedge C_k) + \ldots\pm(c_1\wedge\ldots\wedge c_n|a)\eta(B|a\wedge c_1\wedge\ldots\wedge c_n). \qquad (4.8)$$

The proof, by (4.7) and (4.5), would extend the chapter to a tedious length, and will be omitted.

5. Inference And Inquiry

In the chapter just ended, we have observed the perfect symmetry of Boolean algebra, not only as between conjunction and disjunction but also as between assertions and questions. In this, our final chapter, we consider an extension of this symmetry from a common algebra to a joint calculus, a calculus of the probability of assertions and the bearing of questions.

We conceive bearing, like probability, as a variable relation, capable of every degree between fixed limits. As we call b|a the probability of the inference b on the premise a, so we call B|A the bearing of the inquiry B on the issue A. As an instance we may again borrow from Sir Thomas Browne, taking as issue the question, "What name did Achilles assume when he hid himself among women?" and as inquiry, "What were the time and country of the women among whom Achilles hid?" Or finding a more commonplace example in the toss of a die, we may take as the issue, "What number shows?" and as inquiry, "Is it odd or even?" In both of these instances, it is clear that the bearing is not at either limit but somewhere between the lower limit and the upper: if the inquiry were answered, the issue would be narrowed in some degree, but it would not be closed.

It may be helpful, before proceeding to the axioms and rules of bearing, to examine its limits and compare them, as to likeness and contrast, with the limits of probability.

The upper limit of bearing and that of probability are alike in that both are defined in terms of implication. They differ in that the bearing B|A is maximum when the inquiry B implies the issue A, whereas the probability b|a is maximum when the inference b is implied by the premise a. The case is familiar in respect to probability, and it is easily understood in respect to bearing. To say that B implies A means that A is answered if B is answered: the inquiry is sufficiently comprehensive and pertinent that when it is answered the issue is closed.

The conditions of these two implications can be written as $B = B \wedge A$ and $b = b \vee a$, or alternatively, as $A = A \vee B$ and $a = a \wedge b$. Written either way, they are an instance of a general principle of transformation in the calculus of probability and bearing:

> The exchange of questions with assertions and conjunction with disjunction in any rule of the calculus transforms it into another rule of the calculus.

Hence it follows that the set of all the rules of the calculus is invariant under the two exchanges together. Earlier we referred to "a common algebra" of assertions and questions and "a joint calculus" of probability and bearing. The reason for calling the algebra "common" and the calculus "joint" is now apparent: the algebra is invariant under either exchange separately or both together, the calculus only under both together.

For another instance, let an inference be the disjunction of two assertions which tell nothing in common. Because they can not both be false, they are called an exhaustive pair. Their disjunction is equal to the disjunction of all assertions; therefore, as an inference it is implied by every premise. Thus the disjunction of an exhaustive pair of assertions has maximum probability on every premise.

We have not yet defined an exhaustive pair of questions, but we are accustomed in ordinary discourse to call an inquiry 'exhaustive' if it is exceedingly comprehensive and searching. No inquiry could better deserve this description than the conjunction of two which jointly ask everything. So we define an exhaustive pair of questions as two whose conjunction is equal to the conjunction of all questions. On this definition, the conjunction of an exhaustive pair of questions has maximum bearing on every issue, because it implies every question.

It should be noticed that this property - the conjunction of two being equal to the conjunction of all - by which we have just defined an exhaustive pair of questions, is the same property as we used much earlier to define mutually exclusive assertions. So we are left with no choice for the definition of mutually exclusive questions except the property by which we define an exhaustive pair of assertions. Consequently we call two questions mutually exclusive if their disjunction is equal to the disjunction of all questions, so that the two ask nothing in common.

If the two questions are respectively the inquiry and issue of a bearing, the bearing is plainly minimum, and this meaning of exclusion agrees well with the usage of common discourse.

So in a case at law, the judge will exclude, as having no bearing on the issue before the court, an inquiry which asks nothing in common with the issue. Or, if an inquiry is the disjunction of two questions neither of which asks anything in common with the other, it is an idle inquiry, asking nothing at all, and has no bearing on any issue.

These various statements about assertions and questions, their limits of probability and their limits of bearing, can not be greatly condensed, but they can be more systematically arranged, as follows:

> Two assertions are an exhaustive pair if their disjunction is equal to the disjunction of all assertions; they are mutually exclusive if their conjunction is equal to the conjunction of all assertions; they are complementary if they are an exhaustive pair and mutually exclusive.
>
> Two questions are an exhaustive pair if their conjunction is equal to the conjunction of all questions; they are mutually exclusive if their disjunction is equal to the disjunction of all questions; they are complementary if they are an exhaustive pair and mutually exclusive.

> The probability of an inference is maximum on a given premise if the inference is implied by the premise; it is maximum on every premise if the inference is the disjunction of an exhaustive pair of assertions.
>
> The bearing of an inquiry is maximum on a stated issue if the inquiry implies the issue; it is maximum on every issue if the inquiry is the conjunction of an exhaustive pair of questions.

> The probability of an inference is minimum on a given premise if the inference and the premise are mutually exclusive; it is minimum on every premise if the inference is the conjunction of mutually exclusive assertions.
>
> The bearing of an inquiry is minimum on a stated issue if the inquiry and the issue are mutually exclusive; it is minimum on every issue if the inquiry is the disjunction of mutually exclusive questions.

Let us now proceed to the axioms of the calculus, applying the same principle of transformation by the exchange of assertions with questions and conjunction with disjunction. Thus:

> The probability on a given premise of an inference

which is the conjunction of two assertions is determined by the probability of the first assertion on the given premise and the probability of the second assertion on a premise which is the conjunction of the given premise and the first assertion.

The bearing on a stated issue of an inquiry which is the disjunction of two questions is determined by the bearing of the first question on the stated issue and the bearing of the second question on an issue which is the disjunction of the stated issue and the first question.

The probability of an inference on a given premise determines the probability of its complement on the same premise.

The bearing of an inquiry on a stated issue determines the bearing of its complement on the same issue.

Or expressed as equations:

$$b \wedge c | a = \underline{F}(b|a,\ c|a \wedge b), \qquad \sim b|a = \underline{f}(b|a);$$

$$B \vee C | A = \underline{F}(B|A,\ C|A \vee B), \quad (5.1) \qquad \sim B|A = \underline{f}(B|A). \quad (5.2)$$

For bearing, as for probability, $\underline{F}$ and $\underline{f}$ are unspecified functions of the variables, but they must be consistent with the rules of Boolean algebra. This consistency requires that they satisfy the same functional equations,

$$\underline{F}[\underline{F}(\underline{x},\ \underline{y}),\ \underline{z}] = \underline{F}[\underline{x},\ \underline{F}(\underline{y},\ \underline{z})],$$

$$\underline{f}\ [\underline{f}(\underline{x})] = \underline{x},$$

which we derived in dealing with probability. The proof is the same as before, except that the rule of Boolean algebra (1.3R) replaces (1.3L) in the argument.

Again, as with probability, we have a choice among particular solutions of these functional equations, and again, with no significant loss of generality, we may choose the simplest. So we have:

$$b \wedge c | a = (b|a)(c|a \wedge b), \qquad (b|a) + (\sim b|a) = 1;$$

$$B \vee C | A = (B|A)(C|A \vee B), \quad (5.3) \qquad (B|A) + (\sim B|A) = 1. \quad (5.4)$$

These rules assign to the upper and lower limits of bearing the same numerical values as they assign to those of probability:

Every probability in which the inference is implied by the premise and every bearing in which the inquiry implies the issue is equal to 1.

Every probability in which the inference and premise are mutually exclusive and every bearing in which the inquiry and issue are mutually exclusive is equal to 0.

From the rules of the calculus which were just given, others follow by Boolean algebra. Thus:

$$(b\wedge c|a) + (\sim b\wedge c|a) = c|a,$$

$$(B\vee C|A) + (\sim B\vee C|A) = C|A; \qquad (5.5)$$

$$(b\vee c|a) = (b|a) + (c|a) - (b\wedge c|a),$$

$$(B\wedge C|A) = (B|A) + (C|A) - (B\vee C|A). \qquad (5.6)$$

And for assertions $b_1, \ldots b_m$ and questions $B_1, \ldots B_m$:

$$b_1\vee \ldots \vee b_m|a = \Sigma_i(b_i|a) - \Sigma_{i,j>i}(b_i\wedge b_j|a) + \Sigma_{i,j>i,k>j}(b_i\wedge b_j\wedge b_k|a) - \ldots \pm (b_i\wedge \ldots \wedge b_m|a);$$

$$B_1\wedge \ldots \wedge B_m|A = \Sigma_i(B_i|A) - \Sigma_{i,j>i}(B_i\vee B_j|A) + \Sigma_{i,j>i,k>j}(B_i\vee B_j\vee B_k|A) - \ldots \pm (B_1\vee \ldots \vee B_m|A). \qquad (5.7)$$

These rules of probability were proved in the third chapter. Similar proofs can be given for the rules of bearing, but we are warranted, by the symmetry of the algebra and without further proof, in taking them from the rules of probability by the principle of transformation.

In the third chapter, the definitions of exhaustive and exclusion were extended from a pair of assertions to a set of any number, and were made specific to a given premise. The extended definitions can be so transformed as to define, on a stated issue, an exhaustive set of mutually exclusive questions. Thus:

A set of assertions is exhaustive on a given premise if the premise implies the disjunction of the set.

A set of questions is exhaustive on a stated issue if the issue is implied by the conjunction of the set.

The assertions of a set are mutually exclusive on a given premise if the premise excludes every conjunction of different assertions of the set.

The questions of a set are mutually exclusive on a stated issue if the issue excludes every disjunction of different questions of the set.

In respect to questions, these definitions say, in other words, that the questions of an exhaustive set jointly ask everything which the issue asks, and that no two of a set of mutually exclusive questions ask in common anything which has a bearing on the issue. The consequences of the definitions are familiar insofar as they concern assertions, but for questions they may be unexpected. For assertions and questions together, we find that:

Among assertions which, on a true premise, are an exhaustive set and mutually exclusive, one must be true and all the others false.

Among questions which, on a vain issue, are an exhaustive set and mutually exclusive, one must be vain and all the others real.

What effect this latter consequence might have on the strategy of Chrysippus' dog, though not beyond all conjecture, is a puzzling question. If not the dog, then perhaps we, as "creatures not void of Logick," may find it helpful when, having admitted a vain question to our thought, we need to distinguish the reality from the vanity.

Returning now to the general equations for the probability of a disjunction and the bearing of a conjunction, let us assume mutual exclusion, of the assertions on the premise and of the questions on the issue. The probabilities of the conjunctions of the assertions and the bearings of the disjunctions of the questions all vanish, and the equations become simply:

$$b_1 \vee \ldots \vee b_m | a = \Sigma_i (b_i | a),$$

$$B_1 \wedge \ldots \wedge B_m | A = \Sigma_i (B_i | A). \qquad (5.8)$$

If also both sets are exhaustive,

$$b_1 \vee \ldots \vee b_m | a = 1,$$

$$B_1 \wedge \ldots \wedge B_m | A = 1. \tag{5.9}$$

A further extension of the calculus is indicated by the principles:

> A question answered by an assertion is answered by every implicant of that assertion.
>
> An assertion answering a question answers every implicate of that question.

By the first of these two principles, we were led earlier to define the set of assertions which we called a system. We are led now, by the second principle, to define a system of questions.

> A system of assertions is a set which includes every assertion implying any assertion of the set.
>
> A system of questions is a set which includes every question implied by any question of the set.

The first of the two principles and the first of these two definitions enable us to identify, as we did earlier, the assertions of a system as the set of all the answers to a question. It will be convenient now to call this question the issue of the system. Then, by the second principle, every assertion of the system answers not only the issue but also every implicate of the issue. Since an implicant is the conjunction of all its implicates, and an implicate is the disjunction of all its implicants we are now provided with two convenient definitions: the first, of the issue of a system of assertions; the second, of an assertion which we may call the premise of a system of questions. They are:

> The issue of a system of assertions is the conjunction of all the questions each of which is answered by every assertion of the system.
>
> The premise of a system of questions is the disjunction of all the assertions each of which answers every question of the system.

Systems of both kinds have their irreducible sets and their defining sets. Thus:

> The irreducible set of a system of assertions is the set of all the assertions each of which is an assertion of the system and none of which implies

any assertion of the system except itself.

The irreducible set of a system of questions is the set of all the questions each of which is a question of the system and none of which is implied by any question of the system except itself.

For defining sets of both systems a single definition will suffice:

A defining set of a system of assertions or a system of questions is any set which is included in the system and includes its irreducible set.

The defining set determines the system, and the system in turn determines the issue, if it is a system of assertions, or the premise, if it is a system of questions. When issues or premises are conjoined or disjoined, we have the following theorems for the defining sets:

If assertions $b_1, \dots b_m$ define an issue B, and assertions $c_1, \dots c_n$ define an issue C, then

$$\begin{array}{ccc} b_1 \wedge c_1, & \dots & b_m \wedge c_1, \\ \dots & \dots & \dots \\ b_1 \wedge c_n, & \dots & b_m \wedge c_n \end{array}$$

define $B \wedge C$, and

$$b_1, \dots b_m, \; c_1, \dots c_n$$

define $B \vee C$.

If questions $B_1, \dots B_m$ define a premise b, and questions $C_1, \dots C_n$ define a premise c, then

$$\begin{array}{ccc} B_1 \vee C_1, & \dots & B_m \vee C_1, \\ \dots & \dots & \dots \\ B_1 \vee C_n, & \dots & B_m \vee C_n \end{array}$$

define $b \vee c$, and

$$B_1, \dots B_m, \; C_1, \dots C_n$$

define $b \wedge c$.

The first of these two theorems was proved in the fourth chapter. A similar proof can be given for the second theorem, but a proof in detail is hardly necessary. It is enough to observe that the exchange of assertions with questions and conjunction with disjunction transforms either theorem into the other, and similarly transforms the two definitions, of issue and premise, each into the other. The proof by which the first theorem follows from the first definition becomes, under the same transformation, a proof of the second theorem from the second definition.

By a further use of the same exchange, we may transform the definition of $\eta(D|a)$, which we now call the entropy of the issue D on the premise a, into a definition of $\eta(d|A)$, the entropy of the premise d on the issue A. So we have the two definitions:

$$\eta(D|a) = -\Sigma_i(d_i|a)\ln(d_i|a)$$
$$+\ \Sigma_{i,j>i}(d_i \wedge d_j|a)\ln(d_i \wedge d_j|a) - \ldots$$
$$\pm\ (d_1 \wedge \ldots \wedge d_w|a)\ln(d_1 \wedge \ldots \wedge d_w|a);$$
$$\eta(d|A) = -\Sigma_i(D_i|A)\ln(D_i|A)$$
$$+\ \Sigma_{i,j>i}(D_i \vee D_j|A)\ln(D_i \vee D_j|A) - \ldots$$
$$\pm\ (D_1 \vee \ldots \vee D_w|A)\ln(D_1 \vee \ldots \vee D_w|A). \quad (5.10)$$

Hence there follow equations for the entropies of conjunctions, disjunctions and complements, some as given in the preceding chapter, the rest taken from these by the symmetries of the algebra and the calculus:

$$\eta(B \vee C|a) = \eta(B|a) + \eta(C|a) - \eta(B \wedge C|a),$$
$$\eta(b \wedge c|A) = \eta(b|A) + \eta(c|A) - \eta(b \vee c|A). \quad (5.11)$$
$$\eta(B|a) = \eta(B \vee C|a) + \eta(B \vee {\sim}C|a),$$
$$\eta(b|A) = \eta(b \wedge c|A) + \eta(b \wedge {\sim}c|A). \quad (5.12)$$
$$\eta(B \vee {\sim}C|a) = \Sigma_j(c_j|a)\eta(B|a \wedge c_j) - \Sigma_{j,k>j}(c_j \wedge c_k|a)\eta(B|a \wedge c_j \wedge c_k)$$
$$+ \ldots \pm (c_1 \wedge \ldots \wedge c_n|a)\eta(B|a \wedge c_1 \wedge \ldots \wedge c_n);$$
$$\eta(b \wedge {\sim}c|A) = \Sigma_j(C_j|A)\eta(b|A \vee C_j) - \Sigma_{j,k>j}(C_j \vee C_k|A)\eta(b|A \vee C_j \vee C_k)$$
$$+ \ldots \pm (C_1 \vee \ldots \vee C_n|A)\eta(b|A \vee C_1 \vee \ldots \vee C_n). \quad (5.13)$$

William Gilbert of Colchester wrote in his book, _On the Loadstone_ (Mottelay's translation):

> "Great has ever been the fame of the loadstone and of amber in the writings of the learned: many philosophers cite the loadstone and also amber

> whenever, in explaining mysteries, their minds become obfuscated and reason can no farther go."

In Gilbert's time and for centuries earlier, loadstone and amber (in Latin magnes and electricum) figured as the embodiment of magnetism and electricity. Neither he nor the philosophers of whom he wrote could possibly know how far reason, directing and directed by experiment, would later go in discovering the analogies and connections between magnetic and electric phenomena.

I do not know how far these dualities, of inference and inquiry, probability and bearing, one and another entropy, can be carried. I should be rash indeed if I were to suppose that they could be carried to conclusions so vast and unforeseen as to be at all comparable to those which have rewarded the study of electricity and magnetism. I believe they can be traced, and usefully traced, much farther than I have shown in this essay. But long study has brought me only thus far; and so I end my essay here, still, as I hope, on the safe side of obfuscation.

Notes

The number in brackets following each note is that of the page to which the note refers.

1. Boolean algebra and its application to inductive inference have a long,wide genealogy. Its beginning is briefly described as follows by Hilbert and Ackermann in the introduction to their book Grundzüge der Theoretischen Logik, quoted here in the translation by Hammond, Leckie and Steinhart with the title Principles of Mathematical Logic.

> "The first clear idea of a mathematical logic was formulated by Leibniz. The first results were obtained by A.De Morgan (1806-1876) and G. Boole (1815-1869). The entire later development goes back to Boole. Among his successors, W. S. Jevons (1835-1882) and especially C. S. Peirce (1839-1914) enriched the young science. Ernst Schröder systematically organized and supplemented the various results of his predecessors in his Vorlesungen über die Algebra der Logik (1890-1895) which represents a certain completion of the series of developments proceeding from Boole."

In this development, both necessary and probable inference were included. Thus Leibniz wrote (Nouveaux Essais sur l'Entendement Humain, book 4, ch. 2, Langley's translation):

"Opinion, based on probability, deserves perhaps the name knowledge also; otherwise nearly all historic knowledge and many other kinds will fall. But without disputing about terms, I hold that the investigation of the degrees of probability is very important, that we are still lacking in it, and that this lack is a great defect of our logics."

A like concern was shown by De Morgan, whose Formal Logic was subtitled the calculus of inference necessary and probable, and by Boole, the full title of whose masterpiece was An Investigation of the Laws of Thought, on which are founded the mathematical theories of logic and probabilities. A similar interest was shared by Jevons and Peirce. [119]

2. Different authors have used various signs for conjunction, as x, ·, &. Sometimes it is written without a sign, as ab. Disjunction is usually denoted by **V**. Keeping this sign, I have taken **Λ** for the sign of conjunction, as distinguishing both operations from those of ordinary algebra, and each operation from the other, while suggesting the symmetry between them, which is so striking and pleasing a feature of Boolean algebra, as modified by Jevons, Peirce and Schröder. [119]

3. "Nothing so like as eggs, yet no one on account of this appearing similarity, expects the same taste and relish in all of them".

The quotation is from David Hume, An Inquiry Concerning Human Understanding, Section IV, Part II. He would scarcely have expected to be so quoted in an essay on the logic of probability, a subject in which he was firmly convinced that there was no probability of finding any logic. [120]

4. Boole employed the exclusive disjunction, which he denoted by the sign +, as he denoted conjunction by the sign x. These conventions facilitated (at least in the simpler examples) his symbolic treatment of probability, and enabled him to write:

"From the above investigations it clearly appears, 1st, that whether we set out from the ordinary numerical definition of the measure of probability, or from the definition which assigns to the numerical measure of probability such a law of value as shall establish a formal identity between the logical expressions of events and the algebraic expressions of their values, we shall be led to the same system of practical results.

2dly, that either of these definitions pursued to its consequences, and considered in connection with the relations which it inseparably involves, conducts us, by inference or suggestion, to the other definition. To a scientific view of the theory of probabilities, it is essential that both principles should be viewed together in their mutual bearing and dependence."

In more recent years, I.J. Good, in his Probability and the Weighing of Evidence, has carried this "inference or suggestion" of Boole's almost, if not indeed all the way, to the point of proof.

The change from exclusive to inclusive disjunction was made independently by several authors, of whom Jevons, in his book, Pure Logic: or the logic of quality apart from quantity, was the first. The resulting symmetry between disjunction and conjunction was pointed out by Peirce in an article, "On an improvement in Boole's calculus of logic," Proc. Amer. Acad. Arts and Sci. (1867). (The copy of Jevons' book, inscribed and presented to Peirce, is in the Milton S. Eisenhower Library of The Johns Hopkins University). [121]

5. From the poem, "To his Wife," in The Works of Li Po, the Chinese Poet, done into English Verse by Shigeyoshi Obata. [121]

6. Logicians have recognized several species of implication and have distinguished them by appropriate names. We shall have need only of the kind defined by C. I. Lewis and C. H. Langford in their well known book, Symbolic Logic, and called by them "strict implication". Holding to the one definition, we may, without ambiguity, omit the adjective "strict". [123]

7. La Logique ou l'Art de Penser of Antoine Arnauld, commonly known in English as the Port-Royal Logic, has been printed, since its first publication in 1662, more than fifty times in numerous editions and translations. It is quoted here in the translation by James Dickoff and Patricia James, entitled The Art of Thinking, Port-Royal Logic, published by Bobbs-Merrill in 1964. [129]

8. Many sets of axioms have been proposed for probability. In so short an essay as this, it is clearly impossible to discuss them all, and I shall not attempt even to list their authors, sure as I am that I should overlook some of them. [133]

9. "How many miles to Babylon?"
"Three-score miles and ten."

"Can I get there by candle-light?"
"Yes, and back again."

"If your heels are nimble and light,
You can get there by candle-light." [133]

10. Jaynes has traced the solution of the functional equation (3.2) as far back as Abel (Oeuvres Complètes de Niels Henrik Abel, edited by L. Sylow and S. Lie, 1881). Readers whose scruples permit them to admit the differentiability of the functions may find this, and the other functional equations in this chapter, solved in an article of mine published in 1946 in the American Journal of Physics, or in The Algebra of Probable Inference, published by the Johns Hopkins Press in 1961, but now out of print. Or they may find solutions without the assumption of differentiability in the book by J. Aczel called, in its English version, Lectures on Functional Equations and Their Applications, where they will find also a great deal else of interest and value. [134]

11. Quoted by Jevons in the work cited earlier. [141]

12. In comparison with the attention given to assertions, their character, their varieties and their relations, the neglect accorded to questions throughout most, if not all, of the history of logic and probability seems remarkable. (The Port-Royal Logic is an exception to this generalization.) In a number of texts, the very word 'question' is not even a heading in the index.

Noteworthy, not only for its singularity but for the thoughtfulness and care evident in its composition, is an article entitled "What is a question?" by Felix Klein, which appeared in 1939 in The Monist, 39, 350-364. After citing several writers who defined a question as "a request for information," Klein takes exception to that definition on the ground that it excludes the rhetorical question. He maintains that a question considered as a request for information is not the same as a question considered as a logical entity, and he identifies a question in the latter sense with the expression which other authors have called a "propositional function."

I am inclined to think that the rhetorical question can be accommodated under either definition about as well as under the other. My own preference, avoiding the choice between the two definitions, is to take the word 'question' as undefined, on the ground, perhaps, that the words "What is a question?" will make no sense to anyone who does not know a question when he hears one. With 'assertion' also undefined, we may take these two elements more on a par, and so be better prepared for their mutual symmetry as it emerges in this chapter and the next. [142]

13. Although the names "real" and "vain" are as apt as any that come to mind, it should be noted that questions vain under this definition have sometimes aroused discussion which was anything but trivial. The following instance, remarkable for its influence on the history of physics, is described by Wilson L. Scott in The Conflict between Atomism and Conservation Theory 1644 to 1860. The vain question, which engaged Sir Isaac Newton and for a long time held the attention of some of the finest minds of Europe, was "What laws govern the collisions of perfectly hard bodies?"

Newton and some of the others believed that atoms must have been made perfectly and uniquely hard at the Creation in order to have withstood all their collisions in the ages since. Neither Newton nor the others believed that there were any perfectly hard bodies larger than atoms. Larger bodies, they supposed, were subject to strain and fracture because of the interstices between the atoms of which they were composed, although the atoms themselves underwent no changes, not even the slightest, in their shapes and sizes, however great the stress.

Since the atoms were too small to be seen, the laws governing the collisions of perfectly hard bodies could only be inferred from general laws of nature known in other ways. In the simplest and most crucial instance, two such bodies, conceived as identical spheres, were considered as approaching one another from opposite directions at equal speeds, and colliding along their line of centers.

The symmetry of this scheme was enough to insure that the impact would bring the spheres to a stop, at least momentarily. It insured also, if any motion of the spheres were to be resumed after their collision, that they could move only in a recoil along their line of centers at equal speeds. But recoil was impossible. Being perfectly hard, the spheres must come to a stop as soon as they touched at one point; otherwise they would have to undergo some compression or change of shape. The instant after impact, therefore, would find them at rest, side by side, touching at only a single point, quite unstrained and therefore free of any stress which might set them moving again. And so they would remain at rest.

In this conclusion there was no inconsistency with Newton's laws of motion, and from it there was no escape except by denying the existence of perfectly hard bodies and so calling the question vain. Some indeed from the beginning of the discussion doubted or denied the possibility of perfectly hard bodies. All gave it up at the end, when it was made untenable by the principle of the conservation of energy, which at the outset had been unproved and only imperfectly stated in Leibniz' speculations on _vis viva_.

As not every vain question is trivial, so not every real question is worth asking. The void question, which asks nothing, is sufficiently answered by the void assertion, which tells nothing. So the void question could fairly be called trivial. Yet we are obliged to reckon the void assertion as true, in that it tells nothing false. Therefore we must similarly accept the void question as real, because it can be answered by a true assertion. It is real as asking nothing vain, because it asks nothing at all. [150]

The Following Three Notes were added for the sake of completeness by Robert B. Evans (of the Georgia Tech School of Mechanical Engineering) as a result of his proofreading of this manuscript typed for the M.I.T. press by Ann M. Evans (of the Georgia Tech Engineering Experiment Station) from Professor Cox's original report (which was typed by Thelma Hunter and issued by The Johns Hopkins University Department of Physics):

14. This general solution of (3.2) is obtained on pages 14-16 of Professor Cox's book, _The Algebra of Probable Inference_, The Johns Hopkins Press, Baltimore, Md., 1961. [135]

15. The restricted choice of _g_ required by consistency with (3.3) is obtained on pages 19-22 of _The Algebra of Probable Inference_, op. cit. [136]

16. This entropy expression is derived as a general measure of expected information on pages 44-45 of _The Algebra of Probable Inference_, op. cit. [145]

A NEW APPROACH FOR DECIDING UPON CONSTRAINTS IN THE MAXIMUM ENTROPY FORMALISM

Robert B. Evans

Introduction

The maximum entropy formalism introduced by Jaynes [1] and Tribus [2] is useful for determining the values of Bayesian probabilities when these probabilities are otherwise unknown. This formalism was built upon the earlier information theory contributions of Cox [3], Shannon [4], and several others who contributed indirectly--Elsasser [5] for example. The term "Bayesian" is discussed in the following section.

The question of deciding upon constraints in this formalism will be considered in a notation appropriate for discrete outcomes. Introducing this notation for any exhaustive set of possible outcomes which are mutually exclusive, one has $\Sigma_i p_i = 1$, where p_i denotes the probability of outcome i. The entropy S of any such set is given by $-k\Sigma_i p_i \ln p_i$ where "ln" denotes the natural logarithm and k is a unit conversion constant. Methods for extending this notation to continuous variables and quantum matrices have been treated in detail by Jaynes [6] and others [7], and will not be dwelled upon in this paper.

Using this discrete outcome notation, the maximum entropy formalism is as follows:

Maximize the entropy,

$$S/k = -\Sigma_i p_i \ln p_i$$

subject to the constraints,

$$\Sigma_i p_i = 1$$

$$\Sigma_i p_i g_{ji} = K_j \qquad j = 1,2, ---, m$$

where K_j and g_{ji} are constants and m is at least two less than the number of possible outcomes. Since the constant k may be negative--as was recommend by Gibbs [8] and as occurs in Brillouin's negentropy [9] for example--it is S/k which is a maximum in general, rather than S itself.

Familiar examples of g_{ji} include the energies of microstates in statistical thermodynamics [1,10] and discrete failure times in reliability theory [7].

The following questions naturally arise: Why not use more general constraints,

$$G_j(p_1, p_2 --- p_i --- p_n) = 0 \qquad j = 1,2, ---, m$$

where each G_j could represent any function whatsoever? Could it not be that nonlinear constraints G_j might better characterize the system under consideration? And even if one does restrict the constraints to the linear form $\Sigma p_i g_{ji} = K_j$, might not more general forms of g_{ji} be more appropriate? For example, might not taking g_{ji} to represent microstate forces or pressures instead of microstate energies serve to better characterize thermodynamic systems? Or might not the square root of discrete failure times be a better choice of g_{ji} than the failure times themselves for characterizing the reliability of systems?

Such questions regarding the most appropriate constraints need to be resolved. For until they are, the maximum entropy formalism will continue to appear to many [11] as being little more than an abstract structure of ad hoc representation of results whose rigorous derivation must be made by other means. In this paper, a new approach for deciding upon the most appropriate constraints in the maximum entropy formalism will be considered.

Extension To A More General Principle

This new approach consists of extending the maximum entropy principle into a more general principle, and then determining constraints by requiring consistency with both principles. The more general principle is obtained via a derivation of the maximum entropy principle which is more general than those appearing in the current literature [7,12,13].

The starting point of this derivation is to utilize the familiar extension rule [7,12] (also known as the rule for marginal probabilities [14]) to determine the unknown probabilities p_i as follows:

$$p_i = \Sigma_a\, p_{i|a}\, p_a \qquad (1)$$

where,

$$\Sigma_a\, p_a = 1$$

Here, the slice "|" denotes "given" or "conditional upon"--so that, for example, $p_{i|a}$ represents the probability of outcome "i" conditional upon the proposition "a". The symbol "a" represents mutually-exclusive, hypothetical propositions upon which the conditional probabilities $p_{i|a}$ are either known or may be easily determined.

As mentioned in the opening paragraph of this paper, the probabilities p_i are Bayesian probabilities, which means that they are always conditional upon background or prior knowledge X, so that their values may change (in accordance with the well-known Bayes theorem) when X changes. This is in sharp contrast to non-Bayesian probabilities which represent limiting frequencies assumed to be properties of the natural universe.

For example, if we let the proposition "Today is Henry the 8th's birthday" be represented by the symbol "A", then the probability p_A is a Bayesian probability which is best illustrated as $p_{A|X}$ because of the impact of X. {1}* For most of us in our present state of background knowledge, $p_{A|X} = 1/365$ (not accounting for February 29th). But since King Henry's birthday is a matter of record, one may look up this information to arrive at a state of increased background knowledge X' for which $p_{A|X'}$ is either zero or unity depending upon the date that you happen to read this paper. And in no event can p_A be a non-Bayesian probability, which requires repetition of "A" to support the concept of a limiting frequency--since King Henry's birthday is not a repeatable event. (His mother could well attest to that!)

For a scientifically relevant but similar Bayesian proposition, let "A" represent "$E = mc^2$" while X represents scientific opinion prior to acceptance of Einstein's work and X' represents opinion after its acceptance. Here the change from X to X' is once again simply a matter of record--the difference being that it represents a record of scientifically controlled observations.

In the considerations which follow, specific attention is given to the background, X, so that the extension rule is expressed explicitly as

$$p_{i|X} = \Sigma_a p_{i|aX}\, p_{a|X}$$

where "aX" denotes the conjunction "a and X". {2}

Next, let us stipulate that "a" represents hypothetical states of advanced knowledge such that $p_{i|aX}$ is a definite number for each state "a". Such hypothetical states of advanced knowledge are of course not new. Laplace introduced them to obtain his much criticized "Rule of Succession" [15], and more realistic use of them has since been made by many others [see Refs. 7,12, and 16, for example]. However, in what follows, a completely different class of advanced knowledge states will be employed, as is indicated just before Equation (12) below.

In order to determine the probabilities $p_{a|X}$, one may stipulate that each advanced knowledge state "a" is based upon a hypothetical sequence "I" of observations which is as long as is required to result in the following relation--to as close an approximation as is required by the desired precision of the

*A number in brackets, as appears here, refers to a note in the set of notes collected at the end of this paper.

ensuing results:

$$p_{a|IX} = 1 \tag{2}$$

The extent of the approximation may if desired be expressed explicitly by replacing 1 by 1 - δ, so that for any given percentage error ε in the ensuing results (taken as small as one pleases) there exists a corresponding value of δ.

It should also be pointed out here that each hypothetical advanced knowledge state "a" may in general be obtained from many different hypothetical sequences of observations "I" (but each "I" yields only one "a" in accordance with Equation 2).

Bayes' theorem may now be applied:

$$p_{a|IX} = p_{a|X} \frac{p_{I|aX}}{p_{I|X}}$$

or, in view of Equation (2),

$$p_{a|X} = \frac{p_{I|X}}{p_{I|aX}} \tag{3}$$

Now let the background knowledge X be divided into two parts, X_o and Y, in order to introduce reference levels X_o:

$$X \equiv YX_o \tag{4}$$

where Y denotes the remainder of X. The hypothetical background states X_o will serve as reference levels from which the knowledge content of propositions may be assessed. Substitution of Equation (4) into Equation (3) yields, with A $\equiv$ aY,

$$p_{a|X} = \frac{p_{I|X}}{p_{I|AX_o}} = \frac{p_{I|X_o}}{p_{I|AX_o}} \left(\frac{p_{I|X}}{p_{I|X_o}} \right) \tag{5}$$

In Appendix A it is shown that the ratio $p_{I|X}/p_{I|X_o}$ is a constant C (to as close an approximation as is required by the desired precision of the derivation) with respect to all variations in advanced knowledge states "a" and observation sequences "I":

$$\frac{p_{I|X}}{p_{I|X_o}} = C \tag{6}$$

Thus, in view of Equation (5),

$$p_{a|X} = \frac{p_{I|X_o}}{p_{I|AX_o}} C \tag{7}$$

Equation (7) may be expressed in logarithmic form in order to display the additive quantities which are vital to the completion of this derivation, as pointed out following Equation (12) below:

$$\ln p_{a|X} = -\ln \frac{p_{I|AX_o}}{p_{I|X_o}} + \ln C$$

The additive quantity $\ln (p_{I|AX_o}/p_{I|X_o})$ has been called "essergy" since 1968, because for thermodynamic systems it corresponds to the general essential energy expressions first given by Gibbs [17,18], and because as shown in Appendix B it represents a more general concept than the concepts of "information", and "surprisal" with which it is often identified. In general, the essergy $\mathcal{E}_{i|EX_o}$ with respect to any proposition i is defined by,

$$\mathcal{E}_{i|EX_o} \equiv K \ln \frac{p_{i|EX_o}}{p_{i|X_o}} \quad (8)$$

where $X_o \equiv$ reference level of background knowledge
$E \equiv$ knowledge (about proposition i) based upon reference level X_o
$K \equiv$ unit conversion constant, with units of energy in most thermodynamic applications

In many contexts, the symbols "$|EX_o$" may be dropped without loss of generality--it being understood that they still apply nevertheless--while the symbols "$|X_o$" are replaced by "o", whence with no loss in generality the definition of essergy $\mathcal{E}_i$ may be written as,

$$\mathcal{E}_i \equiv K \ln \frac{p_i}{p_{io}} \quad (9)$$

Applying Equation (8) to Equation (7) yields,

$$p_{a|X} = C\, e^{-\mathcal{E}_{I|AX_o}/K} \quad (10)$$

Each long sequence of observations I is by definition a sequence of N individual observations,

$$I \equiv i_1 i_2 i_3 \text{ --- } i_K \text{ --- } i_N \quad (11)$$

With respect to advanced knowledge "a" in general, the individual observations i_K wouldn't be independent of one another. However, as shown in Appendix C, the class of advanced knowledge states considered in this paper is such that there always exists reference levels X_o from which the individual observations i_K may always be viewed as being independent. With it being understood that X_o represents this type of reference level, one has (from the familiar independent product rule $p_{AB} = p_A p_B$, as shown in Appendix C),

$$\mathcal{E}_{I|AX_o} = \sum_{K=1}^{N} \mathcal{E}_{i_K|AX_o} \tag{12}$$

Equation (12) displays the additive quantities (called "essergy" as pointed out above) which are vital to this derivation. For example, one may apply the familiar variance rule for independent variables (variance of a sum equals sum of the variances) along with Chebyshev's inequality to obtain (as demonstrated in Appendix D) to as close an approximation as one pleases with N taken large accordingly,

$$\mathcal{E}_{I|AX_o}/N = \sum_{K=1}^{N} <\mathcal{E}_{K|AX_o}>/N \tag{13}$$

where $<\mathcal{E}_{K|AX_o}>$ represents the expected value of $\mathcal{E}_{i_K|AX_o}$ which (dropping the "$|AX_o$" for convenience while replacing "$|X_o$" by "o" in accordance with the convention established in Equation 9 above) is given by,

$$<\mathcal{E}_{K|AX_o}> = <\mathcal{E}_K> = \sum_{i_K=1}^{n_K} p_{i_K} \mathcal{E}_{i_K} \qquad K = 1,2 \text{ ---- } N \tag{13a}$$

with,

$$\sum_{i_K=1}^{n_K} p_{i_K} = 1 \qquad K = 1,2 \text{ ---- } N \tag{13b}$$

$$\sum_{i_K=1}^{n_K} p_{i_K o} = 1 \qquad K = 1,2 \text{ ---- } N \tag{13c}$$

Here it is observed that each hypothetical individual observation i_K is understood to be taken from a hypothetical exhaustive set of n_K possible outcomes i_K which are mutually exclusive. Closer approximations of Equation (13) for any given value of N may of course be obtained if desired by applying stronger inequalities or limit theorems--such as Kolmogorov's inequality or the central limit theorem for example [14].

It is convenient at this point to introduce the term "system" in the broad inductive sense employed by Cox [16,19] which includes not only physical systems (such as the systems of statistical thermodynamics) but also may represent <u>any</u> exhaustive set of possible outcomes or sample space (see Appendices B and C for discussion). Using this terminology, each hypothetical sequence I of observations i_K is seen to be made upon a set of N hypothetical systems. It is further stipulated (as discussed in Appendix C) that these systems are sufficiently similar to the original system under consideration (i.e., the original system for which "i" in Equation 1 is one of the possible outcomes) so that the expected essergy $\langle\mathcal{E}\rangle$ for this system is equal to the average expected essergy of the N observations:

$$\langle\mathcal{E}\rangle = \sum_{K=1}^{N} \langle\mathcal{E}_K\rangle/N \tag{14}$$

where,

$$\langle\mathcal{E}\rangle_K = \sum_{i=1}^{n} p_i\ \mathcal{E}_i \tag{14a}$$

with,

$$\sum_{i=1}^{n} p_i = 1 \tag{14b}$$

$$\sum_{i=1}^{n} p_{io} = 1 \tag{14c}$$

while in terms of Equations (8) and (9),

$$\mathcal{E}_i = K \ln\frac{p_{i|AX_o}}{p_{i|X_o}} = K \ln\frac{p_i}{p_{io}} \tag{14d}$$

It will be observed here that the original system has been taken to be a mutually exclusive exhaustive set of n outcomes.

Substitution of Equations (13) and (14) into Equation (10) yields (dropping the "$|X$" from $p_{a|X}$ for convenience),

$$\langle\mathcal{E}\rangle = -\frac{K}{N} \ln \frac{p_a}{C} \tag{15}$$

Equation (15) cannot be written in the same form as Equation (10) (i.e., one cannot merely replace $\mathcal{E}_{I|AX_o}$ by $N\langle\mathcal{E}\rangle$), because of the degree of approximation inherent in Equation (13).

In order to display Equation (15) graphically, let each

advanced knowledge state "a" be indexed by a set of bounded numbers $\{a_1, a_2 a_3, \text{---} a_i \text{---} a_n\}$ with fixed differences Δa_i between all consecutive values of a_i for all i such that $\{\Delta a_i\} \equiv \Delta a_1 \Delta a_2 \Delta a_3 \text{---} \Delta a_i \text{---} \Delta a_n \to 0$, as $N \to \infty$. With this understanding, $p_a / \{\Delta a_i\}$ represents the probability density function on the indices $\{a_i\} \equiv \{a_1, a_2 \text{---} a_i \text{---} a_n\}$. For example, a_i could be represented by N_i/N where N_i denotes the number of times outcomes similar to outcome i occurred in any particular one of the hypothetical observation sequences I--for which case $\Delta a_i = 1/N$. For convenience, let it also be momentarily assumed that $<\mathcal{E}>$ has a unique minimum on the set $\{a_i\}$ subject to whatever mathematical constraints may apply. Then, under this assumption, Equation (15) dictates the variations of $<\mathcal{E}>$ and $p_a/\{\Delta a_i\}$ shown in Figure 1 plotted against any particular one of the indices a_i. As $N \to \infty$, the density function $p_a/\{\Delta a_i\}$ has a very sharp maximum at the values $\{a_i\} = \{a_{im}\}$ corresponding to the assumed unique minimum point of $<\mathcal{E}>$. Hence in Equation (1), the only significant contributions of $p_{i|a}\, p_a$ to the sum $\Sigma_a p_{i|a}\, p_a$ would occur very near the value p_{a_m} of the most probable state a_m of advanced knowledge. Thus the sum in Equation (1), when approximated by a multiple integral on the differentials $\{da_i\}$, yields in the manner of any integral over a Dirac delta function,

$$p_i = p_{i|a_m}$$

Or, replacing "$|a_m$" by "m" for convenience,

$$p_i = p_{im} \tag{16}$$

The next step is to demonstrate that $<\mathcal{E}>$ does indeed have such a unique minimum "m" subject to any constraints which might apply. To do this, first consider a different reference level X_o' which results in a different value C' of the constant C in accordance with the definition of C in Appendix A. Equation (15) then becomes, letting $<\mathcal{E}'>$ represent the expected essergy with respect to X_o',

$$<\mathcal{E}'> = -\frac{K}{N} \ln \frac{p_{a|X}}{C'}$$

Merely changing the reference level cannot alter the value of $p_{a|X}$ so that subtraction of this expression from Equation (15) yields,

$$<\mathcal{E}> - <\mathcal{E}'> = \frac{K}{N} \ln \frac{C}{C'} \tag{17}$$

Equation (17) shows that the essergy difference $\langle\mathcal{E}\rangle - \langle\mathcal{E}'\rangle$ is constant with respect to variations of advanced knowledge states a. This difference is also constant with respect to the sequence length N, since N does not appear in the expressions (14a) and (14d) for the essergy $\langle\mathcal{E}\rangle$. (This of course requires $C^{(1/N)}$ to be constant with respect to N, so that C is seen to be an increasing function of the sequence length, as was pointed out in Appendix A.)

One particularly convenient reference level X_o' is the current background knowledge X, for which,

$$\{p_{io}\} = \{p_{i|X}\} \tag{18}$$

Or in other words, as discussed in Appendix E, for the reference level $X_o' = X$ the reference probabilities p_{io} in Equation (14d) are equal to the solution values $p_{i|X}$ from Equation (1). This result may be utilized by observing that $\langle\mathcal{E}\rangle/K$ corresponds (via Equations 14a and 14d) to the Kullback function [20] for which it is well known that,

$$\begin{aligned} \langle\mathcal{E}\rangle/K > 0 \quad &\text{if} \quad \{p_i\} \neq \{p_{io}\} \\ \langle\mathcal{E}\rangle/K = 0 \quad &\text{if} \quad \{p_i\} = \{p_{io}\} \end{aligned} \tag{19}$$

The proof of Equation (19) may be found in many places--for example Kullback [20], Gibbs [Ref. 17, Eqn. 448], and Tribus [Ref. 7, page 100]. It may be of interest to observe in passing that the expected essergy $\langle\mathcal{E}\rangle$ itself does not correspond to the Kullback function, since K is often negative (see Equation 57 of Gibbs [18] for example).

Now for the reference level $X_o' = X$, the probabilities $\{p_i\}$ must be free to reach the values $\{p_{io}\} \equiv \{p_{i|X_o'}\}$, since these are the solution values $\{p_{i|X}\}$ via Equation (18). But Equation (19) dictates that $\langle\mathcal{E}'\rangle$ has a unique minimum at this point. And finally, Equation (17) dictates that $\langle\mathcal{E}\rangle$ also has a unique minimum at the same point, since $\langle\mathcal{E}\rangle$ and $\langle\mathcal{E}'\rangle$ differ by the constant $(K/N)\ln(C/C')$ at all points--as illustrated in Figure 2. Mathematical constraints will of course have to be present in order for $\langle\mathcal{E}\rangle$ to have this minimum point instead of the unconstrained minimum at $\{p_{io}\} \equiv \{p_{i|X_o}\}$--as illustrated by the dashed curve in Figure 2. Such constraints characterize the system under consideration as well as the reference level X_o. Thus $\langle\mathcal{E}\rangle$ does indeed have a unique minimum point, subject to the constraints which characterize the system and the reference level "o"--as was to be shown. It follows via Equation (16) that the solution values $\{p_i\}$ from Equation (1) correspond to this unique minimum point $\{p_{im}\}$. The following theorem has thereby been proven:

> Theorem of Minimum Essergy: For any system, the actual probabilities $\{p_i\}$ are the ones for which the essergy $\langle\mathcal{E}\rangle/K$ is a minimum subject to the constraints which characterize the system and the reference level "o".

As pointed out above just before Equation (14), the word system is defined here in the broad inductive sense introduced by Cox [16,19] and includes not only physical systems (such as the systems of statistical thermodynamics) but also includes any exhaustive set of possible outcomes or sample space. In the theorem of minimum essergy as just demonstrated, the possible outcomes have been limited to the mutually exclusive case. However, the expected essergy $\langle\mathcal{E}\rangle$ is still well defined even when the propositions are not mutually exclusive, as shown in Appendix B where it is indicated that by using Cox's Boolean algebra transformations, the minimum essergy theorem would appear to remain valid for systems in their most general sense as defined by Cox. More general types of reference levels than those used in the proof (as discussed in Appendix C) may also be permitted, the general criterion for acceptable reference levels being,

$$\langle\mathcal{E}'\rangle = f(\langle\mathcal{E}\rangle) \qquad (20)$$

where f( ) represents any strictly increasing monotonic function. This may be verified via replacing Equation (17) by Equation (20), whereupon it will be observed that the remainder of the proof remains essentially unchanged.

It is convenient to give a name to the type of advanced knowledge states "a" used in this proof in order to distinguish them from the familiar limiting frequency type with which they are compared in Figure 3. The name "macroscopic state" seems appropriate, since for thermodynamic systems they do indeed reduce to the familiar macroscopic states of statistical thermodynamics. This generalization of the term "macroscopic state" is quite in keeping with the generalization of the terms "entropy" and "essergy" (essential energy) already established. Using this term, the main principle embodied in the minimum essergy theorem (as reflected in Equations 15, 16, 18, 19 and 17 or 20) may be stated as follows:

> Principle of Minimum Essergy: The most probable macroscopic state of any system is the one for which the essergy $\langle\mathcal{E}\rangle/K$ is a minimum subject to the constraints which

characterize the system and the reference level "o".

The minimum essergy theorem would not follow from this principle were it not for the sharp maximum exhibited by the probability $p_a/\{\Delta a_i\}$ of macroscopic states as shown in Figure 3. In the words of Jaynes [12] "the most probable state in the N-extension of the system is overwhelmingly more probable than the others." Jaynes makes this statement in connection with his demonstration of the maximum entropy principle. The theorem of minimum essergy constitutes the more general principle which was sought in this section as a generalization of the maximum entropy principle.

Reduction To The Maximum Entropy Principle

It may now be shown that the minimum essergy principle reduces to the maximum entropy principle for a class of systems called "fundamentally symmetrical systems". Such systems by definition are sets of outcomes i for which the fundamental distinguishing features (i.e., the absolute minimum of features required to distinguish any given outcome from any other outcome) are irrelevant. For such systems, it is easily shown that the probabilities $p_{i|z}$ are all equal to each other (where "z" denotes the corresponding fundamental reference state where the only knowledge about the outcomes which one has are the fundamental distinguishing features). This follows, since without these fundamental distinguishing features the outcomes are indistinguishable, and any differences among the values of the probabilities p_i would provide a distinguishing feature in contradiction to the stipulated case--and with "z" irrelevant, the equality among $p_i \equiv p_{i|z}$ remains. Replacing "|z" by "z" for convenience (so that p_{iz} replaces $p_{i|z}$) one has,

$$p_{1z} = p_{2z} = p_{3z} = \text{---} = p_{iz} = \text{---} = p_{nz} \tag{21}$$

To implement Equation (21), one may express the essergy $\langle\mathcal{E}\rangle/K$ from Equation (14a) and (14d) in two parts as follows:

$$\frac{\langle\mathcal{E}\rangle}{K} = \sum_{i=1}^{n} p_i \ln p_{io}^{-1} - \frac{S}{k} \tag{22}$$

where

$$\frac{S}{k} \equiv -\sum_{i=1}^{n} p_i \ln p_i \tag{23}$$

The expression $\sum_{i=1}^{n} p_i \ln p_{io}^{-1}$ for "o" = "z" is, via Equations

(21) and (14b):

$$\sum_{i=1}^{n} p_i \ln p_{iz}^{-1} = \ln p_{iz}^{-1} \sum_{i=1}^{n} p_i = \ln p_{iz}^{-1} \equiv J$$

where J denotes a constant. Thus,

$$\frac{\langle \mathcal{E} \rangle}{K} = J - \frac{S}{k} \qquad (24)$$

While Equation (24) was derived with reference level "o" = "z", it holds for all reference levels consistent with Equation (17), since these will merely alter the constant J by an amount $(K/N)\ln(C/C')$. And for reference levels in general, it will be observed from Equation (20) that $\langle \mathcal{E} \rangle/K$ will always be a strictly decreasing monotonic function of S/k. The theorem and principle of maximum entropy follow as special cases of the theorem and principle of minimum essergy:

> <u>Theorem Of Maximum Entropy</u>: For any fundamentally symmetrical system, the actual probabilities $\{p_i\}$ are the ones for which the entropy S/k is a maximum subject to the constraints which characterize the system.

> <u>Principle Of Maximum Entropy</u>: The most probable macroscopic state of any fundamentally symmetrical system is the one for which the entropy S/k is a maximum subject to the constraints which characterize the system.

The characterization of the reference level "o" need not be stated here (as is required in the minimum essergy principle) since no reference probabilities $\{p_{io}\}$ appear in the expression (23) for entropy.

As shown in Appendix F, the maximum entropy principle applies <u>only</u> to fundamentally symmetrical systems. An example of a fundamentally symmetrical system is the outcomes "1" or "6" from the toss of an honest die. Here the fundamental distinguishing features are irrelevant. An example of a non fundamentally symmetrical system is the outcomes "1" and "6" from the toss of a dishonest die which has the "6" painted on five sides and the "1" on the remaining side. For such a die, one has if the die is otherwise honest,

$$p_{6z} = \frac{5}{6}, \; p_{1z} = \frac{1}{6}$$

Here, the fundamental distinguishing features are not irrelevant. In the light of this example, it would appear that perhaps any physical system may be expressed in terms of a fundamentally symmetrical set of outcomes. For the dishonest die for example, the fundamentally symmetrical set of outcomes constitutes the physical sides of the die, not the numbers "1" and "6". The determination of the fundamentally symmetrical set in continuous variable notation can, as shown by Jaynes, be made by requiring invariance under transformation groups [6] and marginalization [21].

In the following section, it is shown that the simultaneous application of the principles of maximum entropy and minimum essergy dictates that for fundamentally symmetrical systems, all constraints $G_j(p_1, p_2 \text{--} p_i \text{--} p_n)$ must be linear.

The Linear Constraint Theorem {3}

The theorem of maximum entropy as just derived embodies the maximum entropy formalism discussed in the introduction in that one finds the probabilities $\{p_i\}$ by maximizing the entropy,

$$\frac{S}{k} = -\sum_{i=1}^{n} p_i \ln p_i \tag{23}$$

subject to,

$$\sum_{i=1}^{n} p_i = 1 \tag{14b}$$

and subject to any other constraints which characterize the system. One may first consider a single additional constraint which, for the sake of generality, will be assumed to be a nonlinear function of the probabilities $\{p_i\}$:

$$G(\{p_i\}) = 0 \tag{25}$$

Even through this constraint might be highly nonlinear, it nevertheless must be of a form which determines a unique maximum point of S/k, for otherwise, in view of Equation (24), there would not be a unique minimum of the essergy $\langle\mathcal{E}\rangle/K$, as was {3} established in the preceding section. This unique maximum point is given as follows via the method of Lagrange multipliers:

$$p_i = e^{-\lambda_o - \lambda g_i(\{p_i\})} \qquad i = 1,2,\text{----}n \tag{26}$$

where,

$$g_i(\{p_i\}) \equiv \frac{\partial G(\{p_i\})}{\partial p_i} \qquad i = 1,2,\text{----}n \tag{26a}$$

while λ_o and λ are the multipliers for Equations (14b) and (25) respectively.

For convenience, one may denote this unique maximum point by "m" and rewrite the solution (26) as,

$$p_{im} = e^{-\lambda_o - \lambda g_{im}} \qquad i = 1,2,----n \qquad (27)$$

where

$$g_{im} \equiv g_i(\{p_{im}\}) \qquad i = 1,2,----n \qquad (27a)$$

But in accordance with the results established in the preceding section, this solution must also be given by the minimum essergy theorem. And as was pointed out just before the statement of the maximum entropy theorem, the essergy $<\mathcal{E}>/K$ is always a strictly decreasing monotonic function of the entropy S/k.{3} This requires the expression $\Sigma p_i \ln p_{io}^{-1}$ in Equation (22) to be a function f(S/k) of S/k,

$$\sum_{i=1}^{n} p_i \ln p_{io}^{-1} = f(S/k) \qquad (28)$$

such that,

$$\frac{df(S/k)}{d(S/k)} < 1 \qquad (28a)$$

while in view of Equation (19),

$$f(S/k) \geq S/k \qquad (28b)$$

A convenient reference level for the purpose at hand is the current background knowledge X, for which the reference probabilities p_{io} are simply the solution values $p_{i|X}$, as was pointed out following Equation (18). And since the solution values are given by Equation (27), one has for the reference level $X_o = X$,

$$\{p_{io}\} = \{p_{im}\} \qquad (29)$$

Using this reference level, one sets "o" = "m" in Equation (28) to obtain,

$$\sum_{i=1}^{n} p_i \ln p_{im}^{-1} = f(S/k) \qquad (30)$$

One may now substitute the solutions (27) into this expression:

$$\lambda_o + \lambda \sum_{i=1}^{n} p_i g_{im} = f(S/k)$$

For convenience, define

$$f(S) \equiv [f(S/k) - \lambda_o]/\lambda$$

so that one has the following required condition for the reference level $X_o = X$:

$$\sum_{i=1}^{n} p_i g_{im} = f(S) \qquad (31)$$

This condition must coincide completely with the constraint (25), for otherwise it would represent still another constraint--in contradiction to the stipulation that Equation (14b) and (25) are the only constraints thus far imposed. It follows that the content of this condition must remain unchanged if additional independent constraints were to be imposed, since the constraint (25) would remain unchanged in view of the stipulated independence of these additional constraints. For convenience, let these additional independent constraints be of the type which merely specify values for some of the p_i--for example, $p_i = c_1$, $p_2 = c_2$, etc., where c_1, c_2, etc. represents any set of positive constants which is consistent with the prior constraints (14b) and (25). Such specification of p_i would, if $G(\{p_i\})$ is nonlinear, change the values of $g_{im} \equiv g_i(\{p_{im}\})$, since they would change the maximum point $\{p_{im}\}$--as is easily verified by substituting $p_1 = c_1$, $p_2 = c_2$, etc. into Equations (23), (14b), (25), (26), and (27). But this change in g_{im} would alter the content of the condition (31) and therefore of the constraint (25), unless the change is restricted to multiplication of $f(S)$ and each original g_{im} by a common value $h(\{p_{im}\})$ where $h(\ )$ represents an arbitrary function. Hence, in order to allow for all possible such specifications of p_i, the following restriction must be placed upon the form of the function $G(\{p_i\})$:

$$g_i(\{p_i\}) = h(\{p_i\}) g_{im} \qquad i = 1,2,----n \qquad (32)$$

where the quantities g_{im} are the original ones which applied before any such specifications of p_i were made.

The set of equations (32) is equivalent to a functional equation for $G(\{p_i\})$, the solution of which is shown in Appendix G to be given by,

$$G(\{p_i\}) = \sum_{i=1}^{n} p_i g_i - K \qquad (33)$$

where $\{g_i\}$ and K are constants. Thus for any fundamentally symmetrical system, the single constraint (25) must be linear {3}, and as such it may always be written in the form,

$$\sum_{i=1}^{n} p_i g_i = K \qquad (34)$$

Considering the constraint (25) to be any one of a set of m independent constraints which may be treated one at a time, it follows from Equation (34) that all m of the constraints must be linear:

$$\sum_{i=1}^{n} p_i g_{ji} = K_j \qquad j = 1,2,\text{----}m < n-1 \qquad (35)$$

where $\{g_{ji}\}$ and $\{k_j\}$ are constants. If some of these constraints are not independent of each other so that they have to be considered in groups rather than one at a time, then the condition (31) becomes more complex, as shown in Appendix G. Nevertheless the reasoning leading to Equation (32) still applies--whence Equations (33) and (34) still follow for each individual constraint so that the set of equations (35) applies to dependent as well as independent constraints.

It is to be concluded that for any fundamentally symmetrical system, all constraints must be linear, whence they may be written in the form of Equations (35). Of course a linear constraint may always be written in a form which at first sight may appear to be nonlinear. For example, one may take logarithms of both sides of Equation (34) to obtain the following form which in no way alters its actual linear content, even though it gives the appearance of nonlinearity:

$$\ln \sum_{i=1}^{n} p_i g_i = \ln K$$

Also, groups of probability specifications may be written in a form which might at first sight appear to be nonlinear. (It should be noted here that all probability specifications $p_j = c_j$ are embraced by Equations (35) by takeng $K_j = c_j$, $g_{ji} = 0$ for $i \neq j$, and $g_{jj} = 1$.) Consider for example the pair of specifications

$$p_1 = \frac{1}{4},\ p_2 = \frac{1}{2}$$

which may be written as

$$p_1 p_2^2 = \frac{1}{16}\ ,\ p_1 p_2^3 = \frac{1}{32}$$

as long as this second pair of equations is stipulated to be an unbreakable pair. But since this second pair is unbreakable, it contains no actual nonlinearities, since the apparent nonlinearities can be exhibited only by considering each equation separately. So while the set of linear equations (35) may be disguised in what might at first sight appear to be nonlinear forms, the actual content always remains linear. The following theorem has thereby been proven.

> Linear Constraint Theorem: All constraints on the probabilities $\{p_i\}$ of fundamentally symmetrical systems must be linear functions of $\{p_i\}$.

It should be emphasized that this linearity applies only to the probabilities $\{p_i\}$ and not to the quantities $\{g_{ji}\}$. For example, the constraint $\Sigma p_i t_i^2 = K$ complies with the linear constraint theorem because it is linear in $\{p_i\}$--the nonlinearity $g_{ji} = t_i^2$ having no bearing upon the theorem.

It should also be emphasized that it is only f-symmetrical systems (fundamentally symmetrical systems) which must have constraints linear in $\{p_i\}$. Non f-symmetrical systems may have constraints nonlinear in $\{p_i\}$ which can be used as constraints on the minimum of essergy, but not on the maximum of entropy (since the maximum entropy theorem can not apply to non f-symmetrical systems--as was shown in Appendix F).

The Minimum Essergy Formalism

The linear constraint theorem should not be construed as ruling out nonlinear constraints altogether--it merely rules out their use in the maximum entropy formalism, which applies only to f-symmetrical systems. If one is faced with a constraint which is nonlinear in $\{p_i\}$--such as appears to be the case with 1/f noise [22]--then the system of outcomes must be non f-symmetrical, which leaves the investigator with two possible courses of action:

1. Find a transformation to an f-symmetrical system of outcomes, which thereby will transform the constraint into a linear form which can be used in the maximum entropy formalism.

2. Use the nonlinear constraint in the minimum essergy formalism.

The first course of action might be difficult inasmuch as a transformation to an f-symmetrical system of outcomes might be next to impossible. The contributions of Jaynes regarding invariance of $\{p_i|_z\}$ under transformation groups [6] and marginalization [21] should be helpful in this regard. (See Equation 21 regarding $\{p_i|_z\}$ for f-symmetrical systems.) The more promising course of action appears to be to apply the theorem of minimum essergy, which embraces the minimum essergy formalism as follows: To find the actual probabilities $\{p_i\}$,

minimize the essergy

$$\frac{\langle\mathcal{E}\rangle}{K} = \sum_{i=1}^{n} p_i \ln \frac{p_i}{p_{io}} \tag{36}$$

subject to the constraints

$$\sum_{i=1}^{n} p_i = 1 \tag{37}$$

$$G_j(\{p_i\}) = 0 \qquad j = 1,2,{-}{-}{-}{-}m < n-1 \tag{38}$$

In order to use this formalism, an acceptable reference level (one which complies with Equation 20) must be used which yields $\{p_{io}\}$ which characterize any nonlinearities of the constraints being considered. Such an acceptable reference level serves to guarantee that the constrained essergy $\langle\mathcal{E}\rangle/K$ will always have a unique minimum point, as was proven just before the statement of the minimum essergy theorem. There is of course a great deal more to be said regarding the use of the minimum essergy formalism, but it is beyond the scope of this paper to consider the matter further.

Conclusion

The minimum essergy theorem demonstrated in this paper has at least two uses:

1. Determination of probabilities $\{p_i\}$ in systems where the constraints are nonlinear in $\{p_i\}$ (for which case the maximum entropy formalism has been shown not to apply).

2. Determination of constraints in the maximum entropy formalism by requiring consistency with both the theorems of maximum entropy and minimum essergy.

The first use has merely been briefly outlined in the preceding section without any concrete applications although it is anticipated that such applications will be found. The second use has been shown to lead to the linear constraint theorem, which rules out constraints nonlinear in $\{p_i\}$ in the maximum entropy formalism. It is anticipated that requiring consistency with the minimum essergy theorem will lead to still more precise identification of constraints in the maximum entropy formalism when particular classes of systems are considered.

Acknowledgement

Initial groundwork which led to the results reported in this paper was done in close association with Dr. Myron Tribus during the years 1958-1970. {4}

Appendix A: Demonstration That $p_{I|X}/p_{I|X_o}$ Is Constant With Respect To All Variations In "a" And "I"

From Equation (4), $X = YX_o$, whence,

$$\frac{p_{I|X}}{p_{I|X_o}} = \frac{p_{I|YX_o}}{p_{I|X_o}} \qquad \text{(A-1)}$$

and from Bayes' theorem,

$$p_{I|YX_o} = p_{I|X_o} \frac{p_{Y|IX_o}}{p_{Y|X_o}} \qquad \text{(A-2)}$$

The ratio $p_{Y|IX_o}/p_{Y|X_o}$ is a constant C to as close an approximation as desired, since advanced knowledge states "a" are understood to be restricted such that each observation sequence "I" must have approximately the same bearing on one's knowledge Y when viewed from basic reference levels X_o as discussed in Appendix C--whence $p_{Y|IX_o}$ must have approximately the same value for all "I", while the value of $p_{Y|X_o}$ is of course by definition always independent of "I":

$$C \equiv \frac{p_{Y|IX_o}}{p_{Y|X_o}} \qquad \text{(A-3)}$$

The right side of Equation (A-3) will differ from the constant C by an amount δ which varies with "I", but which can be made as small as one pleases by lengthening the sequence I accordingly. That is, only long sequences "I" can yield the condition of each having precisely the same bearing on one's knowledge Y as viewed from reference level X_o, as discussed in Appendix C. The magnitude of δ will also depend upon the particular reference level X_o. For example, if $X_o = X \equiv YX_o$, then $p_{Y|IX_o} = p_{Y|X_o} = 1$, whence $\delta = 0$ in this case regardless of how short the observation sequence "I" might be. With long sequences

"I" such that δ is negligible, the value of the constant C will in general be strongly dependent upon further increasing the length of the sequence "I" (except in the singular case just cited where $X_o = X \equiv YX_o$ in which case C = 1 regardless of the length of the sequence "I") because $p_{Y|X_o}$ is an exceedingly small probability for the basic reference levels X_o used in the development up through Equation (19)--as discussed in Appendix C. It is also possible to introduce more general reference levels for which the right side of Equation (A-3) is not a constant, but for which Equation (20) holds. However, with respect to the development up through Equation (19), it is to be understood that only basic reference levels X_o are used for which Equation (A-3) holds to as close an approximation as desired by lengthening the sequence "I" accordingly. With this understanding, substitution of Equation (A-3) into Equation (A-2) yields,

$$\frac{p_{I|YX_o}}{p_{I|X_o}} = C \tag{A-4}$$

Or, in view of Equation (A-1),

$$\frac{p_{I|X}}{p_{I|X_o}} = C \tag{A-5}$$

Thus, for any given adequate length of the observation sequences "I", the ratio $p_{I|X}/p_{I|X_o}$ is a constant C with respect to all variations in "I" (and therefore in "a") to as close an approximation as is required by the precision of the ensuing results--as was to be shown.

Appendix B: On Essergy As A General Quantity

The quantity "essergy" defined by Equation (9) appears to have originated in statistical mechanics with Gibbs who first demonstrated its importance following his essential energy expression (448) [17], and who presented numerous other essential energy expressions (his Equations 53 and 57 for example [18]). The term "essential energy", may be interpreted either as the "essential aspect or essence of energy" or as "energy in the form essential for the production of power"--the latter being more colloquially understandable in view of its correspondence to the widely used term "useful energy". To see why this quantity

$$\varepsilon_i \equiv K \ln \frac{p_i}{p_{io}} \tag{9}$$

has been called "essergy" since 1968 [23,24,25,26], it may be of interest first to briefly review its more recent history as a measure of information:

The essergy $K \ln(p_i/p_{io})$ is a general form of the Shannon-Wiener information measure (as may be seen from Equation 8)

$$I_{A:B} \equiv \ln \frac{p_{A|B}}{p_A} \qquad \text{(B-1)}$$

named after Shannon [4] and Wiener (1948) by Woodward and Davies [27] who derived it after the manner of Shannon [4] as a measure of information based upon axioms involving probabilities. This measure was called the "surprisal" by Tribus [10] and under this name it has been used extensively by Levine [28,29,30,31] and others [32] to characterize particle collisions and instabilities in chemical reactions.

The expected value of $I_{A:B}$,

$$<I_{A:B}> \equiv \sum_{B}\sum_{A} p_{BA} I_{A:B} \qquad \text{(B-2)}$$

has been shown by Cox to be the entropy of the disjunction of two questions A and B. [16,19] He goes on to derive a complete calculus for systems of questions which is identical in form to probability theory except that "disjunction" is everywhere replaced by "conjunction" and vice versa. Using this Boolean algebra transform, Cox shows that all theorems of probability theory remain valid in the calculus of questions and vice versa [19]. It follows that the minimum essergy theorem derived in this paper holds for systems of questions as well as for systems of propositions or outcomes. (As pointed out before Equation 14, Cox uses the word "system" in its broad inductive sense which includes not only physical systems--such as the systems of statistical thermodynamics--but also includes any system of outcomes or sample space of events, and any system of questions.)

But for systems of questions, essergy is a measure of the depth or broadness of questions and not a measure of information or surprisal--since the terms "information" and "surprisal" have no meaning in the calculus of questions except via the transform to probability theory where one has the information content of propositions (or in equivalent terminology, the surprisal for outcomes or events). And, as shown by Cox (Equations 5.10-5.13 [19]), the concept of entropy also has the same general connotations which applies to expected depth of questions as well as to uncertainty (i.e., expected surprisal [10] or expected information content of propositions). It was with these general connotations in mind that the word "essergy" was introduced in 1968 as a counterpart to Cox's general interpretation of entropy (Cox, private communication, 1967).

It is anticipated that some important (and as yet undiscovered) theorems involving essergy may be easier to prove for systems of questions than for systems of propositions, and that the Cox transformation will thereby provide a powerful tool for deriving such theorems. In any event, the concept of essergy is considerably more general than the concepts of "information" and "surprisal" with which it is often identified.

In macroscopic thermodynamics, it is customary to use terms and symbols such as energy E and mass M when one really means expected energy $\langle E \rangle = \Sigma p_i E_i$ and expected mass $\Sigma p_i M_i$. This practice has been carried over to essergy by Tribus and the writer [33,34,24,25,26,35] whenever the context is non statistical, so that for such applications the essergy $\mathcal{E} \equiv \langle \mathcal{E} \rangle = \Sigma p_i \mathcal{E}_i$ is, from Equation (36), given by

$$\mathcal{E} = K \sum_{i=1}^{n} p_i \ln \frac{p_i}{p_{io}} \qquad (36)$$

A very simple and convenient reference level $\{p_{io}\}$ is the equilibrium distribution as given by the familiar Gibbs grand canonical distribution [10]:

$$-\ln p_i = \frac{PV}{kT} + \frac{E_i}{kT} - \sum_c \frac{\mu_c}{kT} N_{ci}$$

where P, V, and T denote pressure, volume, and temperature respectively while N_{ci} denotes the quantity of component "c" in microstate "i" with μ_c as the corresponding Gibbs potential. Introducing the subscript "o" to denote "reference level", one has,

$$-\ln p_{io} = \frac{P_o V}{kT_o} + \frac{E_i}{kT_o} - \sum_c \frac{\mu_{co}}{kT_o} N_{ci} \qquad (B\text{-}3)$$

Substitution of Equation (B-3) into Equation (36) along with $K = kT_o$ and $S = -k\Sigma p_i \ln p_i$ yields, in the customary notation for non-statistical contexts,

$$\mathcal{E} = E + P_o V - T_o S - \sum_c \mu_{co} N_c \qquad (B\text{-}4)$$

This expression, which corresponds to Gibbs' Equation (53) [18], has been used extensively by Tribus [36,37,38,39,40] and others [25,41,42,43] in studying the optimum design of various types of thermoeconomic systems. In Reference 24, it is shown that for the somewhat more complicated reference level needed for variable T_o, Equation (36) yields exactly the same expression

(B-4) via defining $K = kT_u$ (where T_u is a universal constant with units of temperature), but with T_o now being a variable which applies, for example, to changing weather conditions.

Equation (B-4) is not very useful however, in situations where more than one sink temperature T_o applies simultaneously--such as in the analysis of power plant cooling alternatives (e.g., T_{o1} = air temperature for dry towers, T_{o2} = wet bulb temperature for evaporative towers, T_{o3} = river water temperature, etc.). This nevertheless poses no problem whatsoever for Equation (36) where one needs merely to express the probabilities p_{io} with respect to each different sink temperature T_o, and combine them via the familiar product rule for independent probabilities.

It should also be pointed out that for the general case where exhaustive sets of outcomes i = 1,2,---n are not mutually exclusive, the expected essergy $<\mathcal{E}>$ nevertheless remains quite well defined via merely replacing $-\ln p_i$ by the right side of Equation (9) in Cox's general entropy expression (Equation 8.3 [16] or Equation 4.5 [19]). And since the minimum essergy theorem holds for the extremes of the Cox transformation (i.e., it holds for systems of mutually exclusive questions as well as for systems of mutually exclusive propositions as pointed out earlier in this appendix--these two cases representing the extremes of the Cox transformation [19]), it must hold for the intermediate non mutually exclusive case. Application of the minimum essergy theorem to non mutually exclusive outcomes could perhaps resolve the problem of the equivalence of "orthogonality" in quantum theory to "mutual exclusiveness" in probability theory.

Appendix C: On States Of Advanced Knowledge

As mentioned in Appendix A, the states "a" of advanced knowledge used in the proof given in this paper are restricted to a class which allows the right side of Equation (A-3) to be constant with respect to basic reference levels X_o. And in order to meet the requirement mentioned just before Equation (12), this same class of advanced knowledge states "a" must also satisfy the following condition: From these same basic reference levels, the individual observations i_K in each observation sequence $I = i_1 i_2 i_3 \text{---} i_K \text{---} i_N$ must be viewed as being independent of all preceding observations in the sequence. These two requirements are of course in addition to the fundamental requirement that

$$p_{a|IX} = 1 - \delta \qquad (2)$$

where δ may be rendered as small as one pleases by taking the sequence length N longer accordingly.

A class of advanced knowledge states "a" which meets these three requirements is one where the knowledge is always gained from observation sequences $I = i_1 i_2 --- i_K --- i_N$ made upon an ensembled of systems--where the word "system" denotes any exhaustive set of possible outcomes which are mutually exclusive. (The term "system" is employed here in general inductive sense introduced by Cox [16,19], where it includes not only physical systems--such as the systems of statistical thermodynamics--but also includes systems of possible outcomes or sample spaces of events, as pointed out in Appendix B.) Using this terminology, i_K denotes the outcome observed for the K'th system in the ensembled--that is, for the K'th exhaustive set of mutually exclusive possible outcomes. All of the systems of the ensemble are understood to be similar to the original system under consideration--i.e., the one from which the possible outcome "i" in Equation (1) is taken, And furthermore, this original system is understood to be one of the members of the ensemble.

The systems of such an ensemble may always be looked upon as being independent of each other when viewed from basic reference levels X_o. As viewed from such a reference level, there is absolutely no physical or logical connection via which one outcome could or should influence one's opinion of any other outcome in the ensemble. The following example, though contrived, may serve to illustrate this point: Consider a coin made by a manufacturer of "dishonest" coins, some of which come up mostly heads, others mostly tails, and still others which seem not to "work" as well--some of them even seeming to be "honest". Let the original system be the possible outcomes "heads" and "tails" when this coin is tossed for the very first time after purchase--at which time one has no idea of how it is going to behave. One suitable hypothetical ensemble for this system would be a collection of coins, each from a different manufacturer of similar dishonest coins, <u>each</u> being tossed for the <u>very first time</u>. Such a contrivance is permissible since the ensemble is merely hypothetical, just as all observation sequences $I = i_1 i_2 -- i_K -- i_N$ made upon it are hypothetical--since all advance knowledge states in Equation (1) are always understood to be hypothetical. Still further guarantees of independence between the tosses of these coins may be made by such contrivances as labeling the outcomes "A" and "B" without knowledge of which is "heads" and which is "tails".

Each advanced knowledge state "a" consists of a particular value of the fraction of "A's" (which is stipulated to be the same as the fraction of "heads") for the entire ensemble. (In this regard, it is also stipulated that the number of systems in the ensemble is always several orders of magnitude larger than the number of observations N made upon it.) And since the original system itself is stipulated to be a member of the ensemble without knowing which one it is (due to the interchange of labels "A" and "B" with "heads" and "tails") the probability of heads for this hypothetical state of advanced knowledge "a" is simply equal to the fraction of heads (i.e., of "A's") in the ensemble.

Each sequence of observations $I = i_1 i_2 \text{--} i_K \text{--} i_N$ made on this ensemble (with N understood to be several orders of magnitude less than the number of systems in the ensemble) consists of a sequence where each i_K is either an "A" or a "B". Only sequences "I" which have exactly the same fraction of "A's" as does the advanced knowledge state "a" are paired with "a" in Equation (2). Thus many different sequences "I" may be paired with any given "a"--since many different sequences may have the same fraction of "A's"--but for any given sequence I, there corresponds one and only one advanced knowledge state "a".

If the number of observations N should be increased to the point where it actually equals the number of systems in the ensemble, then obviously $p_{a|IX} = 1$ precisely, since the fraction of "A's" for "I" and "a" would be one and the same for that case. Let us take the mathematical shortcut of considering the number N to be infinite in this case, so that δ approaches zero in the expression $p_{a|IX} = 1 - \delta$ when $N \to \infty$. A similar mathematical shortcut was made by Gibbs [17] when he let the number of mechanical systems in his ensemble "increase indefinitely" to approach "continuous distribution in phase"--a shortcut which he defended as follows: "To avoid tedious circumlocution, language like the above may be allowed, although wanting in precision of expression, when the sense in which it is to be taken appears sufficiently clear."

Finally, let us use this simple "dishonest" coin example to consider the constant "C" define by Equation (A-3):

$$C \equiv \frac{p_{Y|IX_o}}{p_{Y|X_o}} \tag{A-3}$$

Suppose that "Y" represents the knowledge that the coin is one which comes up heads more often than 90% of the time. For this case, "Y" corresponds to the fraction of "A's" being greater than 0.9 in the ensemble. Viewed from the reference level X_o described in the preceding paragraphs, Y is extremely unlikely, since from the symmetry of the ensemble, it corresponds mathematically to the probability of tossing an honest coin N times and getting over 90% heads. The probability $p_{Y|IX_o}$ on the other hand can be made to approach unity as close as one pleases by restricting the advance knowledge states to those with over 90% "A's"--whence only sequences "I" with over 90% "A's" appear in $Y|IX_o$ (since only such sequences are paired with "a" in Equations 2 through 9--in accordance with the procedure established two paragraphs back). Thus restricted, the probability $p_{Y|IX_o}$ would approach unity as $N \to \infty$. More generally, advanced knowledge states "a" may be specified such that all have the same bearing upon "Y", instead of all determining "Y" precisely as in the above example. In this case the sequences "I" all approach this same bearing (i.e., this same probability) as $N \to \infty$. The remainder of the proof that the right side of Equation (A-3) is constant with respect to "a" and "I" (for any given N of course--since $p_{Y|X_o}$ decreases rapidly with increased N as indicated above) is given in Appendix A.

The "dishonest" coin example just given serves to illustrate how the ensemble type of advanced knowledge states "a" used in this paper satisfy the three requirements set forth at the beginning of this appendix. This type of advanced knowledge state is completely different from the limiting frequency type used by Laplace in obtaining the "Rule of Succession" [15]. Going back to the "dishonest" coin example, the limiting frequency type of advanced knowledge states may be characterized by considering hypothetical observation sequences made via hypothetically tossing the same coin over and over--thus arriving at hypothetical limiting frequencies which constitute advanced knowledge states. The probability distributions for such limiting frequency type of advanced knowledge states tend to be much flatter than the distributions for the ensemble typed used in the proof of the minimum essergy theorem--as shown in Figure 3. This flatness simply signifies that the individual observations i_K are not logically independent, as they are for the ensemble type. There is of course a great deal more to be said regarding the details of how ensembles change as the data "Y" is updated--for example by tossing the "dishonest" coin--but it is beyond the scope of this paper to treat the matter further.

As mentioned at the beginning of this appendix, all of the systems of the ensemble are similar to the original system--the criterion of this similarity being that the average expected essergy of the systems in the ensemble is equal to the essergy of the original system. This criterion merely requires that ensembles be formed in a balanced fashion where opposing types of dissimilarities cancel each other out.

Having illustrated how the individual observations i_K in the sequence $I = i_1 i_2 \text{--} i_K \text{--} i_N$ are independent from one another for the ensemble type of advanced knowledge states when viewed from basic reference levels X_o, it follows from the elementary product rule for independent probabilities $p_{AB} = p_A p_B$ that,

$$p_{I|X_o} = \prod_{K=1}^{N} p_{i_K|X_o} \tag{C-1}$$

The probabilities $p_{i_K|aX} \equiv p_{i_K|AX_o}$ are also independent of one another, since with "a" given, the probabilities of all i_K are definite numbers as stipulated in the definition of advanced knowledge states (as given following Equation 1). Thus,

$$p_{I|AX_o} = \prod_{K=1}^{N} p_{i_K|AX_o} \tag{C-2}$$

In view of the definition of the essergy $\mathcal{E}_{i|EX_o}$ given in Equation (8), taking logarithms of Equations (C-1) and (C-2) yields

$$\mathcal{E}_{I|AX_o} = \sum_{K=1}^{N} \mathcal{E}_{i_K|AX_o} \tag{12}$$

Appendix D: Use Of The Variance Rule

For any set of independent "random" variables, the familiar variance rule states that the variance of their sum is equal to the sum of their variances [14]. Or stated in terms of averages, the variance of the average of N variables is equal to 1/N times their average variance. Thus for any finite value of the average variance, the variance of the average approaches zero as $N \to \infty$.

Now Chebyshev's inequality for any "random" variable V is [14]

$$p(|V - \langle V\rangle| \geq \delta) \leq \frac{\sigma^2 (V)}{\delta^2} \tag{D-1}$$

If V is an average of N independent "random" variables, both δ and $\sigma^2(V)/\delta^2$ vanish as $N \to \infty$, as may be illustrated by letting δ equal say $N^{(1/4)}$ and recalling that $\sigma^2(V)$ equals 1/N times the average variance. Thus V approaches $\langle V\rangle$ to within any specified accuracy with a probability which approaches unity as close as one pleases by letting N be accordingly large. Applying this result to the average value of $\mathcal{E}_{i_K|AX_o}$, which by Equation (12) is equal to $\mathcal{E}_{I|AX_o}/N$, it follows that $\mathcal{E}_{I|AX_o}/N$ approaches $\langle\mathcal{E}_{I|AX_o}/N\rangle$ to within any specified range with probability $\to 1$ as $N \to \infty$. Since values of $\mathcal{E}_{I|AX_o}/N$ which depart from $\langle\mathcal{E}_{I|AX_o}/N\rangle$ by more than the specified range are thus so highly improbable, it may be stipulated that any sequences "I" for which this happens may be rejected as not being appropriate for pairing with the advanced knowledge state with which it corresponds--as set forth in Appendix C. In this manner, one obtains to any desired accuracy by taking N accordingly large,

$$\mathcal{E}_{I|AX_o}/N = \langle \mathcal{E}_{I|AX_o}/N \rangle$$

Or in view of Equation (12) and the expectation rule for independent "random" variables (expectation of a sum equals the sum of the expectations), Equation (13) follows.

Appendix E: Concerning The Reference Level $X_o' = X$

As pointed out following Equation (18), for reference level $X_o' = X$, the reference probabilities p_{io} in Eqn. (14d) are equal to the solution values $p_{i|X}$ from Equation (1)--provided of course that such a unique set $\{p_{i|X}\}$ of solution values does indeed exist. If desired, this existence may be taken as a required condition for the minimum essergy theorem--since the theorem will hold only for cases where a solution exists. However, in view of Equations (1), (7), and (A-3), all that is required for the existence of a unique solution $\{p_{i|X}\}$ is the existence of one single reference level X_o for which the quantities in these equations are uniquely determined. When viewed from this vantage point, it would seem extremely doubtful that there could exist a circumstance for which a unique solution $\{p_{i|X}\}$ does not exist, since there are a virtual infinity of different reference levels X_o, any one of which might fulfill this requirement.

Appendix F: On Maximum Entropy And Fundamental Symmetry

Consider any system where there are no constraints other than

$$\sum_{i=1}^{n} p_i = 1 \qquad \text{(F-1)}$$

For this case, it is well known that the maximum entropy point is given by,

$$p_i = \frac{1}{n} \qquad i = 1,2,---n \qquad \text{(F-2)}$$

Consider further that all features except the fundamental distinguishing features (defined just before Equation 21) have been removed. Now if the outcomes were <u>completely</u> indistinguishable, one would also have $p_i=1/n$ for all i, because any inequalities among the p_i would in themselves provide a means for distinguishing among them. It follows that the fundamental distinguishing features are irrelevant, whence the system is fundamentally symmetrical--in accordance with the definition of fundamental symmetry given just before Equation (21).

Now consider any system for which the maximum entropy theorem applies. The fundamental reference level for such a system is by definition the level for which there are neither constraints nor distinguishing features other than the fundamental distinguishing features. Maximizing the entropy for this level thus corresponds to the case just considered, so that once again the system is fundamentally symmetrical. If follows that the maximum entropy theorem applies only to fundamentally symmetrical systems. {5}

Appendix G: Solution For The Function $G(\{p_i\})$

Differentiation of the function $G(\{p_i\})$ yields, in view of Equation (26a)

$$dG(\{p_i\}) = \sum_{i=1}^{n} g_i(\{p_i\})dp_i \tag{G-1}$$

Substitution of the restriction (32) into (G-1) gives,

$$dG(\{p_i\}) = \sum_{i=1}^{n} h(\{p_i\})g_{im}\, dp_i \tag{G-2}$$

The condition for which $G(\{p_i\})$ is constant with respect to $\{p_i\}$ is

$$dG(\{p_i\}) = 0 \tag{G-3}$$

for which

$$\sum_{i=1}^{n} g_{im}\, dp_i = 0 \tag{G-4}$$

since the quantities $h(\{p_i\})$ cancel. Or, since the quantities g_{im} are constant with respect to $\{p_i\}$,

$$d\left(\sum_{i=1}^{n} g_{im}\, p_i\right) = 0 \tag{G-5}$$

Thus $G(\{p_i\})$ is a constant if and only if $\sum_{i=1}^{n} g_{im}p_i$ is constant, so that $G(\{p_i\})$ must depend only upon $\sum_{i=1}^{n} g_{im}p_i$:

$$G(\{p_i\} = f\left(\sum_{i=1}^{n} g_{im}\, p_i\right) \tag{G-6}$$

where f( ) denotes any arbitrary function. Thus, in view of (26a), differentiation of (G-6) with respect to p_i gives,

$$g_i(\{p_i\}) = f'(\sum_{i=1}^{n} g_{im}\, p_i)\, g_{im} \qquad i = 1,2---n \qquad \text{(G-7)}$$

But when the constraint (25) is obeyed, both (G-3) and (G-4) hold. Hence,

$$\sum_{i=1}^{n} g_{im}\, p_i = C \qquad \text{(G-8)}$$

where C is the arbitrary constant of integration. Thus with the constraint obeyed, (G-7) becomes

$$g_i(\{p_i\}) = f'(C) g_{im} \qquad i = 1,2---n \qquad \text{(G-9)}$$

Thus the quantities $g_i(\{p_i\})$ are constant, whence they may be written as,

$$g_i(\{p_i\}) = g_i \qquad i = 1,2---n \qquad \text{(G-10)}$$

Substitution of (G-10) into (G-1) yields,

$$dG(\{p_i\}) = \sum_{i=1}^{n} g_i\, dp_i \qquad \text{(G-11)}$$

which upon integrating gives

$$G(\{p_i\}) = \sum_{i=1}^{n} p_i g_i - K \qquad (33)$$

where -K is the arbitrary constant of integration. Equation (33) is the solution for the constraint (25) when the restriction (32) applies--it being noted once again that the quantities g_i are constants.

For the case of groups of q independent constraints where $q \leq m$, the required condition is the following generalized form of the equation following Equation (30):

$$\lambda_o + \sum_{j=1}^{q} \lambda_j \sum_{i=1}^{n} p_i\, g_{im} = f(S/k) \qquad \text{(G-12)}$$

Here as before, the specification of probabilities p_i would alter the content of this condition, and therefore of the constraints, unless the change resulting from the specifications is restricted to multiplying Equation (G-12) by a common value $h(\{p_{im}\})$. Thus in order to allow for all possible specifications, the restriction (32) applies to each of the constraints in the dependent set. Whereupon each constraint is of the form indicated by Equations (33) and (34).

Notes

The number(s) in brackets following each note is that of the page(s) to which the note refers.

1. In this paper, lower case letters are used to denote numbered propositions, while upper case letters are used for word statements and sets of propositions. [171]

2. In this paper, the conjunction "A and B" of any two propositions "A" and "B" is represented by "AB". [171]

3. All of the statements made in the derivation of the linear constraint theorem apply only to fundamentally symmetrical systems, since the maximum entropy theorem applies only to such systems. For example, in the statement following Equation (25), S/k wouldn't necessarily have a unique maximum if Equation (24) didn't hold--and Equation (24) applies only to fundamentally symmetrical systems. Similarly, in the statements preceding Equation (28), it is only for fundamentally symmetrical systems that $\langle \mathcal{E} \rangle /K$ and S/k are strictly decreasing monotonic functions of each other. [181,182,184]

4. The writer also wishes to acknowledge the support of Dr. S. Peter Kezios and the Administration of the Georgia Institute of Technology in their providing the conditions in which this work could be carried out. [187]

5. It may be of interest to point out that in statistical mechanics, the fundamentally symmetrical set of outcomes corresponds to the use of coordinates which satisfy Liouville's theorm, a generalized form of which has been derived by Robertson [44] based upon "conservation of probabilities." [197]

Bibliography

1. Jaynes, Edwin T., "Information Theory and Statistical Mechanics" Phys. Rev., 106, May 1957, pp 620-630.

2. Tribus, M. "Information Theory as the Basis for Thermostatics and Thermodynamics." Journal of Applied Mechanics. 28: 1-8, March, 1961.

3. Cox, Richard T., "Probability, Frequency, and Reasonable Expectations" Am. J. Phys., 14 January 1946, pp 1-14.

4. Shannon, C.E., "A Mathematical Theory of Communication," Bell System Tech. J. 27, 379, p. 623, 1948.

5. Elsasser, Walter M., "On Quantum Measurements and the Role of Uncertainty Relations in Statistical Mechanics", Phys. Rev., 52, November 1937, pp 987-999.

6. Jaynes, E.T., Syst., Sci., & Cybernetics., Sept., 1968, IEEE.

7. Tribus, Myron. Rational Descriptions, Decisions, and Designs. Pergamon Press, Inc., New York, 1969.

8. Brillouin, L., Science and Information Theory. 2nd ed. Academic press, Inc., New York, 1962, 1st ed., 1956.

9. Gibbs, J.W., "Representation by Surfaces of the Thermodynamics Properties of Substances." (1873). Collected Works Vol. I, Yale U. Press, New Haven, Conn., 1948.

10. Tribus, M., Thermostatics and Thermodynamics. Van Nostrand, Princeton, New Jersey, 1961.

11. Evans, R.A., "The Principle of Minimum Information." IEEE Transactions on Reliability, Vol. R-18, No. 3., August, 1969.

12. Jaynes, Edwin T., Probability Theory in Science and Engineering, Field Research Laboratory, Socony Mobil Oil Company, Inc., Dallas, Texas, (1959).

13. Tribus, M. and Evans, R.B., "The Probability Foundations of Thermodynamics." Applied Mechanics Review, Vol. 16, No. 10, pp 765-769, October 1963.

14. Feller, William. An Introduction to Probability Theory and and its Applications. 2nd. ed. Wiley, New York, 1950.

15. Venn, John. The Logic of Chance: An Essay on the Foundations and Province of the Theory of Probability, Macmillan, 3rd ed. New York. 1888, p 124.

16. Cox, R.T., The Algebra of Probable Inference. Johns Hopkins Press, 1961.

17. Gibbs, J.W., "Elementary Principles in Statistical Mechanics" (1901). The Collected Works, Vol. II. Yale U. Press, 1948.

18. Gibbs, J.W., "On the Equilibrium of Heterogeneous Substances." 1878. The Collected Works, Yale University Press, Vol. 1, p 77, (1948).

19. Cox, R.T., Of Inference and Inquiry: An Essay In Inductive Logic. Dept. of Physics, Johns Hopkins U. (1976). Also, M.I.T. Press, 1978. (Maximum Entropy Formalism Conference Proceedings).

20. Kullback, S., Information Theory and Statistics, Wiley, New York, 1959.

21. Jaynes, E.T., "The Marginalization Paradox," Presented at the 14'th NBER-NSF Seminar on Bayesian Inference, June 1977.

22. Gardner, M., Scientific American, April, 1978.

23. Evans, R.B., "The Formulation of Essergy." Thayer News, Thayer School of Engineering, Dartmouth College, Hanover, N.H. (Fall, 1968).

24. Evans, R.B., "A Proof that Essergy is the Only Consistent Measure of Potential Work (For Chemical Systems)." Ph.D. Thesis. Thayer School of Engineering, Dartmouth College, Hanover, N.H., June 1969.

25. El-Sayed, Y.M. and Evans, R.B., "Thermoeconomics and the Design of Heat Systems." Transactions of the ASME, Journal of Engineering for Power. January 1970. pp 27-35.

26. Evans, R.B., "The Superiority of Essergy Analysis." Discussion of "Economics of Feedwater Heater Replacement." by Fehring T. and Gaggioli, R. Trans. ASME, July 1977, pp 488-489.

27. Woodward, P.M., and Davies, I.L., "Information Theory and Inverse Probability in Telecommunications, Proc. IEEE, 99: Part III, No. 58, pp 34-37. (1952).

28. Bernstein, R.B., and Levine, R.D., J. Chem, Phys. 57, 434 (1972).

29. Ben-Shaul, A., Levine, R.D., and Bernstein, R.B., J. Chem. Phys. 57, 5427 (1972).

30. Levine, R.D. and Ben-Shaul, A., "Thermodynamics of Molecular Disequilibrium. Chemical and Biochemical Applications of Lasers, Vol II. Academic Press, New York, 1977.

31. Levine, R.D. "Maximal Entropy Procedures for Molecular and Nuclear Collisions." Maximum Entropy Formalism Conference Proceedings, M.I.T. Press, 1978.

32. Ross, John. "Relative Stability in Far From Equilibrium Systems." Maximum Entropy Formalism Conference Proceedings, M.I.T. Press, 1978.

33. Tribus M. and McIrvine E., "Energy and Information." Scientific American, Sept. 1971, pp 179-184.

34. Tribus, M. "The Case for Essergy," Journal of Mechanical Engineering, p 75, April, 1975.

35. Evans, R.B., "Essergy," Journal of Mechanical Engineering, January, 1978, p 50.

36. Evans, R.B., Crellin, G. L., and Tribus, M. "Thermoeconomic Considerations of Sea Water Demineralization." 2nd Edition, Principles of Desalination, edited by K. S. Spieler, Academic Press, New York, 1975.

37. Evans, R.B., "Thermodynamic Availability as a Resource and a Tool for System Optimization." (1958). Appendix II of the report by Tribus, M. et al. Thermodynamic and Economic Considerations in the Preparation of Fresh Water From the Sea. Revised September, 1960. University of Calif., Department of Engineering, Los Angeles, Report No. 59-34, 1960.

38. Evans, R.B., A Contribution to the Theory of Thermo-Economics. University of California, Department of Engineering, Report No. 62-36. August, 1962.

39. Evans, R.B. and Tribus, M., "Thermo-Economics of Saline Water Conversion." I&EC Process Design and Development, Vol. 4, No. 2, pp 195-206, April 1965.

40. Tribus, M., and Evans, R.B., "Thermoeconomic Design Under Conditions of Variable Price Structure." Proceedings of the First International Symposium of Water Desalination. Vol. 3, pp 699-716. US. Government Printing Office, Washington, D.C., 1965.

41. Tribus, M., and Evans, R.B., and Grulich, G., "The Use of Exergy and Thermoeconomics in the Design of Desalination Plants." Office of Saline Water, Research and Development Progress Report. Contract No. 14-01-001-928. US. Office of Technical Services, 1966.

42. Berg, C.A., "A Technical Basis for Energy Conservation." Mechanical Engineering, Vol. I, No. 5, pp 30-45, May, 1974.

43. Fehring, T., and Gaggioli, R., "Economics of Feedwater Heater Replacement." Trans. ASME, July 1977, pp 482-488.

44. Robertson, Baldwin. "Application of Maximum Entropy to Non-Equilibrium Statistical Mechanics." Proceedings, Maximum Entropy Formalism Conference, M.I.T., 1978.

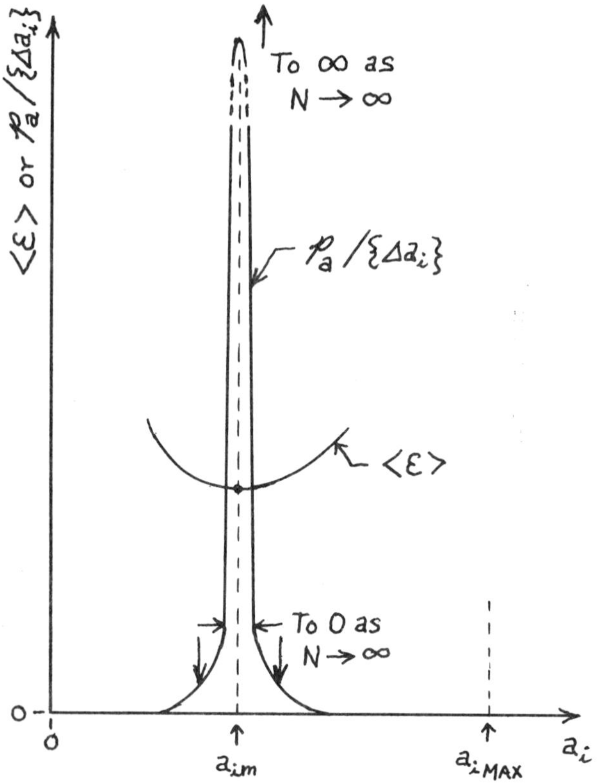

Figure 1. Variation Of $\langle\varepsilon\rangle$ And $p_a/\{\Delta a_i\}$ With Respect To a_i With $\{a_1, a_2, \text{---} a_{i-1}, a_{i+1} \text{---} a_n\}$ Held Constant.

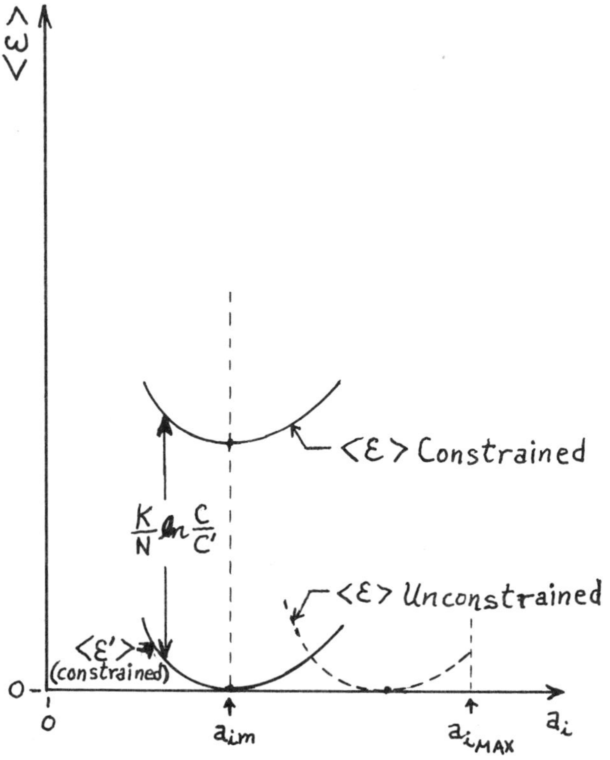

Figure 2. An Illustration Of How Variations Of <ε> And <ε'> Are Separated By The Constant (K/N)ln(C/C').

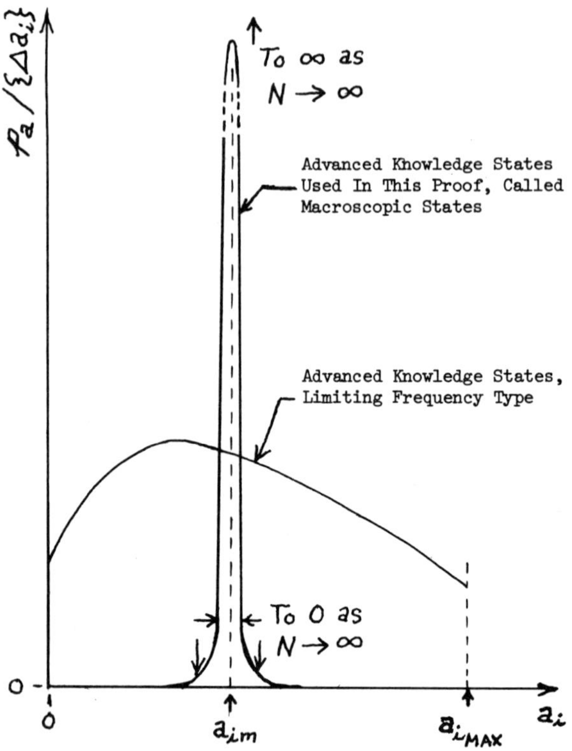

Figure 3. Probability Densities $p_a/\{\Delta a_i\}$ Of Advanced Knowledge States--Comparing The Type Used In This Proof With The Limiting Frequency Type, As Discussed In Appendix C.

AN ALGORITHM FOR DETERMINING THE LAGRANGE PARAMETERS IN THE MAXIMAL ENTROPY FORMALISM*

N. Agmon, Y. Alhassid and R. D. Levine**

This conference discussed a formalism where one needs to determine the maximum of the 'missing information' or entropy function

$$S=-\sum_{i=1}^{n} x_i \ln x_i \tag{1}$$

subject to m constraints

$$\sum_{i=1}^{n} x_i A_{ri} = b_r \quad r=1,\dots,m. \tag{2}$$

Here we consider the practical problem of explicitly determining the distribution $p=(p_1,\dots,p_n)$ which gives the maximum of S subject to the stated constraints. A flow chart, user's guide and a listing of the actual computer program is available from us upon request.

In principle, the problem is solved by the conventional Lagrange multipliers method. The distribution p of maximal entropy is then given by

$$p_i=Z^{-1} \exp(-\sum_{r=1}^{m} \lambda_r A_{ri}) \tag{3}$$

where $Z=\sum_{i=1}^{n} \exp(-\sum_{r=1}^{m} \lambda_r A_{ri})$ is the partition function. To calculate the probability function p one needs yet to solve the set of m coupled, implicit nonlinear equations

$$\langle A_r\rangle-b_r = \sum_{i=1}^{m} p_i A_{ri}-b_r=0 \quad r=1,\dots,m \tag{4}$$

for the Lagrange parameters $\lambda=(\lambda_1,\dots,\lambda_m)$. So, although in principle we know how to solve the maximal entropy problem, in practice it is not so, unless one has an efficient algorithm for solving the set of nonlinear equations 4 for the Lagrange parameters.

Consider equations 3 with a set of 'trial' Lagrange parameters, $\lambda^t=(\lambda_1^t,\dots,\lambda_m^t)$. Then the left-hand side of equation 4 can be written as $f(\lambda^t)$ where $f=(f_1\cdots f_m)$. The desired set of Lagrange parameters λ is the solution of the implicit, nonlinear equation

$$f(\lambda^t) = 0 \quad . \tag{5}$$

Without invoking any special properties of f, one can try some of the standard methods for solving nonlinear equations. For example, the function

$\psi(\lambda^t)$ defined by $\psi(\lambda^t) \equiv \sum_{i=1}^{m} f_i^2(\lambda^t)$ is non-negative and

vanishes if and only if f vanishes. Therefore, one can try to find the minimum of $\psi(\lambda^t)$, using the Newton-Raphson iteration procedure, which converges in principle (i.e. overlooking problems of computer's round-off errors) for any initial guess of λ^t if ψ is a strictly convex function. In the case of the maximal entropy problem, ψ is not convex even for an example as simple as a single constraint given by the first moment (i.e. m=1, $A_{1i}=i$).

Thus, blindly applying a conventional procedure cannot guarantee a solution. The reason is that we have ignored the special properties of the function f of the present case. More explicitly, the Jacobian matrix M of f which is given by

$$M_{rs} \equiv \frac{\partial f_r}{\partial \lambda_s^t} \equiv -\frac{\partial \langle A_r - b_r \rangle^t}{\partial \lambda_s^t} = \langle (A_r - \langle A_r \rangle^t)(A_s - \langle A_s \rangle^t) \rangle^t \tag{6}$$

is a symmetric, positive definite matrix. The first property is immediately transparent from equation 6, and the second property (which is not much harder to show) is proved in (1).

The first property entails the existence of a scalar 'potential function' $F(\lambda^t)$ such that $f=\nabla F$. The second property means that F is strictly convex function (it has a positive definite Hessian M). Therefore, F has a unique global minimum for the λ^t which solves $\nabla F(\lambda^t)=0$. In this manner the problem of solving equation 4 is transformed to finding the minimum of the strictly convex function F. To do that we use the modified Newton-Raphson procedure which is now guaranteed to converge for any initial guess. The algorithm is simple: if for k'th iteration the Lagrange multipliers have the value $\lambda(k)$, $\lambda(k+1)$ is determined as follows: first one chooses a vector u which is directed as closely as possible towards the minimum λ. $u \equiv -M^{-1}\nabla F = -M^{-1}f$ is a better choice than $u = -\nabla F$ because it is the minimum of the second order Taylor expansion of F. Next a number α is chosen such that $\lambda^t(k+1)$, defined by

$$\lambda^t(k+1) = \lambda^t(k) - \alpha u, \tag{7}$$

is the minimum of F along the direction u. It is precisely this determination of α which cannot be accomplished in a conventional iteration method where F is unknown.

Finally, let us mention the explicit form of F (discussed in (2)), namely:

$$F(\lambda^t) = \ln z(\lambda^t) + \sum_{r=1}^{m} \lambda_r^t b_r \quad . \tag{8}$$

(It is easily seen that indeed $-(\nabla F)_r = \langle A_r\rangle^t - b_r$. It should be emphasized that in equation 8 the b_r's are the given 'experimental' averages which are held constant while the λ_r^t's are varied independently. Only at the minimum of F one has that $\lambda^t = \lambda$, the desired Lagrange parameters conjugate to the b's. Then $F(\lambda) = S$ (so that F is an upper bound to the entropy) and the Hessian M is the correlation matrix.

A computer program has been written utilizing the above algorithm. It includes also necessary consistency checks, of which a fuller account is given in (3). The program's source, listing, flow chart and user's guide are available from the authors.

*Work supported by the Office of Scientific Research, USAF, under grant AFOSR 77-3135.

**Permanent address: The Hebrew University of Jerusalem, Jerusalem, Israel.

(1) Y. Alhassid, N. Agmon and R. D. Levine, Chem. Phys. Lett. 53, 22 (1978).

(2) R. D. Levine, J. Chem. Phys. 65, 3302 (1976).

(3) N. Agmon, Y. Alhassid and R. D. Levine, submitted.

INDUCTION AND THE TWO CULTURES

Walter M. Elsasser

The present meeting is on the whole concerned with well-defined technical problems of probability theory. Since, however, I have not been active for some years in this field, I am glad to follow Dr. Tribus' suggestion that I try to delve into the broader implications of our subject, be they called philosophical or cultural. But even after having said this, my title may seem strange, if not mystifying. What does the word induction mean in that context? I am using the term in exactly the sense that mathematicians are accustomed to when they speak of inductive as opposed to deductive probabilities. Later on I shall broaden that use.

The term inductive probability means that one deals with probabilities in a purely observational context: if we are given a die we can throw it many times in order to find out from the results whether the die is "true" or else is "loaded." Ordinarily, mathematicians do not operate in that way; they postulate that the die is true and that hence each face has a fixed probability of one sixth. Such a scheme, based essentially on axioms, is called "deductive".

The use of the term induction in this sense goes back to Francis Bacon. Bacon was born three years before Galilei, and these two men more than any others took science out of theology and put it on an empirical basis. Induction for Bacon means not just the collection of similar instances but also their critical evaluation. This last-named activity can just not be relegated to a computer; hence induction in this sense is far more than a mechanical bringing together of samples.

Bacon's influence in England was tremendous; so one is not surprised that many names in the later history of probability theory are British. I will next jump to a modern renaissance in this field that started in 1921 when a man in Cambridge, England, named John Maynard Keynes published a book with the title, "A Treatise on Probability"(1). As most people know, Keynes went on to other pursuits, and eventually to fame and fortune, but this first technical work of his, written when he was in his thirties, is of great historical significance. Keynes' mode of thinking is squarely within that great British intellectual tradition which early in our century became associated with the names of Whitehead and Russell. There are a number of brilliant names involved in the flowering of science that took place in Europe, in England, and more especially in Cambridge, in the early decades of this century. These men had a common background of thought, and I dare equate their aggregated impact to a major revolution in science.

At this point I can hear some of my hearers say: "What revolution? How come I have never heard of this revolution?" It follows, therefore, that the historians of science must have been asleep. I am here telling you that such a revolution has indeed taken place and that, as you can see for yourself, our community has failed to notice it, at least in these terms. I shall describe this revolution as the triumph of modern induction, where I take the term induction in the Baconian sense, including critical evaluation of the results, not just in the rather trivial, Aristotelian sense of a mere enumeration of similarities.

To understand the nature of the new, inductive mode of thinking, it is best to compare it with the older one. If one had asked a contemporary of Galilei what is the meaning of a mathematical symbol, say _x_, he would have answered that the symbol stands for a _number_. Depending on circumstances, _x_ may denote a variable number, as in the expression for a function, _f(x)_, or else _x_ might be a quite definite number that solves an equation, say _f(x)_ = 0. But in the modern interpretation the emphasis has shifted: toward the end of the nineteenth century mathematicians began to become more and more aware that symbols do not appear in isolation but in huge multitudes connected with each other by formal rules, for instance the order of points along a straight line, or else the connecting operations of algebra such as addition and multiplication.

Out of such ideas there arose gradually the notion of what is called an _abstract structure_, that is to say of a multitude of mathematical symbols related to each other by rules of operation. Many of you might be familiar with this concept but may not realize how new and revolutionary it is in a historical perspective. The term "abstraction" has a Latin root which literally means "pulling away." That means one starts from objects of ordinary experience and by pulling away the most immediate properties one is ultimately left with an "abstract" notion. This is the way the ancient Greeks arrived at their ideas of geometry: one starts with a straight rod and on taking away the thickness of the rod one arrives eventually at the notion of a straight line.

But the view of the modern mathematical thinker is different: if he ever started from ordinary experience, he has left this experience way behind. He deals with multitudes of symbols that are connected to each other by abstract operations; so the abstract structure arises from the mutual relationship of the symbols and their connecting operations. This kind of abstraction can even be explained to a mathematical layman: an architect is able to transform a pile of stones into an arch, and this arch maintains itself through its structure, not by any external means.

In the last 80 years or so mathematicians have given us a vast array of symbol systems of every kind, that are purely abstract structures. There are groups and algebras, there are all the abstract structures of topology, and there is finally the bountiful variety of geometries. Those of you who have had some exposure to mathematics will recognize familiar notions. Now I do not claim to be a mathematician; I think of myself as an empirical scientist who uses the products of the mathematicians' research to describe observations. But to make my point clearer I should bring a few examples to show how widespread the use of this method has become in modern science.

So let me go back to the man I have already quoted, John Maynard Keynes. There is a sentence starting at the bottom of the very first page of Keynes' book which goes as follows: "While it is often convenient to speak of propositions as certain or probable, this expreses strictly a relationship in which they stand to a corpus of knowledge, actual or hypothetical, and not a characteristic of the propositions in themselves." One can hardly express with more clarity the spirit of what I shall from now on simply call modern induction. Probability theory has of course developed since Keynes' times. But since probability has usually been thought of as a branch of mathematics and since mathematics is the science of deduction, mathematicians often don't like to speak of inductive probability; they prefer to call them probabilities of the Bayesian type, Bayes being an 18th century British worthy who developed some of this mathematics. But it is the same thing.

Problems of probability theory are often undetermined. A technical trick that does away with this indefiniteness is expressed by the requirement that the entropy be at a maximum. What this does in practice is that it serves as an extra axiom which makes the deductions unique; so probability statements cease to be mere symbols and yield equations that can be solved and at the end give results that can be expressed by numbers. Jaynes has been successful in applying this method to statistical mechanics, and Tribus has applied the same method to probability and statistics at large and has shown that much of traditional probability theory can be derived in this way in a very elegant manner.

Here we have outstanding examples of the modern inductive method of applying science to bodies of data that start out by being quite messy. We can also see how this method differs from the accustomed one. We do not take a set of observed numbers that we match to a law drawn from our hat of inventiveness. Instead, we apply a somewhat loose abstract structure, but a whole structure, not just parts, to a body of statistical data. We find that this inductive method is far more powerful than any previous methods; so we may well think that it also has a great future. It signifies not the end of an old epoch of science but the beginning of a new one.

But before I come to this, let me mention two more instances of the power of this modern approach. The theory of relativity can, with just a little mental twist, be conceived as a way of using many geometries, not just one, to describe our experiences. It has successfully broken through the 2000-year old prejudice that Euclidean geometry is the "true" geometry describing the "real" world.

Only 4 or 5 years after Keynes published his "Treatise," the theoretical physicist Dirac, also in Cambridge, began his meteoric rise. His idea was to replace the conventional measured numbers in the physics of Galilei and Newton by what he calls q-numbers, what we now call linear operators. Here, the replacement of measured numbers by algebraic symbols defined within their own abstract structure is altogether clear. The rest is history and is known to many of you.

But I should at least mention one other Cambridgeman who lived at about the same time. I have in mind the astronomer, Arthur Eddington. He is one of the great men of science; he was for the theory of the fixed stars what Kepler was for the planets. In his last small book, "The Philosophy of Physical Science" published in 1939 (2) he describes meticulously the modern inductive method. This is, in my personal opinion, the best description of that method and the best introduction to what is called the philosophy of science in existence. Soon thereafter Eddington's tragic decline set in which led first to his writing a baffling work on numerology and not much later to his death in 1944.

At this point I wish to introduce the notion of the two cultures, the second mysterious item of my title. As many of you might know this term was created nearly twenty years ago by the British writer Charles P. Snow, now Lord Snow (3). Snow started his career with a Ph.D. in physics and was working in a scientific research laboratory when the Second World War intervened. As he got involved in various organizational war activities he began to discover his vocation as a novelist and has been successful in this ever since. We all are familiar with some of the examples Snow gives: the successful scientist who is a total barbarian when he steps outside of science, and the passionate humanist who cannot be quick enough to forget the multiplication tables once he is out of high school. But are these not examples of human foibles rather than of a principle of nature. I think they are just that; they do not give us any pointer about the direction in which science is moving.

But they do indicate a rather conspicuous trend in contemporary science. What Snow does is to replace science by the sociology of science, and he is not unique in this. There are other tendencies that show the psychological insecurities of the contemporary scientist; the leaning on sociology and on history is but one of them; and it is widespread. Another is the tendency to see science solely in terms of its practical application. This is so familiar that I need not even describe it.

As against such tendencies one must insist on the dignity of science as a form of creative action. In any such creative activity there are two poles, an internal and an external one. The internal one is the urge to explore and to make discoveries, the external one consists in the existence of conditions that make discovery possible. In this respect the scientist is in the same position as, say, the artist, who would not be an artist if he did not have an innate urge to draw or paint, but he also in the long run needs sponsors, people who pay for the canvases he paints on or erect the wall he decorates. This tension between the internal urge and the external conditions is indispensable for productive people to survive. But as you know, our contemporary society brutally emphasizes the external and tends to crush the internal wherever it can. The telephone whose crude noise has priority over every conversation, is the true symbol of our contemporary social order.

Within this polarity, I feel that I have always been on one side, the side of the internal; in the language of the psychologist, I have always been very much of an introvert. So let me take the liberty to conclude these remarks on a subjective slant, telling you how I have felt and how I myself have reacted to this basic polarization of the scientific effort.

Growing up in Europe I have always been interested in the philosophy of science. I spoke of Eddington's little book on the subject where at a late stage in its development he presents it so well. Eddington incessantly tells us with much emphasis that the chief business of the philosopher of science is to eliminate metaphysical arguments from scientific discourse. This is saying, in the language that I employed before, that nature must be represented by purely abstract structures; the old idea of abstraction as something starting from sense impressions and then "taking something away" must go out of the window. The two great edifices that dominate modern theoretical physics, relativity and quantum theory, have gone as far as one can in eliminating metaphysics in favor of the use of abstract structures for depicting experience.

As a young atomic physicist I was, like all of us, deeply impressed by Niels Bohr although I never had the privilege of working with him personally. Bohr's private and not so secret passion was the application of atomic physics to biology. He came to this in a natural way: his father was professor of physiology at the University of Copenhagen, so Bohr "grew up in the biological laboratory," as he put it. Bohr consistently tried to apply the ideas of quantum theory to biology; he spoke of "generalized complementarity." He has not been as spectacularly successful in this as he was in his physics. But it is surprising that he has had so few followers in his pioneering endeavors. I confess that I am one of the very few people who have followed for many years the direction pointed out by Bohr. Beginning around 1950 I have written a series of articles and three books on the relations of biology to modern physics. The fact that there were so few followers of Bohr is indicative of the havoc that a society set only on respect for the external can produce. But let me not dwell on any lament. Let me end by telling you how I see the connection of physics and biology.

If one tries to understand the relations of these two sciences one runs at once into trouble. One finds that the philosophers have an ancient claim on this area, a claim that they have never relinquished and noisily assert. This claim is based on what I can simply describe as old-fashioned dualism. Since you all know about the dualism of body and soul, or of matter and mind, I do not have to explain this notion. It goes back to pre-historic times and was made into a philosophical axiom or dogma by Descartes around the year 1650. There have been other versions since then, for instance some biologists have developed vitalism but its ideas have been so eminently unsuccessful, that is to say they have so radically eluded expression in operational, scientific terms that vitalism has been abandoned. You can also readily perceive that the notion of the two cultures is just a slightly disguised form of old-fashioned dualism. Mr. Snow has of course long ceased to be an active scientist but I think of the two cultures as a spectacular example of that abuse of language which the scientist must at any costs avoid if he is to remain a scientist.

In dealing with dualism, we come across a very major obstacle. We need not, as Eddington says, seek for hidden remnants of metaphysical thinking; instead we are faced with a mountain topped by a citadel that prohibits our progress. While thinking about this earlier in my life as a physicist it had occurred to me that the method which I have called modern induction is so powerful and so flexible that it ought to be the ideal weapon to drive metaphysics out of the foundations of biology the way it has already been driven from most of physics. The old-fashioned dualism is blatantly non-scientific. There are two qualities in dualism as it is formulated by Descartes, but qualities have no business in science which deals only with relationships. This discrepancy is, I believe, so deep that it cannot be cured and should not be papered over either. Again, I told myself early on that only an utter fool could make a frontal attack on a mountain topped by a citadel. The normal strategy is to circumvent the obstacle. So I tried to find concepts that are specific for biology but do not resist being treated as operational or otherwise formal. After many years of search, I succeeded in the early fifties in discovering one case and have stuck to it since then. I do not think that my discovery exhausts the supply of such ideas; I emphatically deny any such claim.

The concept that I discovered is that of individuality. It has been grossly neglected by scientists and philosophers alike. It has no place in the world of the physicist but there is no biologist who, if you listen to him long enough, will not tell you that no two organisms are ever exactly alike. I use the concept in such a way that it should better be put into the plural, individualities. I think of an organism as a collection of individualities at all levels of its organization. This idea pervades my writing, but at the beginning I was not clear about its significance. I have treated it systematically in my last book "The Chief Abstractions of Biology" which came out in 1975 (4) and is a summary of these ideas.

Bohr has always been searching for a generalization of the physicist's complementarity. The notion of individuality gives us such a generalization. The homogeneity of ordinary, physical objects -- all atoms are exactly alike -- is mutually exclusive with the collection of individualities that form a living thing. I do not claim that this is _the_ generalization of the physicist's complementarity but I do claim to have shown that one can find such a generalization without stepping into the murky world of metaphysics.

Such a quantitative notion as that of entropy can hardly even be spoken of unless one assumes that there is a homogeneous substratum made up of interchangeable atoms. Individuality may be defined as the more or less total absence of homogeneity, and _vice versa_. These terms are not just abstractions to play ball with but they can be made very concrete by connecting them with biological data. So I think one can be quite optimistic about the future of this kind of science. We are not about to run out of intriguing problems and we do not have to go to distant stars or use billion-volt accelerators to find them either.

NOTES

1. Reprinted as Harper Torchbook, TB 557, Harper and Row, New York.

2. Reprinted by the University of Michigan Press, Ann Arbor, Michigan.

3. Reprinted as a Mentor Book, MP 557, The New American Library, New York.

4. Published by North-Holland Publishing Co., Amsterdam; American Elsevier Publishing Co., New York.

RATE-CONTROLLED CONSTRAINED EQUILIBRIUM METHOD FOR TREATING REACTIONS IN COMPLEX SYSTEMS

James C. Keck

1. Introduction

The development of theoretical models for describing the time evolution of complex reacting systems is a fundamental objective of nonequilibrium statistical thermodynamics. It is also of great importance in connection with a variety of practical problems related to combustion, energy conversion, hypersonic aerodynamics, chemical processing, electrical discharges, lasers and biology.

It is generally accepted that the equilibrium state of such systems can be found by "maximizing the entropy subject to the constraints." To do this two types of constraints must be considered. The first are those imposed by the external environment of the system and are under the control of the observer. The volume of the system and its position in a gravitational field are examples of such external constraints. The second are those imposed by the internal structure of the system and the laws of mechanics. Such constraints which are not under the control of the observer and which limit the states accessible to a system on the time scale of interest were termed passive resistance by Gibbs. Semipermeable membranes which segregate species and extremely slow nuclear reactions which imply conservation of the elements are the most familiar examples of passive resistances.

In ordinary thermodynamic calculations involving systems subject to slowly changing external constraints, such passive resistances are set either "on" or "off." The system is assumed to relax through a sequence of shifting equilibrium states determined by the instantaneous values of the external constraints and fixed values of the internal constraints imposed by the passive resistances which are "on," that is by the slow reactions. It is further assumed that all reactions associated with passive resistances which are "off" are sufficiently fast to maintain the shifting equilibrium state. This technique has been used with remarkable success in the design and analysis of a wide variety of practical devices ranging from steam engines and chemical reactors to power plants and refineries.

In cases where the rates of change in external constraints are too high to justify the shifting equilibrium assumption of thermodynamics, as in combustion processes or shock waves, it is necessary to describe the changes in the internal state of a system using a set of rate equations. These must be integrated in conjunction with the equations for the external constraints using a computer. For systems involving a large number of degrees of freedom, this can be a truly formidable task which is further complicated by the fact that most of the reaction rates involved are usually unknown and must be estimated.

To treat such complex reacting systems, Keck and Gillespie (1) have proposed the "rate-controlled partial equilibrium method." This method combines the use of rate equations for both internal and external constraints with the techniques of thermodynamics for determining the constrained equilibrium state. It is based on the assumption that under many conditions it is possible to identify classes of slow reactions which if completely inhibited would prevent the relaxation of the system to complete equilibrium. It is further assumed that fast reactions will equilibrate the system subject to the constraints imposed by the slow reactions and that the system will relax to complete equilibrium through a sequence of constrained equilibrium states at a rate controlled by the slow reactions. Although Keck and Gillespie originally used the term "partial equilibrium" to describe their method, in later work by Galant and Appleton (2) the more accurate term "constrained equilibrium" was substituted.

Prior to the work of Keck and Gillespie, the concept of a "partial equilibrium" state imposed by slow dissociation and recombination reactions was used by Kaskan (3), Schott (4) and Lezberg and Fransciscus (5) to treat the problem of hydrogen-oxygen combustion. It was also used by Bray (6) in his development of the "sudden freezing" approximation for treating rapidly expanding nozzle flows.

As described in companion papers in this volume, variations of the method are also currently being used by Levine (7) to treat reactions in molecular beams, Ross (8) to investigate relative stability in highly nonequilibrium systems and Kerner (9) to investigate speciation in ecological systems.

In the present paper we shall limit our detailed description of the rate-controlled constrained equilibrium method to chemically reacting dilute gas mixtures. The extension of the method to include energy exchange reactions will be obvious. Since the method may truly be regarded as an extension of the familiar techniques of ordinary thermodynamics, we shall begin in section 2 with a review of these techniques as applied to gas mixtures. This will be followed in section 3 by a discussion of the rate-controlled constrained equilibrium method and in section 4 by a brief commentary on conventional methods of treating nonequilibrium gas mixtures. A discussion of some practical applications to combustion will be given in section 5 and the summary and conclusions will be given in the final section.

2. Equilibrium Gas Mixtures

Consider a uniform homogeneous reacting gas mixture of weakly interacting molecules with fixed total energy E in a fixed volume V. Let X_j be the number of moles of species B_j in the gas and a_{kj} the number of atoms of element A_k in a molecule of species B_j. Then the number of moles of element A_k in the gas will be

$$C_k = \sum_{j=1}^{m} a_{kj} X_j \quad : \quad k = 1 \ldots n \tag{1}$$

where m is the number of different species and $n \leq m$ is the number of different elements.

Our problem is to determine the equilibrium state of such a system and, in accord with the general principles of statistical thermodynamics, we shall assume that this can be done by simply maximizing the Gibbs entropy

$$S = -R \sum_i x_i \ \ell n \ x_i \tag{2}$$

subject to the constraints

$$1 = \sum_i x_i \tag{3}$$

$$E = \sum_i \varepsilon_i(V) \ x_i \tag{4}$$

where x_i is the probability that the system is in a microstate (wave function) of energy $\varepsilon_i(V)$ and the summations are over all microstates.

If we further assume that the gas is sufficiently dilute so that the interaction energy between molecules can be neglected, then the total energy of the gas can be approximated by

$$E = \sum_{j=1}^{m} e_j X_j \tag{5}$$

where

$$e_j = \sum_i \varepsilon_{ji}(V) \ x_{ji} \tag{6}$$

is the mean energy of a molecule of species B_j and x_{ji} is the probability that such a molecule will be found in a microstate of energy $\varepsilon_{ji}(V)$ determined only by its internal structure and the volume V of the container. Furthermore, since the molecules have been assumed independent, the probability x_i for the system to be in a state of energy ε_i can be expressed as product of the probabilities x_{ji} for the individual molecules to be in microstates of energy $\varepsilon_{ji}(V)$ and the entropy (2) then takes the form

$$S = -R \sum_{j=1}^{m} X_j \left(\sum_i x_{ji} \ \ln x_{ji} + \ln X_j - 1 \right) \tag{7}$$

Maximizing this expression subject to the constraints

$$1 = \sum_i x_{ji} \quad : \quad j = 1 \ldots m \tag{8}$$

$$C_k = \sum_{j=1}^{m} a_{kj} X_j \quad : \quad k = 1 \ldots n < m \tag{9}$$

$$E = \sum_{j=1}^{m} \sum_i \varepsilon_{ji}(V) \ x_{ji} \ X_j \tag{10}$$

we obtain for the equilibrium state

$$\ln x^\circ_{ji} = -\alpha_j - \beta \ \varepsilon_{ji}(V) \tag{11}$$

$$\ln X^\circ_j = \alpha_j - \sum_{k=1}^{n} \gamma_k \ a_{kj} \tag{12}$$

where α_j, β, and γ_k are the Lagrange multipliers conjugate to the constraints (8) - (10)

Substituting x_{ji} from (11) back into the normalization condition, (8) we obtain

$$e^{\alpha_j} = \sum_i e^{-\beta\varepsilon_{ji}(V)} = Q_j(\beta,V) \tag{13}$$

where $Q_j(\beta,V)$ is the Boltzmann partition function for the species β_j.

Given the partition function all the other thermodynamic properties of the individual molecules can be determined as functions of β and V from the well known formulae of statistical thermodynamics. In particular, the temperature is

$$T = (R\beta)^{-1} \tag{14}$$

and the chemical potential is

$$\mu_j = RT \ \ln X_j \ / \ Q_j(T,V) \tag{15}$$

To determine the equilibrium composition of the gas, we may use (13) to eliminate α_j in (12). This gives

$$\ln X^\circ_j = \ln Q_j(\beta,V) - \sum_{k=1}^{n} \gamma_k \ a_{kj} \tag{16}$$

which in turn may be substituted into (9) to obtain a set of n equations

$$C_\ell = \sum_{j=1}^{m} a_{\ell j}\ Q(\beta,V)\ \exp(-\sum_{k=1}^{n} \gamma_k\ a_{kj}) \quad : \ell = 1\ldots n \tag{17}$$

which can be solved numerically using the methods developed by Warga (10), Zeleznik and Gordon (11) and Agmon, Alhassid and Levine (12) to obtain the n unknown Lagrange multipliers

$$\gamma_k = \gamma_k(T,V,C_1\ldots C_n) \quad : \quad k = 1\ldots n \tag{18}$$

Substituting (18) and (13) into (11) and (12) and using (14), we obtain the energy distribution

$$x_{ji} = x^\circ_{ji}(T,V,C_1\ldots C_n) \tag{19}$$

and composition

$$X_j = X^\circ_j(T,V,C_1\ldots C_n) \tag{20}$$

as functions of T,V, and $C_1\ldots C_n$. This completely determines the equilibrium state of the system.

<u>2.1 Thermodynamics of Gas Mixtures</u> In the above analysis, which may be properly termed <u>thermostatics</u>, we have considered only the final equilibrium state in which a gas mixture subject to fixed external constraints will be found after an infinitely long period of time. We now wish to extend the analysis to include situations in which the external constraints may be functions of time. The science of <u>thermodynamics</u> enables us to do this by introducing the assumption that, if the changes in the external constraints are sufficiently slow, a system such as the gas mixture under consideration will relax through a sequence of quasistatic states which remain close to equilibrium at all times. In a gas, this condition is supposed to be maintained by molecular collision processes whose relaxation time τ_c is short compared to the characteristic time for a change in the constraints, e.g. $\tau_c \ll \tau_V = (d\ell nV/dt)^{-1}$. Under these conditions, the time dependent energy distribution and composition can be obtained from (19) and (20) in the form

$$x_{ji}(t) = x^\circ_{ji}(T(t),\ V(t),\ C_1\ldots C_n) \tag{21}$$

and

$$X_j(t) = X_j^\circ(T(t),V(t),C_1\ldots C_n) \tag{22}$$

Shifting equilibrium calculations carried out using equations (21) and (22) are widely used in thermodynamics, aerodynamics and chemistry and are familiar to most scientists and engineers. They clearly describe reversible processes since the energy distribution and composition are functions only of the instantaneous values of the constraints, in this case temperature and volume.

3. Rate-Controlled Constrained Equilibrium

In thermodynamic calculations of the type considered in the previous section, it is usually assumed that the only constraints on the allowed composition of a reacting gas mixture are those imposed by nuclear reactions which are extremely slow at ordinary temperatures and imply conservation of the elements. However, as has already been observed in the Introduction, there are many situations of practical interest in which it is possible to identify classes of slow chemical or energy exchange reactions which if completely inhibited would prevent the relaxation of the system to complete equilibrium. Some examples of such reactions important in high temperature chemistry and combustion are given below.

Passive Resistance	Constrained Property
Nuclear Reactions	Moles of Elements
Ionization Reactions	Moles of Ions
Dissociation Reactions	Moles of Gas
Nitrogen Fixing Reactions	Moles of Fixed Nitrogen
CO Oxidation Reactions	Moles of CO_2
T-V Transfer Reactions	Vibrational Temperature
V-V Selection Rules	Vibrational Quanta

A common feature of all these reactions is that, under the conditions of interest, their characteristic relaxation times are long compared to the relaxation times for the large majority of binary energy transfer or particle exchange reactions occurring in a reacting gas mixture.

The fundamental hypothesis of the rate-controlled constrained equilibrium method is that the fast reactions will equilibrate the system subject to the constraints implied by the slow reactions and that the approach to complete equilibrium will proceed through a sequence of constrained equilibrium states at a rate controlled by the slow reactions.

3.1 Constrained Equilibrium Gas Mixtures Except for the inclusion of additional constraints

$$C_k = \sum_{j=1}^{m} a_{kj} \, X_i \quad : \quad k = n+1 \ldots n+r \tag{23}$$

of the same form as (9) imposed by the rate-controlling reactions, the calculation of the constrained equilibrium composition may be carried out in exactly the same manner as in the equilibrium case. Depending on the choice of the thermodynamic variables which are held fixed, one may then either

1) Maximize the Entropy

$$S = -R \sum_{j=1}^{m} X_j \left(\sum_i x_{ji} \, \ell n \, x_{ji} + \ell n \, X_j - 1 \right) \tag{24}$$

subject to the constraints

$$1 = \sum_i x_{ji} \quad : \quad j = 1 \ldots m \tag{24}$$

$$C_k = \sum_{j=1}^{m} a_{kj} \, X_j \quad : \quad k = 1 \ldots n+r \tag{25}$$

$$E = \sum_{j=1}^{m} \sum_i \varepsilon_{ji}(V) \, x_{ji} \, X_j \tag{27}$$

2) Minimize the Helmholtz Free Energy

$$A(T,V) = \sum_{j=1}^{m} X_j \left(e_j(T) - T \, s_j(T,V,X_j) \right) \tag{28}$$

subject to the constraints

$$C_k = \sum_{j=1}^{m} a_{kj} \, X_j \quad : \quad k = 1 \ldots n+r \tag{29}$$

3) Minimize the Gibbs Free Energy

$$F(T,p) = \sum_{j=1}^{m} X_j \left(e_j(T) + RT - T \, s_j(p,T,X_j) \right) \tag{30}$$

subject to the constraints

$$C_k = \sum_{j=1}^{m} a_{kj} \, X_j \quad : \quad k = 1 \ldots n+r \tag{31}$$

In all cases, one obtains the result

$$\ell n \, X_j^c = \ell n[Q_j(T)] + \ell n \, V - \sum_{k=1}^{n+r} \gamma_k \, a_{kj} \tag{32}$$

where $[Q_j(T)] = Q_j(V,T)/V$ is the partition function per unit volume for molecules of species B_j.

From a computational point of view the most convenient independent variables are T and V or T and p. For the case where T and V are the independent variables, substitution of (32) back into the constraints (29) then gives the set of n+r equations

$$C_\ell = \sum_{j=1}^{m} a_{\ell j} \exp(-\sum_{k=1}^{n+r} \gamma_k a_{kj} - A_j(V,T) - 1) : \ell = 1 \ldots n+r \qquad (33)$$

where

$$A_j(T,V) = - \quad (\ell n[Q_j(T)] - \ell n\, V + 1) \qquad (34)$$

which may be solved for the n+r unknowns

$$\gamma_k = \gamma_k(T,V,C_1 \ldots C_{n+r}) \quad : \quad k = 1 \ldots n+r \qquad (35)$$

For the case where T and p are the independent variables, the situation is slightly more complicated since equation (32) can not be solved explicitly for X_j^c in terms of T and p after the equation of state

$$pV = RT \sum_{j=1}^{m} X_j^c \qquad (36)$$

has been introduced to eliminate V. It has therefore been found convenient by Warga (10) to introduce an additional unknown X_o and an additional equation

$$X_o = \sum_{j=1}^{m} X_j^c \qquad (37)$$

to the original set. Substitution of (32) into (37) and the constraints (31) then leads to the set of n+r+1 equations

$$C_\ell = X_o \sum_{j=1}^{m} a_{\ell j} \exp(-\sum_{k=1}^{n+r} \gamma_k a_{kj} - F_j(pT)) : \ell=1 \ldots n+r \qquad (38)$$

and

$$1 = \sum_{j=1}^{m} \exp(-\sum_{k=1}^{n+r} \gamma_k a_{kj} - F_j(pT)) \qquad (39)$$

where

$$F_j(T,p) = - \quad (\ell n[Q_j(T)] + \ell n(p/RT)) \qquad (40)$$

which may be solved for the n+r+1 unknowns

$$\gamma_k = \gamma_k(T,p,C_1\ldots C_{n+r}) \quad : \quad k=1\ldots n+r \tag{41}$$

and

$$X_o = X_o(T,p,C_1\ldots C_{n+r}) \tag{42}$$

It should be noted that equation (37) which gives the total moles of gas does not constitute an additional constraint on the system but is merely an algebraic identity introduced for mathematical convenience.

Substituting (35) or (41) back into (32) we obtain the constrained equilibrium composition of the gas for the two cases considered.

$$X_j = X_j^c(T,V,C_1\ldots C_{n+r}) \tag{43}$$

or

$$X_j = X_j^c(T,p,C_1\ldots C_{n+r}) \tag{44}$$

as functions of T and V or T and p and the n+r constraints.

In concluding this discussion it may be observed that the same procedures can be used in conjunction with any other free energy function to treat cases where the energy states of the molecules may be affected by gravitational, electromagnetic, or other body forces.

3.2 Thermostatics of Constrained-Equilibrium Once the composition of the gas mixture has been determined, all its other thermodynamic properties may easily be calculated. In particular, the chemical potential and Gibbs free energy can be expressed

$$\mu_j \equiv RT\,\ell n\, X_j/Q_j = \sum_{k=1}^{n+r} \lambda_k\, a_{kj} \tag{45}$$

and

$$F \equiv \sum_{j=1}^{m} \mu_j\, X_j = \sum_{k=1}^{n+r} \lambda_k\, C_k \tag{46}$$

in which we have introduced the parameter

$$\lambda_k = -R\,T\,\gamma_k \tag{47}$$

It follows from (46) that

$$\mu_j = (\partial F/\partial X_j)_{Tp\underline{X}} \tag{48}$$

and

$$\lambda_k = (\partial F/\partial C_k)_{Tp\underline{C}} \tag{49}$$

Also from (45), (46) and (31)

$$\sum_{j=1}^{m} X_j \, d\mu_j = \sum_{k=1}^{n+r} C_k \, d\lambda_k \tag{50}$$

and

$$\sum_{j=1}^{m} \mu_j \, dX_j = \sum_{k=1}^{n+r} \lambda_k \, dC_k \tag{51}$$

Substituting (51) into the Gibbs equation

$$T\,dS = dE + p\,dV - \sum_{j=1}^{m} \mu_j \, dX_j \tag{52}$$

we obtain

$$T\,dS = dE + p\,dV - \sum_{k=1}^{n+r} \lambda_k \, dC_k \tag{53}$$

It can be seen from equations (46), and (48) - (53) that the parameters λ_k play precisely the same role with respect to the constraints C_k as the chemical potentials μ_i play with respect to the mole numbers X_i. We may therefore define λ_k as the *generalized chemical potential* conjugate to the constraint C_k. The important distinction between the λ's and the μ's is that, since the number of constraints n+r is in general much smaller than the number of species m, it is far more efficient to determine the composition in terms of the λ's than the μ's.

For processes at constant T and p, the Gibbs free energy must be a minimum at equilibrium and since the constraints C_k must be linearly independent, it can be seen from (49) that, unless dC_k vanishes identically in any allowed process, the corresponding λ_k must be zero at equilibrium. Thus, in general, a condition for chemical equilibrium will be

$$\lambda_k^\circ = 0 \quad : \quad dC_k \neq 0 \tag{54}$$

<u>3.3 Dynamics of Constrained Equilibrium</u> Following the lead of

thermodynamics, we now wish to consider the situation in which the constraints C_k on the composition of a gas mixture may be slowly varying functions of time. We assume that energy exchange reactions have equilibrated so that the internal degrees of freedom of the molecules can be characterized by a single temperature T(t), which may also be a slowly varying function of time. We further assume that changes in the composition of the gas are the result of chemical reactions of the type

$$\sum_{j=1}^{m} \nu_{ji}^{+} B_j \underset{R_i^-}{\overset{R_i^+}{\rightleftarrows}} \sum_{j=1}^{m} \nu_{ji}^{-} B_j \tag{55}$$

where R_i^+ and R_i^- denote the forward and reverse rates of the reaction R_i and ν_{ji}^+ and ν_{ji}^- are the corresponding stoichiometric coefficients. Finally we assume that the rate equations for the individual species are of the form

$$\dot{X}_j = \sum_i \nu_{ji}(R_i^+ - R_i^-), \tag{56}$$

where

$$\nu_{ji} = (\nu_{ji}^- - \nu_{ji}^+), \tag{57}$$

and that the reactions rates are given by the phenomenological equations

$$R_i^{\pm}(T,V,\underline{X}) = k_i^{\pm}(T)\ V \prod_j [X_j]^{\nu_{ji}^{\pm}}, \tag{58}$$

where $k_i^+(T)$ and $k_i^-(T)$ are rate constants for the forward and reverse reactions and $[X_j] \equiv X_j/V$ is the concentration of species B_j.

At equilibrium the gas must be in a stationary state and thus the left hand side of (56) must vanish. A *sufficient* condition for this is the *detailed balancing condition*

$$R_i^+ = R_i^- = R_i^\circ \tag{59}$$

If we further assume, consistant with the laws of mechanics and thermodynamics, that the forward and reverse rates of a reaction can only vanish simultaneously at equilibrium and that the equilibrium composition of a reacting gas mixture must be independent of the magnitude of the reaction rates provided there is at least one reaction whose rate is not identically zero, then (59) becomes a *necessary* condition. This proof of necessity, which to the author's knowledge has not been given before, is of considerable fundamental importance. It is also of practical importance since in actual calculations of rate processes only a small fraction of the possible reactions can be included and, for these, the condition *must* be imposed.

Substituting (58) into (59) we obtain the rate quotient law

$$k_i^+/k_i^- = \prod_{j=1}^{m} [Q_j]^{\nu_{ji}} = K_i(T) \tag{60}$$

where $K_i(T)$ is the equilibrium constant of the reaction R_i. This law must hold under constrained-equilibrium as well as equilibrium conditions as long as the internal degrees of freedom of the individual molecules are in equilibrium at the temperature T.

<u>3.3.1 Relaxation Times</u> The characteristic times τ_{Bj} and τ_{Ri} for the relaxation of a species of B_j or reaction R_i to equilibrium may be defined by

$$\tau_{B_j}^{-1} = |\partial \ln \dot{X}_j / \partial t|_{TV}^{\circ} \tag{61}$$

and

$$\tau_{R_i}^{-1} = |\partial \ln(R_i^+ - R_i^-) / \partial t|_{TV}^{\circ} \tag{62}$$

Substituting (56) and (58) into (61) and (62) and using (57) and (59) we find

$$\tau_{B_j}^{-1} = (X_j^\circ)^{-1} \sum_i \nu_{ji}^2 R_i^\circ \tag{63}$$

and

$$\tau_{R_i}^{-1} = R_i^\circ \sum_j \nu_{ji}^2 (X_j^\circ)^{-1} \tag{64}$$

These equations, which also apply under constrained equilibrium conditions, are useful for determining which species are close to equilibrium and which reactions are rate-controlling.

<u>3.3.2 Rate Equations for Constraints</u> Differentiating the equations for the constraints with respect to time gives

$$\dot{C}_k = \sum_{j=1}^{m} a_{kj} \dot{X}_j \quad : \quad k = 1 \ldots n+r \tag{65}$$

from which we may eliminate $\dot{X}_j$ using (56) to obtain the rate equation for the constraints

$$\dot{C}_k = \sum_i b_{ki}(R_i^+ - R_i^-) \quad : \quad k = 1 \ldots n+r \tag{66}$$

where

$$b_{kj} = \sum_{j=1}^{m} a_{kj} \nu_{ji} = \sum_{j=1}^{m} a_{kj}(\nu_{ji}^- - \nu_{ji}^+) \tag{67}$$

is the change in constraint C_k due to reaction R_i. Assuming elements are conserved in all reactions, we obtain

$$b_{ki} = 0 \quad : \quad k = 1 \ldots n \tag{68}$$

which by virtue of (66) gives

$$C_k(t) = C_k \quad : \quad k = 1 \ldots n \tag{69}$$

Using the rate-controlled constrained equilibrium composition given by (32) to evaluate the reaction rates $R_i^{\pm}(T,V,\underline{X})$ in (66), we obtain a set of r first order differential equations of the form

$$\dot{C}_k(t) = \dot{C}_k(T(t),V(t),C_1 \ldots C_n,C_{n+1}(t) \ldots C_{n+r}(t)) : k=n+1 \ldots n+r \tag{70}$$

for the remaining unknown constraints. Given the initial values of T,V and $C_1 \ldots C_{n+r}$ these equations can be integrated in conjunction with the equations of state and the conservation equations of thermodynamics and fluid mechanics to determine the evolution of the system. In simple cases this can be done analytically but, in general, numerical methods must be employed. In this connection it should be noted that, except for the addition of the rate equations for the time dependent constraints, a constrained-equilibrium calculation may be carried out in exactly the same manner as a shifting equilibrium calculation.

3.3.3 Degree of Disequilibrium A useful parameter for measuring the departure of a reaction from equilibrium is the ratio of the reverse to forward reaction rates

$$R_i^-/R_i^+ = (k_i^-/k_i^+) \prod_{j=1}^{m} [X_j]^{\nu_{ji}} \tag{71}$$

Substituting the constrained-equilibrium composition (32) into (71) and using the rate quotient law (60), we find under constrained-equilibrium conditions

$$R_i^-/R_i^+ = (k_i^-/k_i^+)\prod_{j=1}^{m}[Q_j]^{\nu_{ji}} \exp(-\sum_{k=1}^{n+r}\gamma_k b_{ki}) = \exp(-\sum_{k=1}^{n+r}\gamma_k b_{ki}) \tag{72}$$

The *degree of disequilibrium* may be defined as

$$\ln R_i^-/R_i^+ = -\sum_{k=1}^{n+r}\gamma_k b_{ki} \tag{73}$$

Using the definition (45) and (43), (73) may also be written

$$\ln R_i^-/R_i^+ = (RT)^{-1}\sum_{k=1}^{n+r}\lambda_k b_{ki} = (RT)^{-1}\sum_{j=1}^{m}\mu_j \nu_{ji} \tag{74}$$

From (74) it can be seen that a condition for chemical equilibrium is

$$\sum_{k=1}^{n+r}\lambda_k b_{ki} = \sum_{j=1}^{m}\mu_j \nu_{ji} = 0 \tag{75}$$

It also follows from (74) that for any set of reactions which change only a single constraint, say C_ℓ, we have

$$b_{\ell i}^{-1} \ln R_i^-/R_i^+ = (RT)^{-1}\lambda_\ell = -\gamma_\ell \tag{76}$$

This relation can be useful for evaluating the validity of the constrained equilibrium assumption under practical conditions.

3.3.4 Entropy Production Using the rate equations (66) for the constraints, the Gibbs equation (53) can be written in the form

$$T\dot{S} = \dot{E} + p\dot{V} + T\sigma_R \tag{77}$$

where

$$\sigma_R = -\frac{1}{T}\sum_{k=1}^{n+r}\lambda_k \dot{C}_k = -\frac{1}{T}\sum_i(\sum_{k=1}^{n+r}\lambda_k b_{ki})(R_i^+ - R_i^-) \tag{78}$$

is the entropy production due to chemical reactions. Substituting (74) into (78), we obtain the familiar positive definite form

$$\sigma_R = R \sum_i (R_i^+ - R_i^-) \ln(R_i^+/R_i^-) \geq 0 \tag{79}$$

Thus it can be seen that, for a gas in a fixed volume V passing through a sequence of constrained equilibrium states of fixed energy E, the entropy production is always greater than or equal to zero and is zero if and only if $R_i^+ = R_i^-$ for all reactions. It follows that such a gas will relax unconditionally to a unique stable equilibrium state determined only by the values of E and V and the fixed constraints $C_1 \ldots C_n$. It can also be clearly seen from the Gibbs equation written in the alternative forms

$$\dot{A} = - S \dot{T} - p \dot{V} - T \sigma_R \tag{80}$$

or

$$\dot{F} = - S \dot{T} + V \dot{p} - T \sigma_R \tag{81}$$

that, for systems at fixed T and V or T and p, the Helmholtz free energy A(T,V) or Gibbs free energy F(T,p) must always decrease and will be a minimum at equilibrium. These results can be generalized to include systems subject to external constraints in addition to volume and internal rate-controlled constraint due to slow energy exchenge reactions or selection rules.

4. Conventional Treatments of Nonequilibrium

Before discussing the applications of the constrained-equilibrium method to a few practical combustion problems, we should like to consider briefly two of the conventional methods for treating chemical reactions in nonequilibrium systems with which we shall wish to compare it.

4.1 Integration of Complete Set of Rate Equations

In principle, the most accurate method of determining the evolution of a homogeneous chemically reacting system is to integrate a complete set of rate equations

$$\dot{X}_j = \sum_{j=1}^{m} \nu_{ji}(R_i^+ - R_i^-) \quad : \quad j = 1 \ldots m-n \tag{82}$$

sufficient in number to determine all the unknown species concentrations given the concentrations of the n elements. For systems such as those encountered in combustion involving 50 or more interesting species and thousands of potentially significant reactions, this is an extremely difficult task and truncation of both the species and reaction lists are necessary, expecially if the chemical kinetics are to be combined with a flow calculation. The problem with truncation is that, unless great care is used, the omission of an important intermediate species or fast reaction can result in serious errors. Furthermore, most of the rate constants needed to integrate even a modest set of rate equations are unknown and must be estimated. Thus, unless the corresponding reactions are at or close to equilibrium the results will depend strongly on the estimated rate constants and one is probably better off making the constrained equilibrium assumption in the first place. In view of this, it seems obvious that the best method of carrying out an "exact" calculation would be to include all interesting species, but limit the number of rate equations to those for which the reaction rates are known and use the constrained equilibrium assumption to determine the remaining unknowns.

4.2 Steady State Approximation To reduce the number of differential equations which must be integrated, the steady state approximation is frequently used. This involves the assumption that for certain species, say B_1 to B_ℓ,

$$0 \approx \sum_i \nu_{ji}(R_i^+ - R_i^-) \quad : \quad j = 1 \ldots \ell \tag{83}$$

Equation (83) may then be solved in conjunction with the reduced set of m-n-ℓ differential equations

$$X_j = \sum_j \nu_{ji}(R_i^+ - R_i^-) \quad : \quad j = \ell+1 \ldots m-n \tag{84}$$

Although this can result in considerable simplification of the calculations, the steady state species must be identified and the other problems discussed above still remain. In addition, a more subtle problem which must be faced is that it can no longer be proved, as for the constrained equilibrium approximation or a full set of rate equations, that the entropy production will be non-negative. This could, in principle, lead to a Second Law violation or failure of the system to reach a unique stable equilibrium point. Whether this is serious in practice will depend on the particular situation involved however.

5. Applications in Combustion

The rate-controlled constrained equilibrium method has been applied with considerable success to the problem of predicting CO and NO formation in the combustion products from internal combustion engines and burners. These species, which are important air pollutants, are formed at relatively high concentrations at temperatures T ≥ 2000 °K and their removal is controlled by strongly temperature dependent reactions many of which involve trace species such as N atoms as collision partners. For these reasons, chemical kinetic models of CO and NO formation can easily involve 30 or more "significant" species. A typical list of such species in the order of their abundance in the equilibrium combustion products of a stoichiometric mixture of isooctane and air at a temperature of 2000°K and pressure of 1 atmosphere is given below.

N_2	H_2O	CO_2	CO	O_2	NO	H_2	OH	H	O
NO_2	N_2O	HO_2	H_2O_2	HNO_2	HNO	NH_3	NH_2	HCO	HNCO
NH	N	H_2CO	O_3	HCN	NCO	HNO_3	NO_3	H_2N_2	N_2O_3
CN	CH_3	CH_4	...						

Clearly the integration of a full set of rate equations necessary to determine the concentrations of so many species would be a formidable task. In addition, when one also considers the fact that for such a species list it is easily possible to identify upwards of 1000 potentially significant reactions and that for the overwhelming majority of these the rate constants are unknown, one might well be discouraged from even attempting it. This is, of course, just the situation in which the constrained equilbrium method can be of most value.

The problem has also been treated using the conventional methods discussed in the preceeding section. In all the cases considered severe truncation of the species and reaction rate list has been used to simplify the calculations. Comparisons between these methods and the constrained equilibrium method will be made in the discussion below.

5.1 CO Freezing in a Steady Flow Burner The constrained equilibrium method has been used by Morr and Heywood (13) to interpret their measurements of CO concentrations in a steady flow cylindrical burner. This work illustrates in a dramatic way both the effect of using different constraints and the effect of combining constraints.

The experimental geometry is shown schematically in the upper part of Figure 1. Below it are shown the mean temperature profile along the axis of the burner and the experimental and theoretical results. It can be seen that, upstream of the heat exchanger where the temperatures are high, the measured CO mole fractions, shown as points, agree well with the calculated equilibrium values, shown as a dashed line. However, downstream of the heat exchanger where the temperatures are low, the measured CO levels far exceed the equilibrium value, giving clear evidence for the freezing of CO oxidation reactions.

To predict the CO levels in this highly non-equilibrium region, Morr and Heywood carried out a 14 species constrained equilibrium calculation involving two probable constraints. The first was that on the total number of moles of gas

$$M = N_2 + H_2O + CO_2 + CO + O_2 + \dots \tag{85}$$

controlled by the dissociation and recombination reactions

$$R_{11} \quad : \quad M + H + H \rightleftarrows H_2 + M \tag{86}$$

$$R_{12} \quad : \quad M + H + OH \rightleftarrows H_2O + M$$

$$R_{13} \quad : \quad M + H + O_2 \rightleftarrows HO_2 + M$$

$$R_{14} \quad : \quad M + OH + OH \rightleftarrows H_2O_2 + M$$

and governed by the rate equation

$$\frac{dM}{dt} = \sum_{i=1}^{4} (R_{1i}^{-} - R_{1i}^{+}) \tag{87}$$

The results obtained when this constraint was used alone are shown by the curve marked $(CO)_M$ in Figure 1. Although the $(CO)_M$ levels are substantially higher than the $(CO)_{eq}$ levels they are still far below the measured values.

The second constraint which was tried was that on the moles of CO controlled by the CO oxidation reactions

$$R_{21} \qquad OH + CO \rightleftarrows CO_2 + H \tag{88}$$

$$R_{22} \qquad O_2 + CO \rightleftarrows CO_2 + O$$

$$R_{23} \qquad NO_2 + CO \rightleftarrows CO_2 + NO$$

$$R_{24} \qquad M + O + CO \rightleftarrows CO_2 + M$$

$$R_{25} \qquad M + H + CO \rightleftarrows HCO + M$$

and governed by the equation

$$\frac{d\ CO}{dt} = \sum_{i=1}^{5} (R_{2i}^{-} - R_{2i}^{+}) \tag{89}$$

The results obtained when this constraint was used alone are shown by the curve marked $(CO)_{CO}$ in Figure 1 which gives CO levels which substantially exceed the measured values.

The effect of imposing both constraints simultaneously is shown by the remaining curve marked $(CO)_{M,CO}$. Although this curve is slightly lower than the measured values the agreement was felt to be well within the combined uncertainty of the experimental measurements and the rate constants used. Additional constraints could in principle have been employed to improve the agreement but no further *useful* information would have been obtained.

The reason for the difference between the $(CO)_{CO}$ curve and the $(CO)_{M,CO}$ curve is due to the effect of the total moles constraint which greatly increases the number of free radicals, especially OH, in the gas which in turn increases the rate of the most important CO oxidation reaction R_{21}.

5.2 NO concentrations in an I.C.E. A theoretical investigation of NO formation and removal in an internal combustion engine (I.C.E.) has been carried out by Keck and Gillespie (1). One of the objectives of this work was to compare the results of a constrained equilibrium calculation with a corresponding steady state calculation and the results are shown in Figure 2. In the constrained equilibrium calculations 14 species and two constraints were again considered. The first constraint was that on "fixed nitrogen."

$$NO_F = NO + NO_2 + NH + N \tag{90}$$

controlled by the reactions

$$\begin{array}{ll} R_{31} & N + NO \leftrightarrow N_2 + O \\ R_{32} & N + NH \leftrightarrow N_2 + H \\ R_{33} & NO + NO \leftrightarrow N_2O + O \\ R_{34} & NH + NO \leftrightarrow N_2O + H \\ R_{35} & N_2 + NO + NO \leftrightarrow N_2O + N_2O \end{array} \tag{91}$$

and governed by the rate equation

$$\frac{d\,NO_F}{dt} = 2\sum_{i=1}^{5}(R_{3i}^{-} - R_{3i}^{+}) \tag{92}$$

The mole fractions of NO, O and N obtained in a calculation in which this was the only constraint are shown by the solid curves in Figure 2. The freezing of nitrogen fixing reactions can clearly be seen by comparison of the $(NO)_N$ and $(N)_N$ with the corresponding equilibrium values shown by the short dashed curve. By way of contrast the $(O)_N$ levels remain close to their equilibrium values.

The second constraint considered was that on total moles M previously discussed. As indicated by the "Note" in Figure 2 the addition of this constraint had no effect on the single constraint curves indicating that dissociation and recombination reactions were in equilibrium throughout. This was checked by a single constraint calculation involving only total moles and found to be the case.

The corresponding steady state calculations were based on the model of Lavoie, Heywood and Keck. The major combustion reactions were assumed to be in equilibrium and the extended Zeldovich mechanism

$$R_{z1} \quad N + NO \rightleftarrows N_2 + O \tag{93}$$

$$R_{z2} \quad N + O_2 \rightleftarrows NO + O$$

$$R_{z3} \quad N + OH \rightleftarrows NO + H$$

was used to describe the kinetics of N and NO. The rate equations for these species were

$$\frac{dN}{dt} = R_{z1}^{-} + R_{z2}^{-} + R_{z3}^{-} - R_{z1}^{+} - R_{z2}^{+} - R_{z3}^{+} \tag{94}$$

$$\frac{d\,NO}{dt} = R_{z1}^{-} - R_{z2}^{-} - R_{z3}^{-} - R_{z1}^{+} + R_{z2}^{+} + R_{z3}^{+} \tag{95}$$

and the steady state assumption $dN/dt = 0$ was made for N. As can be seen from the long dashed curves in Figure 2, the steady state NO concentration $(NO)_{SS}$ was lower than the constrained equilibrium concentration $(NO)_N$ early in the cycle when it was being formed and lower later in the cycle when it was being removed. It can also be seen that the N atom concentration behaved in just the opposite way. These discrepancies are due to the fact that the reactions R_{z2} and R_{z3} included in the steady state model were not sufficiently fast to maintain relative equilibrium between the fixed N species N and NO. Which of the two calculations gives the better approximation is difficult to say without additional information. The list of species and reactions included in the steady state calculation is of course highly truncated and parallel reactions would tend to drive the fixed nitrogen system closer to a constrained equilibrium situation. In any case, as can be seen in Figure 3 which shows average NO concentration in the exhaust of an I.C.E. as a function of fuel/air ratio, either calculation predicts the experimental results within the expected accuracy. The steady state predictions shown by the dashed curve are based on unpublished work by Keck and Gillespie. The constrained equilibrium predictions shown by the solid curve were obtained by Delichatsios and Keck (14). The experimental points are from Heywood, Mathews and Owen (15).

5.3 Degree of Disequilibrium in I.C.E. Combustion Products

Considerable insight into the validity of the constrained equilibrium approximation as applied to NO and CO formation and removal in internal combustion engines can be gained from the results of a 13 species 17 reaction "exact" calculation carried out by Newhall (16). The species included were the first twelve in the list above plus N, the twenty second.

Of particular interest is the degree of disequilibrium calculated for 14 reactions during the expansion stroke and shown in Figure 4. It can be seen that the 5 exchange reactions involving the C-H-O system and the dissociation reaction $M+N_2O \rightarrow N_2+O+M$ involving the most abundant species N_2 were in equilibrium throughout. The 5 remaining dissociation reactions and the 3 exchange reactions involving NO all exhibit highly non-equilibrium behavior, however. At first sight it is not obvious that the behavior of most of the non-equilibrium reaction is related in any simple way. In fact, only 2 constraints, one on total moles M and another on NO, are necessary to obtain an excellent approximation to the behavior of the system and the combinations of the Lagrange multipliers necessary to do this are given in the figure for the various reactions. If desired, a perfect fit can be obtained with 2 additional constraints on N and H.

These results indicate that the previously discussed constrained equilibrium calculations of NO concentrations in I.C.E. by Keck and Gillespie could have been improved by taking the constraint to be NO rather than fixed nitrogen. They also cast considerable doubt on the use of the steady state approximation for N. As seen in Figure 2, this predicts N concentrations less than equilibrium during expansion whereas Figure 4 shows N concentrations greater than equilibrium. However, because of the severe truncation of the species and reaction rate lists in the "exact" calculations even these conclusions must be regarded as tentative. In a general way one would expect parallel reactions to drive the system closer to a constrained equilibrium thus improving the accuracy of this method relative to both the "exact" and steady state methods.

6 Summary and Conclusions

The rate-controlled constrained equilibrium method of treating reactions in complex systems has been reviewed and its application to the treatment of chemically reacting gas mixtures considered in detail. For systems with a large number of degrees of freedom, the method offers a number of advantages over the conventional techniques of integrating the full set of rate equations necessary to determine its state: 1) Since in general the number of constraints necessary to specify the state of a complex system to a specified degree of accuracy is very much smaller than the number of degrees of freedom of the system, there are fewer differential equations to integrate. This greatly simplifies numerical calculations. 2) Only the rate constants for the fastest rate-controlling reactions are needed to carry out a calculation and these are the most likely to be known. 3) Since they enter explicitly as constraints, the conservation laws are satisfied at all times and are not subject to round-off errors during integration. 4) The entropy production for a system relaxing through a sequence of constrained equilibrium states is greater than or equal to zero at all times and the system approaches a unique stable equilibrium point. This is not necessarily true for calculations in which the steady state approximations has been used. 5) The accuracy of the calculations can be systematically improved by the addition of constraints one at a time and they become "exact" when the number of constraints equals the number of degrees of freedom. 6) The method can easily be generalized to include constraints imposed by slow energy exchange reactions, selection rules, body forces and diffusion. 7) As indicated by the work of Kerner (9) on ecological cycles, the method can be applied to extremely complex systems.

The only obvious disadvantage of the method is that the constraints must be identified. Even this can be an advantage, however, because it forces one to *think* about the problem before embarking on elaborate calculations !

Acknowledgement The author would like to acknowledge his indebtedness to the late Professor Joseph H. Keenan for valuable insight into the structure of classical thermodynamics gained over a period of years from his published works and in conversations with him. He is also indebted to Drs. John P. Appleton,

Serge Galant and George N. Hatsopoulos for many interesting and stimulating discussions concerning the concept of constrained equilibrium.

Bibliography

1) Keck, J.C., and Gillespie,D., Combustion and Flame, 17, 237 (1971)

2) Galant, S., and Appleton,J.P., M.I.T. Fluid Mech. Lab. Publ. 73-6.

3) Kaskan, W.E., Combustion and Flame, 3, 229, 286 (1958)

4) Schott, G.L., J. Chemical Phys., 32, 710(1960)

5) Lezberg, EA., and Fransciscus, L.C., AIAA J., 1,2077 (1963)

6) Bray, K.N.C., J. Fluid Mech., 6, 1(1959)

7) Levine, R.D., This volume

8) Ross, John, This volume

9) Kerner, E.H., This volume

10) Warga, J., J. Soc. Indust. Appl. Math., 11, 594 (1963)

11) Zeleznik, F.J., and Gordon, S., Ind. Eng. Chem., 60, 27(1968)

12) Agmon, N., Alhassid, L., and Levene, R.D., Submitted to J. Comput. Phys. 1978.

13) Morr, A.R., and Heywood, J.B., Acta Astronautica, 1, 949 (1974)

14) Delichatsios, M., and Keck, J., ACS Symposium Preprint, 20, 105(1975)

15) Heywood, J.B., Mathews, S.M., and Owen, B., SAE preprint 710011(1971)

16) Newhall, H.D., Twelfth Symposium (International) on Combustion, P. 603, The Combustion Institute (1969)

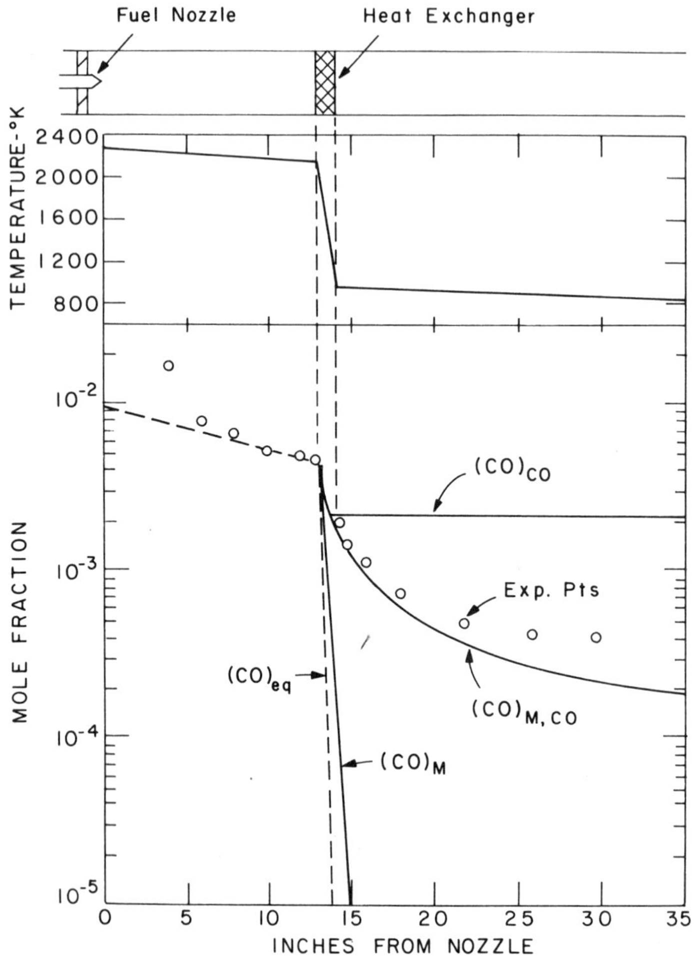

Figure 1. Temperature and CO profiles in a steady flow cylindrical burner.

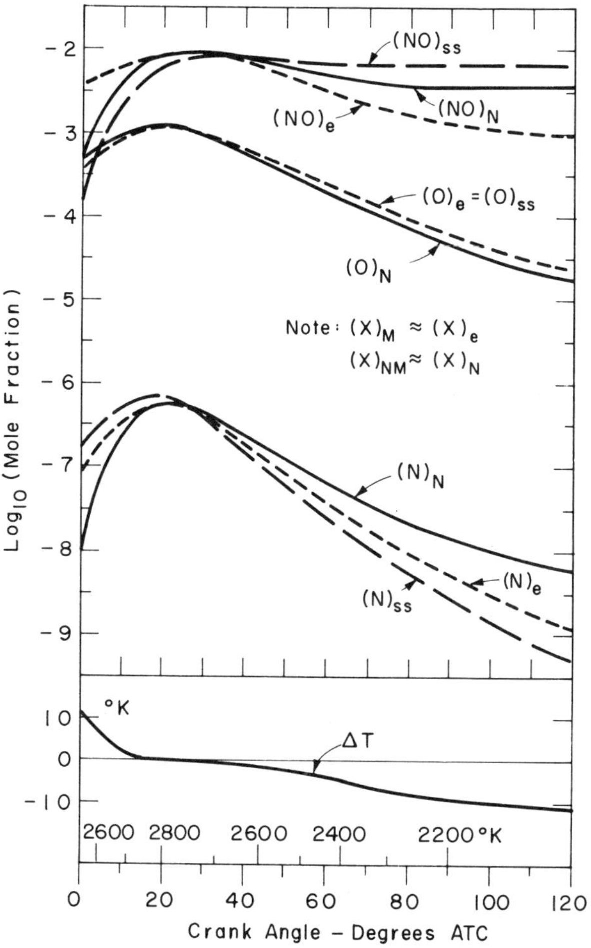

Figure 2. Comparison of equilibrium, steady state and constrained equilibrium NO, N and O mole fractions during the power stroke of an I.C.E.

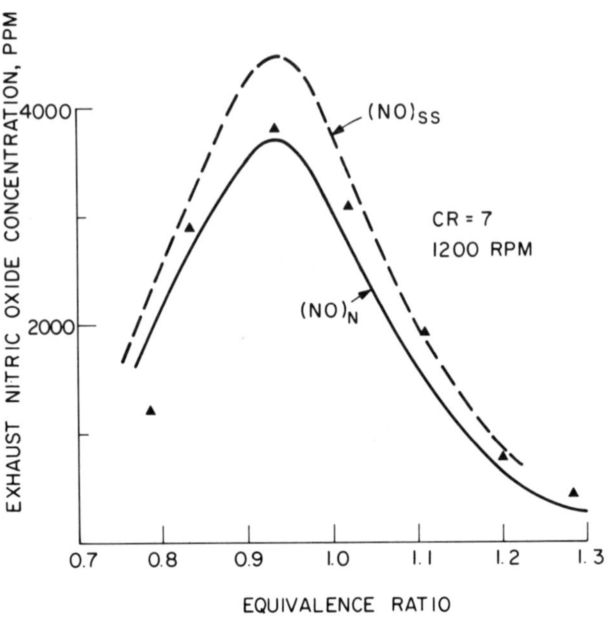

Figure 3. Comparison of predicted and measured average NO concentration in the exhaust of an I.C.E. as a function of fuel/air equivalence ratio.

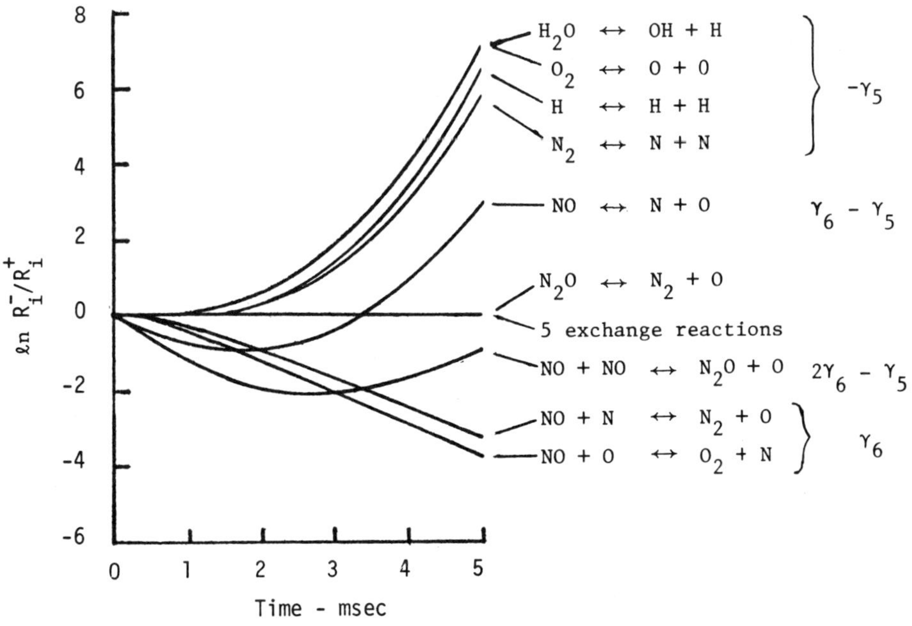

5 exchange reactions: $CO + OH \leftrightarrow CO_2 + H$

$OH + O \leftrightarrow O_2 + H$; $OH + H \leftrightarrow H_2O + H$

$OH + H \leftrightarrow H_2 + O$; $OH + OH \leftrightarrow H_2O + O$

Time dependent constraints: M ; NO

Lagrange multipliers γ_5; γ_6

Figure 4. Degree of disequilibrium, ln R_i^-/R_i^+, for 14 reactions in the combustion products of an I.C.E. as a function of time after top dead center.

MAXIMAL ENTROPY PROCEDURES FOR MOLECULAR AND NUCLEAR COLLISIONS

R. D. Levine

Introduction

Statistical mechanics of disequilibrium is concerned with the motion of systems in phase space. The thrust of this paper is that it is possible to regard the study of binary collisions as an effective probe of such motions. In so doing one not only obtains a useful route for the analysis of collision phenomena (1-6) but also an improved insight into statistical mechanics and the procedure of maximal entropy (7-13).

Collision experiments consist essentially of two stages (14) both of which are relevant to our discussion. The first is the selection of the initial state before the collision. By employing different experimental arrangements it is possible to exercise considerable control over this selection. In particular, it is possible to prepare an initial state that is well localised in phase space (This is typically the case in computational studies where the collision is described through a numerical solution of the equations of motion). It is also possible however to prepare a less sharply defined state, up to the limit of a uniform distribution in phase space. Following the initial state selection the system evolves under its own Hamiltonian (i.e. a collision is taking place). The second stage is that of final state analysis: once the colliding particles are well separated, (i.e. non-interacting), their distribution in phase space is determined.

The entire collision experiment consists therefore in the determination of the change in the distribution in phase space as a result of a single binary encounter. If the initial state corresponds to a uniform distribution in phase space, then microscopic reversibility implies that the final state will also be uniform. For the more typical case of a selected (i.e. non-uniform) initial state, the final distribution need not be uniform and hence serves as a probe of the dynamics. The purpose of our studies (1-6) was to explore the following question: Can one describe the collision by specifying a set of constraints that prevent the final distribution from being uniform in phase space? In other words, can one describe the final state as a distribution of maximal entropy subject to constraints. We contend that the answer is in the affirmative and shall bring both theoretical and empirical results in support of the argument. Specifically, given an initial state and the Hamiltonian of the system it can be shown (6,15) possible to specify a set of constraints such that the exact solution of the Liouville equation (both after and during the collision) is given as a distribution of maximal entropy subject to these constraints. In addition it is found (1-5) that experimental final state distributions are often remarkably well represented as distributions of maximal entropy subject to very few (one, two) constraints.

An Example (4) The present point of view provides a unified framework for describing disequilibrium on both the microscopic and the macroscopic levels. Pending the technical discussion below, figure 1 shows typical results. The final, nascent vibrational state distribution of CO molecules produced in a single collision, $O + CS \rightarrow CO(v') + S$ is an example of 'micro-disequilibrium' as discussed in the previous paragraph. The triangles are experimental results (16) while the dashed line is the distribution of maximal entropy subject to a single constraint (4), namely the mean vibrational energy of the CO molecules. (Contrary to one's immediate reaction, such a constraint does not necessarily imply a Boltzmann distribution. The reason is that the total energy is conserved during an elementary collision and this need be taken into account (1-4)). When the same reaction is allowed to take place in the bulk, the newly formed, vibrationally excited, CO molecules undergo subsequent collisions with other molecules, leading to a relaxation of the nascent vibrational distribution. Any such partially relaxed distribution is an example of a macroscopic system in disequilibrium. The open circles show a measured distribution (17) and its representation (dashed line) as a distribution of maximal entropy subject to constraints (4). Finally, at the end of the relaxation, the CO molecules attain thermal equilibrium. The characterisation of the Boltzmann thermal distribution by the maximal entropy procedure is, of course, well known. In the rest of this paper we center attention on distributions following a single collision and refer to other sources (4,18-20) for our results on macroscopic systems in disequilibrium.

Differences Having discussed the description of a binary collision in statistical mechanical terms one should also note the following differences. The most essential is the number of degrees of freedom. Even through we are dealing with collisions of composite projectiles, the number of degrees of freedom of our system is small; often of order 1 or less (i.e. <10) in contrast to the very high numbers (e.g. 10^{23}) characteristic of macroscopic systems. This has important implications regarding the number of constraints required for an exact description. Next, there is the nature of the dynamics. One is dealing here with a single, isolated collision. The experimental arrangement insures that the collision partners are detected as soon as they emerge from their mutual interaction and before any subsequent relaxation can take place. It is for this reason that the final state will not be an equilibrium distribution and hence can be analysed to determine the constraints implied by the dynamics. Finally, one should note that any actual experiment consists in exceedingly many independent repetitions of the same collision. Hence, an ensemble point of view is quite appropriate. It is not necessary to distinguish between probabilities and frequencies (21), i.e. the probabilities determined by the procedure of maximal entropy do admit of an objective interpretation as frequencies (2-4). In other words, the distribution of maximal entropy can be interpreted *a là* Boltzmann (22) and Planck (8) as the (overwhelmingly (2,3)) most probable distribution subject to the specified constraints. This does not preclude a subjective interpretation (11, 21, 23, 24) of probabilities as reflecting a state of knowledge but is not dependent on such an interpretation.

Outline The next section forges the link between the maximal entropy procedure and the equations of motion points of view and provides a simple example. Complete details and additional worked out examples are available elsewhere (6,15). Next, several illustrations of data analysis are provided. Many additional examples are available (1-5), including a compilation of the work done during the first five years (5) and a 'guidebook' for would be practitioners (25).

Theory

The purpose of this section is to determine the constraints that characterise (via the procedure of maximal entropy) the exact solution of the Liouville equation for the collision. Given the Hamiltonian and the constraints that characterise the initial state it is possible to proceed to determine the required constraints by purely algebraic means. At the end of this algebraic stage, one has determined the functional form of the solution of the Liouville equation throughout the collision. Explicitly, and using the quantal, density oeprator, $\rho(t)$ (6,12,13,26) to describe the system at the time t,

$$\rho(t)=\exp[-\sum_{r=0}^{m}\lambda_r(t)A_r]. \qquad 1.$$

Here the m+1 operators A_r ($A_0 \equiv I$) are the constraints and the λ_r's are the Lagrange parameters, whose magnitude is defined via the m+1 implicit equations

$$\langle A_r(t)\rangle \equiv \mathrm{Tr}[A_r \rho(t)] \qquad 2.$$

where $\langle A_r(t)\rangle$ is the mean value of A_r at the time t. In the second stage it is shown that the implicit set of equations 2 can explicitly be solved, given the form of the initial state (i.e. $\rho_{in} \equiv \rho(t \to -\infty)$) and the Hamiltonian. A simple example is worked out in detail. Finally, two simple general results are noted. First, and as is well known in practice (1-5), similar collisions can be characterised by the same set of constraints. Second, an explicit expression for the branching ratio (27) is derived. A brief discussion of approximation procedures concludes this section.

The Initial State In formal collision theory it is customary to choose the initial, pre-collision, state of the system as an eigenstate of the Hamiltonian, H_0, of the non-interacting molecules. Such a choice corresponds to the complete specification of the initial state and has, thus far, seldom been realized in collision experiments. An incomplete (i.e.partial) state selection of the reactants is the rule rather than the exception. There is however a common feature to the actual and the idealized experiment. In both, the initial state is selected so as not to undergo transitions in the absence of the collision. The initial state is stationary under the time evolution governed by H_0. We now argue (6) that a stationary initial state can be specified by the maximal entropy procedure subject to constraints that are constants of the motion.

A normalized stationary density operator ρ is a function of (say n) constants of the motion $\rho = \rho(c_1,\ldots,c_n)$. Consider another (normalized) density operator, say σ, which gives precisely the same results as ρ for the distribution of the constants $(c_1,\ldots,c_n)$ of the motion. Explicitly, for any function $f(c_1,\ldots,c_n)$ of the constants of the motion

$$\mathrm{Tr}\{\rho f(c_i)\} = \mathrm{Tr}\{\sigma f(c_i)\}. \qquad 3.$$

In particular, since ρ (and hence $\ell n\rho$) is a function only of the constants of the motion

$$\mathrm{Tr}\{\rho \ell n\rho\} = \mathrm{Tr}\{\sigma \ell n\rho\}. \qquad 4.$$

Any two normalized density operators ρ and σ satisfy the inequality (28)

$$S[\sigma] = -\mathrm{Tr}\{\sigma \ell n\sigma\} \le -\mathrm{Tr}\{\sigma \ell n\rho\} \qquad 5.$$

with equality if and only if $\sigma \equiv \rho$. Using equation 4 to evaluate the right-hand side of the inequality,

$$S[\sigma] = -\mathrm{Tr}\{\sigma \ln \rho\} \leq -\mathrm{Tr}\{\rho \ln \rho\} = S[\rho] \qquad 6.$$

Among all the states that agree on the distribution of n constants of the motion, the stationary state, $\rho = \rho(c_1, \ldots, c_n)$, is the (unique) state of maximal entropy. The state ρ is unique because equality in (2.10) holds if and only if $\sigma \equiv \rho$. This quantum mechanical proof is an extension of the proof by Gibbs (29) that a classical distribution in phase space which is a function of the energy has a higher entropy than 'any other distribution in which the distribution in energy is unaltered'.

As in equilibrium statistical mechanics (7) the number of the constants of the motion can be dramatically reduced by noting that an initial state in a binary collision necessarily corresponds to two independent systems. Moreover, the description need be invariant under a Galilean transformation (i.e., the collision is essentially the same whether viewed in the laboratory or in the center of mass system of coordinates since the interaction is only dependent on the relative position of the colliding particles.)

Another route to the characterisation of ρ_{in} is provided by the argument that the initial state is a result of a preparation procedure and hence must be a state of maximal entropy subject to those variables that were specified during its selection (12,13,15). Finally, one can always adopt the position that the procedure of maximal entropy is the most reasonable induction that can be made for an incompletely specified initial state (30).

The Variational Principle The central formal result of this section can now be stated. It is that an initial state of maximal entropy evolves into a state of maximal entropy throughout the collision. More explicitly, consider a set of initial states, all consistent with the same set of constraints. Let ρ_{in} be that particular initial state of maximal entropy (subject to the given set of constraints) and let $\rho(t)$ be the exact solution of the equation of motion that evolved from the initial state ρ_{in}. To other possible initial states, $\tilde{\rho}_{in}$, there are similarly corresponding solutions, $\tilde{\rho}(t)$, of the equation of motion. Then

$$S[\tilde{\rho}(t)] \leq S[\rho(t)] \qquad 7.$$

with equality if and only if $\tilde{\rho}_{in} \equiv \rho_{in}$. The proof is immediate, and quite trivial. The state of maximal entropy is unique and since all initial states must be consistent with the constraints

$$S[\rho_{in}] \geq S[\tilde{\rho}_{in}] \qquad 8.$$

with equality if and only if $\tilde{\rho}_{in} \equiv \rho_{in}$. Exact solutions of the equations of motion do not change their entropy (28). Hence

$$S[\rho_{in}] = S[\rho(t)] \qquad S[\tilde{\rho}_{in}] = S[\tilde{\rho}(t)]. \qquad 9.$$

Combining equations 9 and 8 leads unavoidably to the desired conclusion, equation 7.

The formal result, equation 7, shows that among all the contenders, $\rho(t)$ has the highest entropy. It remains to be shown that $\rho(t)$ can be explicitly determined by the procedure of maximal entropy and, on practical grounds, to specify the constraints that need be imposed in the determination of $\rho(t)$.

The Surprisal To determine the functional form of $\rho(t)$ throughout the collision it is convenient to work in terms of the surprisal, $-\ell n\rho(t)$. The problem is thus to identify the operators A_r that appear in the surprisal,

$$-\ell n\rho(t) = \sum_r \lambda_r(t)A_r. \qquad 10.$$

Towards this end one notes (6,15) that the surprisal itself satisfies the Liouville equation of motion

$$-i\hbar\partial\ell n\rho(t)/\partial t = [H,-\ell n\rho(t)]. \qquad 11.$$

Since $\rho_{in} \equiv \rho(t\rightarrow-\infty)$ is given, and the Hamiltonian is assumed known, the constraints are given as that set of operators such that equation 10 is the general solution of equation 11.

The Constraints Introducing the form 10 into the equation of motion for the surprisal

$$i\hbar \sum_{s=0}^{m} (\partial\lambda_s/\partial t)A_s = \sum_{s=0}^{m} \lambda_s[H,A_s] \qquad 12.$$

one notes that 10 is indeed the desired solution provided that each member of the set of operators s=0,...,m, satisfy the condition

$$[H,A_s] = i\hbar \sum_{r=0}^{m} A_r g_{rs}. \qquad 13.$$

Here the g_{rs} are (possibly complex) numerical coefficients.

Substituting for $[H,A_s]$ from equation 13 into equation 12 and invoking the requirement that the operators A_r be linearly independent (so that one can compare the coefficients of A_s on both sides) we obtain

$$\partial\lambda_r/\partial t = \sum_{s=0}^{m} g_{rs}\lambda_s. \quad 14.$$

The equation of motion for the density operator has thus been converted to a set of coupled equations of motion for the Lagrange parameters. The number of coupled equations equals the number of constraints.

The boundary conditions on the equations of motion are determined by the requirement that the initial state ρ_{in}, be a state of maximal entropy subject to constraints specified by the experiment. Explicitly,

$$-\ell n\rho_{in} = \sum_{r=0}^{m} \lambda_r^{in} A_r \quad 15.$$

The operators A_r that appear in equation 15 are thus known to begin with. Starting with this set of constraints one proceeds (if necessary) to add additional operators until the condition 13 obtains for any operator in the (augmented) set. The set of (sufficient) constraints is that set of operators that is closed under commutation with the Hamiltonian. The boundary condition on the $\lambda_s(t)$'s is $\lambda_s^{in}=0$ if A_s is not included in the set of constraints required to specify ρ_{in}. Otherwise, λ_s^{in} is determined by the experiment (i.e. by the particular initial state under consideration). An open question is under what conditions on the Hamiltonian will the set of constraints be finite. At this point, however, we already know of several examples where the eigenstates of H_0 are countably infinite, yet where a finite set of constraints does exist (see, e.g. below).

The procedure just outlined has an obvious conceptual similarity to the close coupling method of collision theory. There too one replaces a complicated equation of motion (The Schrödinger equation) by a set of coupled (but simpler) equations. The present procedure has the advantage that one can solve directly for the evolution of the initial state of interest (which need not be a pure state) and that the number of coupled equations (required to obtain an exact solution) will often (but not always) be far smaller. Similarly, the present procedure is derivable from (the maximal entropy) variational considerations (cf. (5)) and hence approximations are readily formulated. In particular, and as in the close coupling procedure, one can truncate the set of equations 13.

For a uniform initial state the only constraint initially present is the identity, which commutes with H. Thus, as expected, the final state is also uniform, as is indeed $\rho(t)$ throughout the collision. In general, we recover the known implication of equilibrium statistical mechanics: whenever ρ_{in} is a state of maximal entropy subject to operators that commute with H it will remain unchanged due to collisions.

For a macroscopic system, the condition 13 does not provide a practical procedure since, typically, the number of terms will be unrealistically large and alternative routes must be devised (13,31-33). For a system of a few degrees of freedom, one can however consider an actual determination of a set of constraints via 13 (6,15) and a very simple example is now considered.

<u>An Example</u> Perhaps the simplest collision problem is that of a structureless particle with a harmonic oscillator (34), with their relative motion being treated classically. The Hamiltonian of the oscillator is simply

$$H_0 = \hbar\omega a^{\dagger}a \qquad 16.$$

while the particle-oscillator interaction is represented as

$$V(t) = f(t)(a+a^{\dagger}). \qquad 17.$$

Here $a^{\dagger}$ and a are the usual creation and annihilation operators for the oscillator. f(t) is a function of time representing the approach and departure of the projectile so that $f(t)\to 0$ before $(t\to-\infty)$ and after $(t\to+\infty)$ the collision.

As an initial state consider a thermal distribution of oscillator states,

$$\rho_{in} = \exp(-\beta H_0-\lambda), \qquad 18.$$

$$\exp(\lambda) = \mathrm{Tr}\{\exp(-\beta H_0)\}. \qquad 19.$$

Two constraints (the identity operator and H_0) are thus necessary to specify the initial state. Of the two, only H_0 will fail to commute with $H=H_0+V$,

$$[H,H_0] = [V,H_0] = \hbar\omega f(t)[(a+a^{\dagger}),a^{\dagger}a] = \hbar\omega f(t)(a-a^{\dagger}). \qquad 20.$$

Two new operators, a and $a^{\dagger}$, are obtained on first application of equation 13. We now proceed to check the commutation of these operators with H

$$[H,a] = \hbar\omega[a^\dagger a,a] + f(t)[a^\dagger,a] = -\hbar\omega a - f(t)$$

$$[H,a^\dagger] = \hbar\omega a^\dagger + f(t). \qquad 21.$$

No new operators are obtained in equation 21. Hence, the set of four operators I, $a^\dagger a$, a and $a^\dagger$ is closed under commutation with H. It follows that for any initial state of the form 18, the exact solution of the Liouville equation at any time t during (or after) the collision must be of the form

$$\rho(t) = \exp[-\beta(t)H_0-\gamma(t)a^\dagger-\gamma^*(t)a-\lambda(t)]. \qquad 22.$$

The functional form of $\rho(t)$ is as far as one can go with purely algebraic manipulations. To determine the magnitude of the Lagrange parameters one must solve the equations of motion 14. Using equations 13, 14, 20 and 21, we find

$$\partial\beta(t)/\partial t = 0$$

$$\hbar\partial\gamma(t)/\partial t = -i\omega\gamma(t)+if(t)\beta(t) \qquad 23.$$

$$\hbar\partial\lambda(t)/\partial t = if(t)[\gamma^*(t)-\gamma(t)].$$

The boundary conditions implied by the form 18 of ρ_{in} are $\beta^{in}=\beta$, $\gamma^{in}=0$, $\lambda^{in}=\lambda$. Hence

$$\beta(t) = \beta$$

$$\gamma(t) = -\exp(-i\omega\Delta t/\hbar)\alpha(t,t_0)\beta \qquad 24.$$

$$\lambda(t) = \lambda + |\alpha(t,t_0)|^2\beta.$$

Here $\Delta t=t-t_0$ where t_0 is a time in the remote past when $f(t_0)=0$ and

$$\alpha(t,t_0) = -i\int_{t_0}^{t} f(\tau)\exp(i\omega\tau/\hbar)d\tau. \qquad 25.$$

As $t\to+\infty$, $f(t)\to 0$ and α attains a constant value. It is possible to avoid the rapidly oscillating term in $\gamma(t)$ by working in the interaction picture (15).

The solution is now complete. For any initial state of the

form 18, we have determined not only the form of the solution but also the magnitude of the Lagrange parameters. While the magnitude of the parameters clearly reflects the details of the specific system, (i.e. the oscillator frequency ω and the form of the force term f(t), cf. equation 25), the set of constraints is independent of these details. Indeed, it is evident from the general scheme 13 that the same set of constraints would obtain irrespective of the magnitude of any numerical parameters in the Hamiltonian. Similar systems, which have the same operators appear in their Hamiltonian and differ only in the coefficients in front of these operators will thus be characterised by the same set of constraints (15).

Equations of Motion The complete solution of the scattering problem requires not only the specification of the constraints but also the determination of the Lagrange parameters. In the previous example this was implemented by integrating the linear equations of motion 14 subject to their boundary conditions. It is possible however to generalise this procedure in order to emphasise the following point: The Lagrange parameters at any time t are linear combination of the initial Lagrange parameters with coefficients that are independent of the particular initial state of interest. The proof is immediate.

Consider the Lagrange parameters arranged as a column vector $\underset{\sim}{\lambda}(t)$. Equations 14 can thus be written as

$$\partial\underset{\sim}{\lambda}(t)/\partial t = \underset{\sim}{g}(t)\underset{\sim}{\lambda}(t) \qquad 26.$$

where the matrix $\underset{\sim}{g}(t)$, defined via the closure relation 13, is independent of the magnitude of the Lagrange parameters of the initial state. One can thus define a matrix $\underset{\sim}{G}(t,t_0)$ by the linear differential equation

$$\partial\underset{\sim}{G}(t,t_0)/\partial t = \underset{\sim}{g}(t)\underset{\sim}{G}(t,t_0) \qquad 27.$$

and the boundary condition

$$\underset{\sim}{G}(t_0,t_0) = \underset{\sim}{I} \qquad 28.$$

such that

$$\underset{\sim}{\lambda}(t) = \underset{\sim}{G}(t,t_0)\underset{\sim}{\lambda}(t_0). \qquad 29.$$

A similar result obtains for the mean values of the constraints. Specifically, if the mean values are arranged as a row vector then

$$\langle \underset{\sim}{A}(t)\rangle = \langle \underset{\sim}{A}(t_0)\rangle[\underset{\sim}{G}(t,t_0)]^{-1} \qquad 30.$$

A simple proof is provided by the time invariance of the entropy (cf. equation 9)

$$\langle \underset{\sim}{A}(t)\rangle \underset{\sim}{\lambda}(t)=S(t)=S(t_0)=\langle A(t_0)\rangle \underset{\sim}{\lambda}(t_0). \qquad 31.$$

On using equation 29 in 31 we verify equation 30. It follows that the final values of the constraints after the collision are given as linear combination of their initial values before the collision, with coefficients that are independent of the particular initial state. When the Hamiltonian is known, the required coefficients, i.e. the G matrix, can be computed by first determining the g matrix using 13 and then solving the linear equation 27, subject to the boundary condition 28. Clearly neither the equation of motion for G nor its boundary conditions require any input concerning the magnitude of the initial values of the Lagrange parameters.

Sum Rules We refer to the relations 30 as a 'sum-rule' since they determine the mean (i.e. sum over states) value of the constraints. In particular, for the final state after the collision

$$\langle A_r\rangle_{out} = \sum_s F_{sr}\langle A_s\rangle_{in} \qquad 32.$$

where F is the inverse of the G matrix. Such sum rules have been effectively employed as routes to the determination of the Lagrange parameters of final states (35,36) and an example is shown in the next section. The sum rule used there is the same type as the one that obtains for the example of the forced harmonic oscillator. Of the four operators (I, $a^{\dagger}$ a, a and $a^{\dagger}$) only two (I and $a^{\dagger}a$) have non-vanishing averages for a thermal initial state. Hence

$$\langle E_{vib}\rangle_{out} = a\langle E_{vib}\rangle_{in} + b \qquad 33.$$

where a and b are constants which have the same magnitude irrespective of the value of $\langle E_{vib}\rangle_{in}$ (b is multiplied by $\langle I\rangle_{in}\equiv 1$). It should however be noted that the same sum rule obtains also for more realistic Hamiltonians (see the example in (6)). What distinguishes these cases is the number of constraints and hence the existence of additional sum rules.

Branching Fractions It is often of interest to determine the fraction of collisions that proceed to form products in a specified range of states. (The different ranges may correspond to distinct arrangements, etc (27)). An immediate implication of the present formalism is that such fractions are directly proportional to the corresponding partition functions, i.e. to the effective volume in phase space available to the products.

The partition function for the entire system is defined, as usual, by

$$Q_{out} \equiv \mathrm{Tr}\{\exp[-\sum_{r=1}^{m} \lambda_r^{out} A_r]\}. \quad 34.$$

In 34 the trace is over all final states. Consider however a trace over a restricted group 'a' of final states, denoted as Tr_a. One can then define the appropriate partition function, Q_{out}^a by

$$Q_{out}^a \equiv \mathrm{Tr}_a\{\exp[-\sum_{r=1}^{m} \lambda_r^{out} A_r]\}. \quad 35.$$

If the groups a are an exclusive and exhaustive set,

$$Q_{out} = \sum_a Q_{out}^a . \quad 36.$$

The fraction p_a of products in group a is defined by

$$p_a \equiv \mathrm{Tr}_a(\rho_{out}) = \mathrm{Tr}_a\{\exp[-\sum_{r=1}^{m} \lambda_r^{out} A_r]\}/Q = Q_a/Q \quad 37.$$

and the branching ratio is given by

$$\Gamma_{ab} \equiv p_a/p_b = Q_a/Q_b . \quad 38.$$

The results 37 and 38 were previously (27) derived directly from the principle of maximal entropy without constructing the distribution of final states.

Approximations An approximate solution is obtained by failing to include all the constraints. It is indeed possible to express the exact density operator as a sum of a density operator of maximal entropy plus a correction term (13,31-33). When the set of constraints is closed under commutation with the Hamiltonian (i.e. when 13 obtains) the correction term vanishes (33). Otherwise, one can use such a resolution in order to systematically improve any approximate set of constraints.

Another practical point concerns the freedom to transform to a set of equivalent constraints (15). Since only $\rho(t)$ has operational significance one can always introduce a non-singular matrix $\underset{\sim}{w}$ such that an equivalent set of constraints and Lagrange parameters is given by

$$C_r = \sum_s w_{sr} A_s \qquad 39.$$

$$\gamma_r = \sum_r w^{-1}_{rs} \lambda_s . \qquad 40.$$

Such a transformation leaves the surprisal, $-\ln\rho(t)$, invariant. The matrix $\underset{\sim}{g}$ for the set C_r is now given by (cf. 13)

$$\underset{\sim}{\bar{g}} = \underset{\sim}{w}^{-1} \underset{\sim}{g} \underset{\sim}{w} \qquad 41.$$

and, since the constraints are operators in the Schrödinger picture (i.e., time independent)

$$\underset{\sim}{\bar{G}} = \underset{\sim}{w}^{-1} \underset{\sim}{G} \underset{\sim}{w} . \qquad 42.$$

In particular, one can sometimes bring the G matrix to a triangular form. This, and other rigorous reduction procedures have been discussed (15). A particularly important result is the specification of the minimal number of coupled equations that need be solved. Proceeding to approximations, one can choose w such that one constraint makes a leading contribution (i.e. that the magnitude of one term $\gamma_r\langle C_r\rangle$ makes a dominant contribution to the entropy, cf. equation 31).

Summary Given the Hamiltonian and an initial state characterised by the procedure of maximal entropy, it was shown possible to (a) Define a set of constraints such that the exact solution of the Liouville equation throughout the collision is a density operator of maximal entropy subject to these constraints and (b) Provide an explicit procedure for the computation of the Lagrange parameters or of the mean values of the constraints throughout the collision in terms of their initial values.

Surprisal Analysis

Surprisal analysis is the empirical application of the procedure of maximal entropy towards the characterisation of the final state of the collision. A few selected examples are discussed in this section. To simplify the description we shall proceed to the limit of quasi-classical mechanics. In other words, we shall retain the quantisation of the internal degrees of freedom but regard the constraints as classical (and hence, commuting) observables. The probability of any final quantum state is thus written as

$$P(\gamma)=\exp[-\sum_{r=0}^{m} \lambda_r A_r(\gamma)] \qquad 43.$$

Here $A_r(\gamma)$ is the magnitude of the observable A_r for the state γ and $A_0(\gamma)=1$.

The Prior Distribution Typically one does not measure the fully resolved final quantum state distribution. Rather, final states are collected into groups. If the magnitude of the constraint is the same for all states of the group a then 43 can be written as

$$P(a)=n_a\exp[-\sum_{r=0}^{m} \lambda_r A_r(a)] = P^O(a)\exp[-\sum_{r=0}^{m} \lambda_r' A_r(a)]. \qquad 44.$$

Here, n_a is the number of quantum states in the group a and $P^O(a)=n_a/\sum_a n_a$ (so that λ_0 has been redefined in the second line of 44). By construction $P^O(a)$ is the probability of the group a in the absence of any constraint (i.e. when the λ_r's=0). It is thus referred to as the 'prior' distribution. By construction again, the prior distribution obtains when (at a given total energy) all final quantum states are equally probable. The prior distribution is the distribution of maximal entropy subject to conservation of probability and of good quantum numbers.

The computation of a prior distribution consists, therefore, in the counting of the number of quantum states in the group of interest. Explicit results for examples of practical interest are available (25).

Surprisal Analysis The purpose of the analysis is to find a compact representation of the surprisal in the form (cf. 44)

$$-\ln[P(a)/P^O(a)] = \lambda_0 + \sum_{r=1}^{m} \lambda_r A_r(a) \qquad 45.$$

retaining as few terms as possible. An efficient numerical procedure has been described (37). Remarkably often however a single constraint will already suffice, in which case the Lagrange parameter can be determined as a slope in a plot of the surprisal vs. the appropriate observable as shown in figure 2.

Example: Population Inversion Figure 2 shows a surprisal analysis for the final vibrational state distribution in the

$$F + D_2 \rightarrow DF(v') + D$$

reaction. The prior distribution (open circles) is a monotonically decreasing function of the vibrational energy of DF. The reason is clear. The more energy is deposited in the non-degenerate vibrational mode of DF, the less energy is available to populate the degenerate rotational states of DF and the states of relative motion of DF and D. The actual experimental results (38,39) (solid symbols) are in qualitative discord with the prior expectations. Indeed, they show population inversion with the v=3 state being more populated than the lower states. Despite the considerable deviance of the observed from the prior distribution, the surprisal is very well represented by a single constraint

$$I(f_{v'}) = -\ln[P(f_{v'})/P^{o}(f_{v'})] = \lambda_0 + \lambda_v f_{v'} \quad . \qquad 46.$$

Here $f_{v'} = E_{v'}/E$ is the fraction of the total energy E deposited in the DF vibration, the primes denote a final state and $\lambda_v=-5.7$. The negative value of λ_v reflects the initial rise of $P(f_{v'})$ with $f_{v'}$, and is thus characteristic of population inversion. In other words, the observed mean vibrational excitation of DF exceeds the value computed for the prior distribution.

While we do not know the magnitude of $\langle f_{v'} \rangle$ *a-priori* and hence need to determine λ_v as the slope of the surprisal plot, we do not know *a-priori* that probability is conserved and hence that $\langle 1 \rangle = 1$. The magnitude of λ_0 can thus be determined in terms of λ_v and the normalisation condition, i.e.

$$\exp(\lambda_0) = \sum_{v'} P^{o}(f_{v'})\exp(-\lambda_v f_{v'}) .$$

That λ_0 can be regarded as a function of the other Lagrange parameters, is, of course, a general result of the maximal entropy formalism.

There are many other examples of a linear vibrational surprisal plot (1-5), the O + CS reaction shown in figure 1 being one, and the dashed line labelled 'nascent' in figure 1 is just the $P(f_{v'})$ distribution defined implicitly by equation 46. There is however a family of reactions for which a single constraint does not suffice and two constraints are definitely required to account for the final vibrational state distribution (40).

Example: Surprisal Analysis of Computational Studies For sufficiently simple collisions it is now possible to obtain numerically exact solutions of the Schrödinger equation. Figure 3 shows an analysis (4) of such results for rotational energy transfer in collisions of a He atom with HX molecules in a definite initial rotational state j (41). The only spread in the initial state is in that the 2j+1 'magnetic' quantum numbers (m_j) of the HX molecules are taken to be equally probable.

The variable employed is the fraction Δg_R,

$$\Delta g_R = (E_{j'} - E_j)/E, \qquad 47.$$

of the total energy E converted into rotational excitation. The surprisal,

$$I(\Delta g_R) = -\ln[P(j'|j)/P^{\circ}(j'|j)] \qquad 48.$$

is found to be a linear function of Δg_R. In particular, different $j \rightarrow j'$ transitions with a similar value of Δg_R (e.g. $2 \rightarrow 3$ and $0 \rightarrow 2$) are seen to have a similar surprisal.

A linear dependence of the surprisal of the final rotational state distribution on Δg_R was also found for the numerical quantum mechanical solution for the simplest, $H+H_2 \rightarrow H_2+H$ (42) exchange reaction. It was furthermore found (43) that the same was true throughout the collision.

Example: Surprisal Synthesis The empirically observed simple functional form of the surprisal suggests that a possible simple route to the computation of the final state in the collision is to determine the surprisal directly. The actual distribution is then obtained by 'inverting' the definition of the surprisal,

$$P(a) = P^{\circ}(a)\exp[-I(a)]. \qquad 49.$$

The theory of the previous section showed that this is indeed possible provided one knows the Hamiltonian. Unfortunately, only too often it is precisely the Hamiltonian that is unknown. Indeed, collision experiments are typically conducted towards trying to determine the nature of the interaction. There is however another route. All that we really need are the mean values of the constraints in the final state. One way to obtain these is by using sum rules (35,36). Figure 4 shows results (35) for inelastic collisions that were derived in such a fashion. The bars in the midpanel are the surprisals derived from the data (44,45). The lines show the surprisal computed using a sum rule (35). From these lines one obtains the distributions shown as dots in the top panel.

Example: Nuclear Collisions An illustration of an application to the nuclear rearrangement collision (46)

$$^{16}O + {}^{6}L_i \rightarrow {}^{20}Ne + D$$

is shown in figure 5. (47) This also documents an analysis of the angular distribution of the collision products. The experimental results are quite deviant from the prior angular distribution (48) shown as a dashed line. Yet a surprisal based on a momentum transfer constraint ((47), continuous line) accounts well for the angular distribution of this and similar nuclear reactions.

Interpretation of Surprisal Analysis The procedure of surprisal analysis seeks to represent the observed frequencies f_a by theoretical probabilities P(a) of the functional form

$$P(a)=P^{o}(a)\exp[-\sum_{r=0}^{m} \lambda_r A_r(a)]. \qquad 50.$$

The magnitudes of the m+1 Lagrange parameters λ_r which appear in 50 are determined by minimising the average deviance of the surprisal of the actual data from its assumed form

$$H \equiv \sum_a f_a\{\ell n[f_a/P^{o}(a)] + \sum_{r=0}^{m} \lambda_r A_r(a)\} = \min . \qquad 51.$$

To show that such a determination of the Lagrange parameters is precisely that implied by the procedure of maximal entropy one proceeds as follows. The variables in equation 51 are the λ_r's . Hence the minimising condition can be written as

$$\partial H/\partial \lambda_r = 0, \ r = 0,1,\ldots,m. \qquad 52.$$

Using the definition, equation 51, of H, one finds that the optimal values of the m+1 Lagrange parameters are given through the solution of m+1 implicit equations

$$\langle A_r \rangle \equiv \sum_a f_a A_r(a) = \sum_a P(a)A_r(a), \ r = 0,1,\ldots,m \qquad 53.$$

where the P(a)'s depend on the λ_r's as in 50. These equations are however just the familiar condition that the magnitude of the expectation of A_r (the right-hand side of equation 53) be set equal to its average value.

The set of distributions $Q(a)$ which are normalised and are consistent with the given magnitude of the average value of the m constraints is defined by

$$\langle A_r \rangle = \sum_a Q(a) A_r(a) \quad r = 0,1,\ldots,m. \qquad 54.$$

The procedure of surprisal analysis identifies that particular member of the above set (the distribution P, cf. equation 53) which is of minimal average deviance from its prior distribution. The distribution of final quantum states corresponding to the form 50 is given by equation 43. Since the condition 53 can also be written as

$$\langle A_r \rangle = \sum_\gamma P(\gamma) A_r(\gamma) \qquad 55.$$

it follows that the final quantum state distribution with the λ_r's determined via 51 is indeed the distribution of maximal entropy consistent with the magnitude of the constraints. The proof is immediate. H can also be written as

$$H = \sum_\gamma f_\gamma \, \ell n[f_\gamma / P(\gamma)] \geq 0 \qquad 56.$$

where the summation is over quantum states. Using 43, 55 and the definition of entropy

$$S[P] = - \sum_\gamma P(\gamma) \ell n P(\gamma) \qquad 57.$$

it follows from 56 that

$$H = \sum_r \lambda_r \langle A_r \rangle - S[f] = S[P] - S[f] \geq 0, \qquad 58.$$

with equality iff $P(\gamma) \equiv f_\gamma$.

Finally, a very important practical point is that there is no inherent harm in including as many observables as one wishes in expressing $P(a)$ in the form 50. Any particular observable, say A_s, whose average value does not constrain the distribution will be identified by the procedure and will lead to $\lambda_s \equiv 0$. To show this note that the function H is a concave function of the λ_r's with a unique minimum at the distribution $P(a)$ defined by equation 53 (37). But the observable A_s has been introduced such that $\langle A_s \rangle = \sum P(a) A_s(a)$ even when λ_s is set equal to zero. Hence the solution of 52 will be $\lambda_s \equiv 0$.

Concluding Remarks

The procedure of maximal entropy was shown to offer a novel formulation of the exact dynamics of collision processes as well as providing a particularly convenient framework for empirical analysis. The procedure views the collision as a mapping from an initial to a final distribution in phase space. It seeks to determine the constraints that prevent the final state distribution from being as chaotic as possible. The theory identifies these constraints in terms of the Hamiltonian and the constraints required to specify the initial state. It also provides an explicit route to the determination of the Lagrange parameters. The empirical surprisal analysis seeks to identify the constraints by a fit to the data. It owes its success to the very frequent situation where one particular constraint is quite dominant, so that with a single term in the suprisal it is possible to offer a good account of the observations.

This work was supported by the Office of Scientific Research, USAF, under grant 77-3135.

References

1. Levine, R. D., Bernstein, R. B. 1974. Acc. Chem. Res. 7: 393-400.

2. Bernstein, R. B., Levine, R. D. 1975. Adv. At. Mol. Phys. 11:215-297.

3. Levine, R. D., Bernstein, R. B. 1975. Dynamics of Molecular Collisions, ed. W. H. Miller, Part B, pp. 323-364. New York: Plenum.

4. Levine, R. D., Ben-Shaul, A. 1977. In Chemical and Biochemical Applications of Lasers, ed. C. B. Moore, II: 145-197. New York: Academic Press.

5. Levine, R. D. 1978. Ann. Rev. Phys. Chem. 29: 59-92.

6. Alhassid, Y., Levine, R. D. 1977. J. Chem. Phys. 67:4321-39.

7. Gibbs, J. W. 1902. Elementary Principles in Statistical Mechanics. New Haven: Yale University Press.

8. Planck, M. 1914, The Theory of Heat Radiation. New York: Blakiston.

9. Elsasser, W. M. 1937. Phys. Rev. 52: 987-99.

10. Stratonovich, R. L. 1955. Sov. Phys. J.E.T.P. 1: 426-31.

11. Jaynes, E. T. 1957. Phys. Rev. 106: 620-630.

12. Wichman, E. H. 1963. J. Math. Phys. 4: 884-90.

13. Robertson, B., 1966. Phys. Rev. 144: 151-161.

14. Levine, R. D., Bernstein, R. B. 1974. Molecular Reaction Dynamics, Oxford: Clarendon Press.

15. Alhassid, Y., Levine, R. D. 1978. Phys. Rev. In press.

16. Hancock, G., Morley, C., Smith, I.W.M. 1971. Chem. Phys. Lett. 12: 193-198.

17. Tsuchiya, S., Nielsen, N. Bauer, S. H. 1973. J. Phys. Chem. 77: 2455-2463.

18. Procaccia, I., Shimoni, Y., Levine, R. D. 1976. J. Chem. Phys. 65: 3284-3301.

19. Levine, R. D. 1976. J. Chem. Phys. 65: 3302-3315.

20. Kafri, O., Ben-Shaul, A., Levine, R. D. 1975. Chem. Phys. 10: 367-392.

21. Tribus, M. 1969. Rational Descriptions, Decisions and Design. Oxford: Pergamon Press.

22. Boltzmann, L. 1964. Lectures on Gas Theory. Berkeley: Univ. of California.

23. Cox, R. T. 1946. Am. J. Phys. 14: 1-14

24. Good, I. J. 1950. Probability and the Weighing of Evidence. London: Griffin.

25. Levine, R. D., Kinsey, J. L. 1978. Atom-Molecule Collision Theory: A Guide for the Experimentalist ed: R. B. Bernstein, New York: Plenum.

26. Fano, U. 1956. Rev. Mod. Phys. 29: 74-94.

27. Levine, R. D., Kosloff, R. 1974. Chem. Phys. Lett. 28: 300-304.

28. Tolman, R. D. 1938. The Principles of Statistical Mechanics, Oxford: Clarendon Press.

29. Theorem 1 in ref. 7.

30. Jaynes, E. T. 1965. Am. J. Phys. 33; 391-398.

31. Kawasaki, K., Gunton, J. D. 1973. Phys. Rev. A8: 2048-64.

32. Grabert, H. 1977. Z. Phys. B27: 95-9.

33. Oppenheim, I., Levine, R. D. 1978.

34. Carruthers, P., Nieto, M.N. 1965. Am. J. Phys. 33: 537-42.

35. Procaccia, I., Levine, R. D. 1975. J. Chem. Phys. 63: 4261-4279.

36. Procaccia, I., Levine, R. D. 1976. J. Chem. Phys. 64: 808-817.

37. Alhassid, Y., Agmon, N., Levine, R. D. 1978. Chem. Phys. Lett. 53: 22-26.

38. Perry, D. S., Polanyi, J. C. 1976. Chem. Phys. 12:419-431.

39. Berry, M. J. 1973. J. Chem. Phys. 59: 6229-6253.

40. Korsch, H. J., Levine, R. D. 1978. Chem. Phys.

41. Collins, L. A., Lane, N. F. 1976. Phys. Rev. A14:1569-75.

42. Wyatt, R. E. 1975. Chem. Phys. Lett. 34: 167-169.

43. Nesbet, R.K. 1976. Chem. Phys. Lett. 42:197-201.

44. Ennen, G., Ottinger, Ch. 1974. Chem. Phys. 3:404-17.

45. Thompson, D. L. 1974. J. Chem. Phys. 60: 455-62.

46. Ogloblin, A. A. 1970. Vth Intl. Con. Nuclear Reactions Induced by Heavy Ions, p. 231. Amsterdam: North Holland; Middleton, R., Rosner, B., Pullen, D. J., Polsky, L. 1968. Phys. Rev. Lett. 20: 118-21.

47. Alhassid, Y. 1978. Ph.D. Thesis. Jerusalem: Hebrew Univ.

48. Faist, M.B., Levine, R. D., Bernstein, R. B. 1977. J. Chem. Phys. 66:511-523.

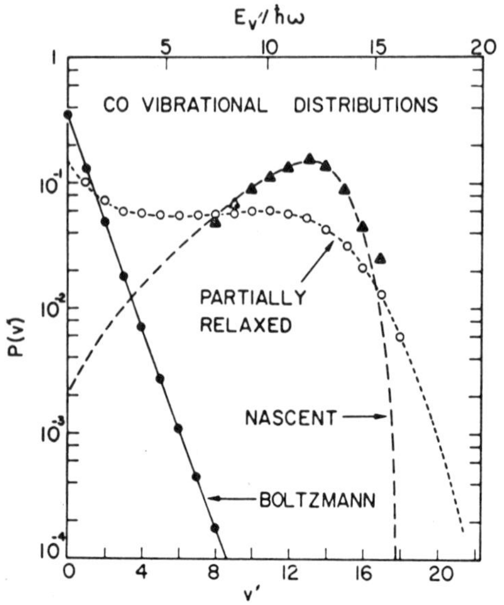

1. Three distribution of CO vibrational states (logarithmic scale) vs. the vibrational energy (top scale) or the vibrational quantum number (bottom scale. The prime denotes a final state distribution.) The nascent distribution is observed (triangles (16)) following an isolated O + CS → S + CO(v') collision. The partially relaxed distribution is observed (open circles (17)) when the same reaction is carried out in the bulk, leading to a partial relaxation of the nascent distribution by secondary collisions. The thermal (Boltzmann) distribution is shown as solid circles. The lines are the distributions determined (4) via the procedure of maximal entropy subject to the appropriate constraints.

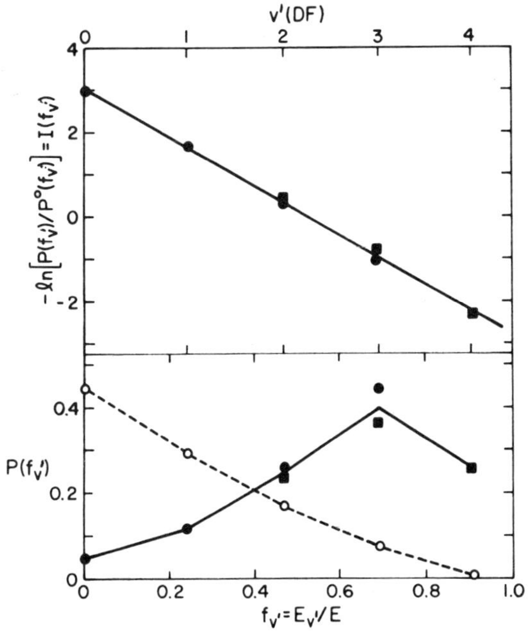

2. Surprisal analysis of the nascent DF vibrational state distribution from the F + $D_2 \rightarrow$D + DF(v′) reaction using reagents in thermal equilibrium (38,39). Bottom panel: Experimental distributions (squares: determined via chemiluminescence (38), circles: determined via a chemical laser (39)) and the prior distribution (open circles), vs. $f_{v'} = E_{v'}/E$ the fraction of total energy E in the DF vibration. (The final quantum number scale is on top). Top panel: the surprisal vs. $f_{v'}$.

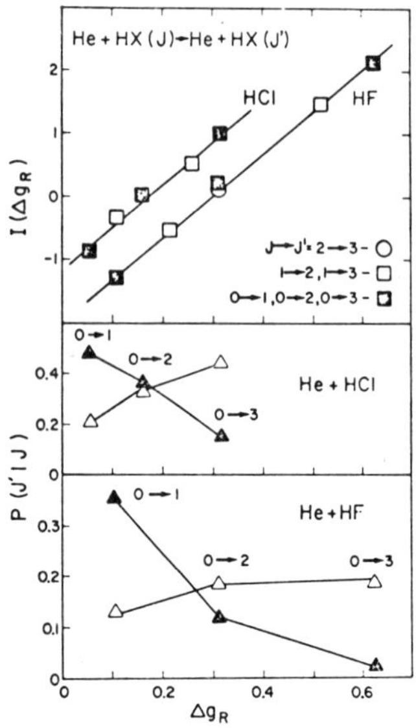

3. Surprisal analysis of final rotational state distributions determined via a numerical solution of the Schrödinger equation for He + HX(j) collisions, at a given total energy (4). Bottom panels: computed ((41), solid triangles) and prior (open triangles) distributions for He + HX (j=0) vs $\Delta g_R=(E_{j'}-E_j)/E$. Top panel: Surprisal vs. Δg_R.

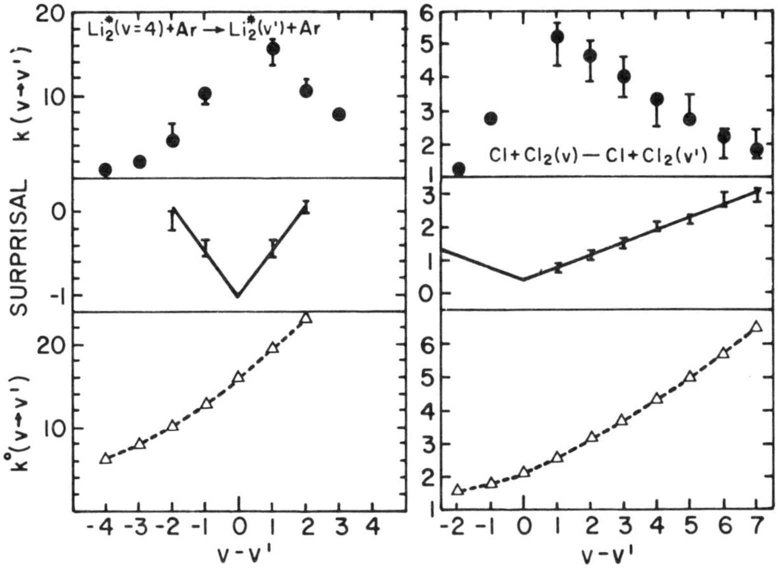

4. Surprisal analysis and surprisal synthesis for vibrational final state distributons ininelastic collisions (35). Bottom panels: prior distributions vs. the change in the vibrational quantum number. top panels: Left-measured ((44), bars) distribution in the Li_2^* (v=4) + Ar → Li_2^* (v′) + Ar collision. Right-classical trajectory computations ((45), bars) for the Cl + Cl_2 (v=7) → Cl + Cl_2(v′) collision. (Degrees of freedom not explicitly selected were in thermal equilibrium) Note that for both examples the distribution is a-symmetric about the elastic (v=v′) peak. The surprisal of the data is shown (as bars) in the mid-panels. Also shown is the surprisal computed (35) using a sum rule (continuous line). From this synthetic surprisal, the final distribution can be determined (using equation 49). The results are shown as dots in the top panels.

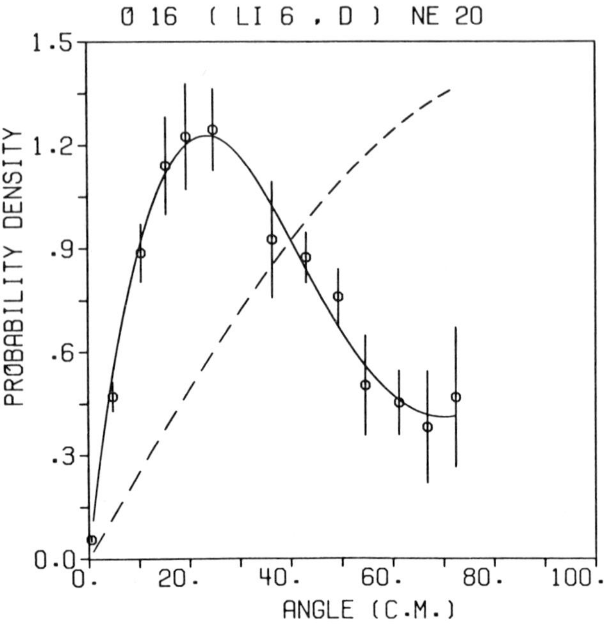

5. The final angular distribution of the products in a nuclear reaction (47). The experimental results ((46), bars) are compared to a fit (continuous line) of the distribution in the form of equation 50, where the constraints are the longitudinal and transverse components of the vector momentum transfer. The prior angular distribution (48) is shown as a dashed line.

THE SPECIAL ROLE OF MAXIMUM ENTROPY IN THE APPLICATION OF "MIXING CHARACTER" TO IRREVERSIBLE PROCESSES IN MACROSCOPIC SYSTEMS

C. Alden Mead, Chemistry Department, University of Minnesota, Minneapolis, Minnesota 55455

I. Introduction: Mixing Character and Mixing Distance

The "Principle of Increasing Mixing Character" was first expounded in a 1975 paper by Ruch,[1] and the idea has been further developed since then by several authors,[2-8] including Lesche, Ruch, Schlögl, and the present author. In a somewhat different form, some of the ideas had been anticipated by Uhlmann.[9] The basic mathematical structure which is involved had been developed originally for a completely different purpose by Ruch and Schönhofer.[10] In work which is now largely forgotten, Muirhead[11] had studied basically the same structure in the early years of the present century.

The principle has been formulated in various ways. Here we will briefly sketch the formulation of ref. 5, not because this is necessarily superior to the others, but because some choice must be made, and this author feels more at home with this formulation. Of course, this formulation was not created out of nothing, but owed much to refs. 1 and 3.

We say that a probability distribution $\{\overline{p}\}$ is more mixed (or has greater mixing character) than $\{p\}$ if it is possible to reach $\{\overline{p}\}$ from $\{p\}$ by some combination of the following two types of steps: (i) "reversible" processes, in which probabilities are merely permuted among the states; and (ii) "two-state irreversible" processes, in which a state of lower probability gains probability at the expense of a more probable state. When this relation holds, it is denoted by

$$\hat{\mathcal{S}}\{\overline{p}\} > \hat{\mathcal{S}}\{p\} . \tag{1}$$

Here $\hat{\mathcal{S}}$ is a symbol for the mixing character; as we shall see, it is not reducible to any single quantity characteristic of the distribution. A necessary and sufficient condition for (1) to hold is that

$$\mathcal{S}_\Phi\{\overline{p}\} \geq \mathcal{S}_\Phi\{p\} \tag{2}$$

for all convex functions $\Phi(x)$ (functions with Φ'' never greater than zero), where

$$\mathcal{S}_\Phi\{p\} = \Sigma_\alpha \Phi(p_\alpha) , \tag{3}$$

with the sum going over all states of a member of the ensemble. If $\Phi(x) = -x \ln x$, then $\mathcal{S}_\Phi$ is just the entropy. For increase in mixing character, however, all the functions $\mathcal{S}_\Phi$ must increase (or at least not decrease). The functions $\mathcal{S}_\Phi$ are called "mixing homomorphic" or "entropy-like" functions. The principle of increasing mixing character states that, if the p's are understood as microcanonical ensemble probabilities, the ensemble members being isolated systems, the mixing character actually does increase in allowed irreversible processes. It has been shown[1,3,5] that this principle holds under rather general conditions. For the type of ensemble considered, basically the same type of "Ansatz" that enables one to prove increase of entropy also leads to increasing mixing character. The principle may be thought of as a stronger form of the second law.

For a system which is not isolated, but is in contact with a heat bath,[3,5] the principle takes the form of "decreasing mixing distance" from the equilibrium distribution:

$$\hat{\mathcal{F}}\{\overline{p}\} < \hat{\mathcal{F}}\{p\}$$

for which it is necessary and sufficient that

$$\mathcal{F}_\Psi\{\overline{p}\} \leqslant \mathcal{F}_\Psi\{p\}$$

for all concave functions $\Psi(x)$, with

$$\mathcal{F}_\Psi\{p\} = \Sigma_\alpha w_\alpha \Psi(p_\alpha/w_\alpha) \; , \tag{4}$$

where the w_α are the equilibrium probabilities. The functions $\mathcal{F}_\Psi$ are called "mixing distance homomorphic" or free-energy-like" functions. It is easy to show[3,5] that decrease of mixing distance for the system of interest is the same as increase of mixing character for the supersystem composed of the system of interest plus the heat bath.

It has also been shown that one need not consider all concave functions in (4); it is sufficient to consider only the "angle" or "hinge" functions

$$\Psi_\lambda(x) = (x-\lambda)h(x-\lambda) \; , \tag{5}$$

where h(x) is the step function, and the corresponding free-energy-like functions

$$\mathcal{F}_\lambda\{p\} = \Sigma_\alpha w_\alpha \Psi_\lambda(p_\alpha/w_\alpha) \; , \tag{6}$$

It is felt that the use of this principle in studying macroscopic irreversible processes has both the strengths and the weaknesses of classical thermodynamics. Its strength lies in its generality, and its independence of assumptions about the detailed dynamics of the system. Its weakness is its inability to yield complete information on what will happen. Like thermodynamics, it only restricts the direction of an irreversible process, though it does restrict the direction more severely.

II. Role of Maximum Entropy

In the approach we are taking, the entropy appears as only one of an infinite number of state functions that are requited to increase in an irreversible process. Nevertheless, if one wishes to apply the principle to experimental observations, it is necessary to assign the entropy a special role.

Experimentally, in observing an irreversible process, one measures certain macroscopic parameters v_1, v_2, ... (denoting the deviation of the system from its equilibrium state) as functions of time, and it is the task of theory to make predictions about their behavior. The principle of increasing mixing character, if it holds, makes predictions about the behavior of the ensemble probabilities. Thus, to use the principle to make predictions based on a measurement of the v's only, one must have some way of going from a knowledge of the v's (interpreted as ensemble averages) to a knowledge of the p's, and this is not unique. Of course, this problem is common to all of irreversible thermodynamics, and is not a special property of our approach.

One's first impulse for getting around this problem is to maximize the mixing character subject to the constraints of the v's having the prescribed average values. Unfortunately, this does not work: there is no distribution which simultaneously maximizes all mixing homomorphic functions subject to this kind of constraint. In this extremity, appropriately in view of the theme of this conference, we are saved by the prescription of Jaynes.[12] In the absence of further knowledge, we choose to be "maximally noncommittal," which means in effect that we maximize the entropy subject to the constraints imposed by our knowledge of the v's. This leads to the "accompanying canonical distribution"

$$p_\alpha = \Xi^{-1} w_\alpha \exp(\gamma \cdot v_\alpha) \qquad (7)$$

where by definition

$$\gamma \cdot v \equiv \Sigma_j \gamma_j v_j$$

and

$$\Xi = \Sigma_\alpha w_\alpha \exp(\gamma \cdot v_\alpha) .$$

In this distribution, the thermodynamic conjugate variables γ_j, of course, are assigned values in such a way as to produce the measured values of the v's as ensemble averages.

For macroscopic systems, it is known that the overwhelming majority of distributions compatible with the measured v's differ only slightly from (7). Thus, if one knows only the v's, one could scarcely base predictions of the future on any other assumption than (7). If the resulting predictions prove to be badly wrong, one is forced to conclude that the v's alone provide an inadequate description, and one measures more variables (perhaps including correlations between v's) until predictions based on (7) with the augmented set of v's are confirmed. As long as (7) is not satisfied, a phenomenological description of the behavior of the system based on the instantaneous v's alone is impossible.

For systems that show a clear separation between short-time irreversible and long-time irreversible behavior, the use of (7) has also been justified[6] by means of a transformation due to Mori.[13] According to this approach, the actual distribution can be considered to be obtainable from (7) by application of a Liouville time-displacement operator for a time which is long on the short-time scale but short on the long-time scale, and to have the same mixing character as (7).

As we shall show in the following sections, the assumption (7) enables one to extract nontrivial predictions about irreversible processes from the requirement of decreasing mixing distance, i.e., from the requirement that all functions (6) must decrease. In the next section, we discuss the so-called η-theorem, which gives the form of this requirement in cases where (7) can be assumed to hold.

III. The η-Theorem

The η-theorem has been previously proved in refs. 5 and 6, but both proofs are slightly more restrictive than necessary. In ref. 5, the result is restricted to "many-parameter" systems, while in ref. 6 it is assumed that the v's are all independent, so that the use of additional v's which are correlations between other v's is ruled out. The proof which we sketch below differs from the earlier ones only in minor details, but is free of these restrictions.

If (7) holds, a small change in the probability distribution is given by

$$\begin{aligned}\delta p_\alpha &= p_\alpha\{v_\alpha\cdot\delta\gamma - \Xi^{-1}\delta\Xi\}\\ &= p_\alpha\{v_\alpha - \Xi^{-1}\Sigma_\beta w_\beta \exp(\gamma\cdot v_\beta)v_\beta\}\cdot\delta\gamma\\ &= p_\alpha(v_\alpha - \langle v\rangle)\cdot\delta\gamma\ ,\end{aligned}\tag{8}$$

where <> denotes an average. With the help of (8), the requirement that all the functions (6) must decrease takes the form

$$\delta\mathcal{F}_\lambda = \Sigma_\alpha h(\frac{p_\alpha}{w_\alpha} - \lambda)p_\alpha(v_\alpha - \langle v\rangle)\cdot\delta\gamma \leqslant 0\ .\tag{9}$$

This must hold for all positive λ. Because of (7), however, we have

$$\begin{aligned}h(\frac{p_\alpha}{w_\alpha} - \lambda) &= h(\gamma\cdot v_\alpha - L)\ ,\\ L &= \ell n(\Xi\lambda)\ .\end{aligned}\tag{10}$$

Inserting (10) into (9), and taking the Fourier transform of the step function, we obtain

$$\delta\mathcal{F}_\lambda = \frac{1}{2\pi i}\Sigma_\alpha\int_{-\infty}^{\infty}\frac{e^{i\kappa(\gamma\cdot v_\alpha - L)}d\kappa}{\kappa}\,p_\alpha(v_\alpha - \langle v\rangle)\cdot\delta\gamma \leqslant 0\ ,\tag{11}$$

where the contour goes slightly below the pole at the origin. Taking the sum over α first, and making use of (7) again, we can transform (11) into

$$\delta\mathcal{F}_\lambda = \frac{\Xi^{-1}}{2\pi i}\int_{-\infty}^{\infty}\frac{\Xi_{i\kappa}e^{-i\kappa L}\left(\langle v\rangle_{i\kappa} - \langle v\rangle\right)\cdot\delta\gamma}{\kappa}\,d\kappa \leqslant 0\ ,\tag{12}$$

where $\Xi_{i\kappa}$ and $\langle v\rangle_{i\kappa}$ are the values that Ξ and <v> would take on if all the γ's were multiplied by a common factor (1+iκ). Note that there is now no singularity at the origin, so the path may be distorted freely.

The part of the integrand in (12) that varies rapidly with κ is

$$\Xi_{i\kappa}e^{-i\kappa L}\ .$$

Accordingly, we seek its saddle point. Setting

$$\Xi_{i\kappa}e^{-i\kappa L} = e^{\sigma},\quad \sigma = \ell n\,\Xi_{i\kappa} - i\kappa L\ ,$$

we have at the saddle point

$$\sigma' = i(\gamma\cdot\langle v\rangle_{i\kappa} - L) = 0 \ ,$$
$$\gamma\cdot\langle v\rangle_{i\kappa} = L \ . \qquad (13)$$

According to (10), L is real and may be positive or negative, extending over the entire possible range of $\langle\gamma\cdot v\rangle$. If we set $\kappa = -i\eta$, with η real, then (13) takes the form

$$\gamma\cdot\langle v\rangle_\eta = L \ , \qquad (14)$$

which will always have a solution for real η. At the saddle point we have for the second derivative

$$\sigma'' = -\Sigma_j\gamma_j(G_{jk})_\eta\gamma k \ ,$$

where

$$G_{jk} = \frac{\partial^2 \ell n\Xi}{\partial\gamma_j\partial\gamma_k} = \langle v_j v_k\rangle - \langle v_j\rangle\langle v_k\rangle \ .$$

In all these expressions, the subscript η means that quantities are calculated from the distribution (7), but with all γ's multiplied by the factor $(1+\eta)$. The second derivative is always negative, and is proportional to the size of the system, becoming infinite in the thermodynamic limit. This part of the integrand therefore behaves like a δ-function near the saddle point, so a path through the saddle point effectively picks out the value of the rest of the integrand at that point. When inherently positive factors are dropped, (12) now becomes

$$\frac{1}{\eta}(\langle v\rangle_\eta - \langle v\rangle)\cdot\delta\gamma \lesssim 0 \ . \qquad (15)$$

This is the so-called η-theorem. It must hold for all η in the "physical range," i.e., the range of η which corresponds, via (10) and (14), to the physically allowed range of $\gamma\cdot v$. In practice, as we shall see, the physical range is the range in which all η-subscripted quantities take on values consistent with their definitions, e.g., temperatures remain positive.

Equation (15) must, as we have said, hold for all values of η in the physical range, so each such value gives a different inequality that must be satisfied. To make the connection with traditional thermodynamics, consider the case where η is very close to zero. For this case we have

$$\langle v_j\rangle_\eta - \langle v_j\rangle = \eta\Sigma_k\gamma_k G_{kj} \ ,$$

so (15) becomes

$$\Sigma_{jk}\gamma_k G_{kj}\delta\gamma_j = \Sigma_n \gamma_m \delta v_k \lessgtr 0 ,$$

which is just the usual requirement of decrease of free energy. The limit $\eta \to 0$ thus reproduces the second law as we know it. By considering other values of η, we obtain the further requirements imposed by decrease of mixing distance. In the applications made so far, it has been found convenient to concentrate on the inequalities obtained by letting η approach the edge of its physical range, as these often turn out to be simple and easy to interpret. In the next section, we consider two such examples.

IV. Two Applications

A. Substance Exchanging Heat and Matter with Surroundings.

Here we follow closely the treatment of Schlögl, ref. 6. Consider a substance exchanging energy (U) and particles (N) with its surroundings. The v's in this case are

$$v_1 = U - U_0 ; \tag{16}$$
$$v_2 = N - N_0 , \tag{17}$$

with the corresponding γ's given by

$$\gamma_1 = -\frac{1}{T} + \frac{1}{T_0} ; \tag{18}$$
$$\gamma_2 = \frac{\mu}{T} - \frac{\mu_0}{T_0} . \tag{19}$$

Here the subscript 0 denotes the equilibrium value in each case. Multiplying (18) by $(1+\eta)$, we can express η in terms of the temperature T_η as follows:

$$-\frac{1}{T_\eta} + \frac{1}{T_0} = -\frac{1}{T} + \frac{1}{T_0} + \eta\gamma_1 ;$$
$$\eta = \left(\frac{1}{T} - \frac{1}{T_\eta}\right)\gamma_1^{-1} . \tag{20}$$

Inserting this into (19), we obtain

$$\left(\frac{\mu}{T}\right)_\eta - \frac{\mu_0}{T_0} = \frac{\mu}{T} - \frac{\mu_0}{T_0} + \eta\gamma_2 ;$$
$$\left(\frac{\mu}{T}\right)_\eta = \frac{\mu}{T} + \left(\frac{1}{T} - \frac{1}{T_\eta}\right)\frac{\gamma_2}{\gamma_1} , \tag{21}$$

where η itself has now been eliminated. We assume that the substance is one for which the chemical potential μ is allowed to take on all values, positive and negative. In this case, the physical range is determined by the requirement that T_η be positive. Equation (15) becomes

$$\frac{1}{\eta}[(U_\eta-U)\delta\gamma_1 + (N_\eta-N)\delta\gamma_2] \lesssim 0 \ . \tag{22}$$

The simplest requirements are obtained by considering the edges of the physical range, i.e., $T_\eta \to \infty$, $T_\eta \to 0$. For the first of these, we have

$$\eta^{-1} \approx - T_\eta \gamma_1 \ ,$$
$$U_\eta >> U, \ |U_\eta \delta\gamma_1| >> |(N_\eta-N)\delta\gamma_2| \ ,$$

so (22) simply becomes

$$\text{sgn}\ \delta\gamma_1 = - \text{sgn}\ \gamma_1 \ . \tag{23}$$

For $T_\eta \to 0$, we must distinguish two cases, according to whether (γ_2/γ_1) is positive or negative. If it is positive, both U_η and N_η go to zero, and we have

$$\text{sgn}(U\delta\gamma_1+N\delta\gamma_2) = - \text{sgn}\,\gamma_1 = - \text{sgn}\,\gamma_2 \ . \tag{24}$$

If the ratio (γ_2/γ_1) is negative, both U_η and N_η will in general be finite. A simple special case is when $U_\eta = 0$, while N_η takes on a saturation value N_s, independent of the γ's. This is the case, for instance, if the particles are hard spheres. For this case, we have

$$\text{sgn}[U\delta\gamma_1-(N_s-N)\delta\gamma_2] = - \text{sgn}\,\gamma_1 = \text{sgn}\,\gamma_2 \ . \tag{25}$$

The restrictions imposed on the motion of the system of the system by (23-25) are illustrated graphically in figure 1, where at each point the motion must be within the angle described by the two arrows.

B. Concentration and Correlation in Simple Landau Model.

This work is contained in a recently-completed Ph.D. thesis of my student Sin-Chong Park.[14] It is an example of a case where a quantity is considered along with correlations involving the same quantity. In the simple Landau model for a substance well above its critical point, the functional whose maximum determines the most probable concentration distribution and correlation function is

$$\Lambda = \int\{-\tfrac{1}{2}ac^2(r) - b[\nabla c(r)]^2 + \gamma(r)c(r)\}d^3r + \tfrac{1}{2}\iint c(r)\Gamma(r,r')c(r')d^3rd^3r', \tag{26}$$

where c is local concentration, with the equilibrium uniform concentration subtracted off, γ is its thermodynamic conjugate, and Γ is the thermodynamic conjugate of the two-point correlation function. a and b are positive constants. Maximizing (26), we obtain

$$(a-b\nabla^2)c(r) = \gamma(r) + \int\Gamma(r,r')c(r')d^3r' ,$$

or in Fourier-transformed representation

$$[a + bk^2 - \Gamma(k)]c(k) = \gamma(k) . \tag{27}$$

Solving, we obtain for c(k)

$$c(k) = \frac{\gamma(k)}{a + bk^2 - \Gamma(k)} , \tag{28}$$

and for the correlation function G(k):

$$G(k) = \frac{\partial c(k)}{\partial\gamma(k)} = \frac{1}{a + bk^2 - \Gamma(k)} . \tag{29}$$

In this problem our v's are the c(k), G(k), and the γ's are $\gamma(k)$, $\Gamma(k)$. When the $\gamma(k)$ and $\Gamma(k)$ are zero, we have the equilibrium situation: c = 0 (uniform concentration) and

$$G(k) = G_0(k) \equiv (a+bk^2)^{-1} . \tag{30}$$

Starting from a nonequilibrium situation, the time development of c describes diffusion, while that of G represents the fact that the correlation function may also change with time. As preparation for examining the consequences of (15) for this case, we need to go through a few manipulations. From (29) and (30), we have

$$\Gamma = G_0^{-1} - G^{-1} , \tag{31}$$

where it is to be understood from now on that the quantities listed are functions of k. With the aid of (29), (30), (31), and some manipulation, we find

$$G_\eta = \frac{1}{G_0^{-1} - (1+\eta)\Gamma} = \frac{GB}{B - \eta} , \tag{32}$$

where

$$B = \frac{G_0}{G - G_0} \quad . \tag{33}$$

From (32), we also find

$$G_\eta - G = \frac{G_\eta}{B - \eta} \quad . \tag{34}$$

From (28) and (32), we obtain for c_η:

$$c_\eta = (1+\eta)\gamma G_\eta = (1+\eta)c \frac{G_0}{G} = \frac{(1+\eta)cB}{B - \eta} \quad ; \tag{35}$$

$$c_\eta - c = \frac{\eta(1+B)c}{B - \eta} \quad . \tag{36}$$

For the differentials of γ and Γ we have

$$d\Gamma = dG/G^2 \quad ; \tag{37}$$

$$d\gamma = \frac{dc}{G} - \frac{cdG}{G^2} \quad , \tag{38}$$

Inserting (34), (36), (37), and (38) into (15), we obtain

$$\int \frac{k^2 dk}{B - \eta} \left\{ \left[1 - \frac{(1+B)c^2}{G} \right] dG + (1+B)cdc \right\} \lessgtr 0 \quad . \tag{39}$$

The physical range of η is determined by the fact that G_η must be positive everywhere. In regions where B is positive, $G > G_0$, this means that η must be less than the minimum value of B, which occurs where $B^{-1} = (G-G_0)/G$ has its maximum. Similarly, if B takes on negative values for any range of k, η is not permitted to take on negative values greater in absolute value than the smallest absolute negative value of B. Summarizing, if B takes on both positive and negative values, then the physical range of η is determined by

$$(B-)_{max} < \eta < (B+)_{min} \tag{40}$$

where $(B-)_{max}$ and $(B+)_{min}$ are respectively the largest (least negative) negative value and the smallest positive value taken on by B. Note that B does not pass through zero when it changes sign, but has a singularity. If B is positive everywhere, or negative everywhere, there is only one bound to the physical range.

In this case too, it turns out that one obtains simple results by considering inequalities obtained when η approaches the edge of its physical range. To see how this happens in this case, let $\eta = (B+)_{min} - \varepsilon^2$, where ε will be allowed to become arbitrarily small, and consider the region near the minimum of positive B, assumed to occur at $k = k_m$, where B can be approximated by

$$B = (B+)_{min} + \alpha^2 q^2 ,$$

where $q = k - k_m$. In this region, the factor $(B-\eta)^{-1}$ in the integrand of (39) behaves like

$$\frac{1}{\alpha^2 q^2 + \varepsilon^2} \approx \frac{\pi}{\alpha\varepsilon}\,\delta(q) .$$

For sufficiently small ε, therefore, the integral (39) is dominated by the region arbitrarily close to k_m, and the inequality becomes:

$$\left\{1 - \frac{(1+B_m)c_m^2}{G_m}\right\}\dot{G}_m + (1+B_m)c_m\delta c_m < 0 , \tag{41}$$

where the subscript m here means that the quantities are evaluated for $k = k_m$. In the special case $c = 0$, where only the correlations are changing with time, this means that the correlation must always be decreasing at the point where its positive deviation from equilibrium, as measured by $(G-G_0)/G_0$, is greatest. Similar considerations apply to the region around $(B-)_{max}$. The situation for $c = 0$ is illustrated graphically in figure 2, where the curve showing deviation from equilibrium as a function of k is shown being squeezed toward the equilibrium value from both above and below, as if by a vise. It can also be shown[14] that consideration of the quartic terms in Λ, omitted in (26), does not change this basic conclusion.

V. Discussion

The difference between the consequences of mixing character and entropy increase may be visualized as follows: If we have a system described by n macroscopic observables, then a small change in the system is denoted by a vector in an n-dimensional space. The requirement of entropy increase singles out one direction along which the component of this vector must be positive, i.e., it limits the vector to a half-space. If mixing character is also required to increase, each inequality limits this motion vector to a different half-space, and it is finally limited to the intersection of all of them. Thus, its direction is much more closely specified by mixing character than by entropy alone.

We have seen that the mixing character principle, together with the Ansatz (7), leads in simple systems to simple, easily interpreted results. If (7) is not satisfied in the course of an irreversible process, how would we expect the experimental results to differ from the predictions of (15)? I don't know the answer to this for sure, but one can speculate as follows: Suppose we create a nonequilibrium state by imposing external forces which are then suddenly turned off, or by a T-jump or similar method. In this case, (7) will be obeyed, at least initially. Suppose that the irreversible process fails to obey (7), so that after a short time the mixing distance has decreased, but (7) is no longer obeyed, and therefore also not (15). Imagine that we now reimpose external forces, so that the v's are held at their new values. This means that (7) is imposed again, with new γ's. The mixing distance relative to the equilibrium state without the external forces will not necessarily decrease further in this process, but the free energy will certainly decrease, and we have seen that this corresponds to (15) with $\eta = 0$. It seems reasonable to assume, therefore, that (15) will continue to hold for a range of η around zero, but not necessarily for large η. The effects of deviation from (7), therefore, are probably to be sought in failure of (15) for large values of η.

A final word about the relation between mixing character and entropy, from the point of view of the present work: The mixing character principle, if true, gives considerably more detailed information about the physical behavior of macroscopic systems than does the entropy principle. There thus exists the possibility that the entropy principle, considered as a statement about the direction of irreversible processes, may be supplanted by the mixing character principle, for which it is necessary but not sufficient. Considered as a prescription for how to be maximally noncommittal in the absence of complete information, however, the entropy principle retains its unique status, and seems indeed to be indispensable for applications of the mixing character principle. Thus, this work seems to confirm once again the wisdom of the Jaynes approach.

Acknowledgement
Acknowledgement is made to the donors of the Petroleum Research Fund, administered by the American Chemical Society, for support of this research.

Bibliography

1. E. Ruch, Theor. Chim. Acta 38, 167 (1975).

2. B. Lesche, J. Math. Phys. 17, 427 (1976).

3. E. Ruch and A. Mead, Theor. Chim. Acta 41, 95 (1976).

4. F. Schlögl, Z. Physik B25, 411 (1976).

5. C. A. Mead, J. Chem. Phys. 66, 459 (1977).

6. F. Schlögl and C. A. Mead, Ann. Phys. (N.Y.) (in press).

7. E. Ruch, R. Schranner, and T. H. Seligman, J. Chem. Phys. (in press).

8. E. Ruch and B. Lesche, J. Chem. Phys. (in press).

9. A. Uhlmann, Wiss. Z. Karl-Marx-Univ. Leipzig Math. Naturwiss. Reihe 21, 421 (1972); 22, 139 (1973).

10. E. Ruch and A. Schönhofer, Theor. Chim. Acta 19, 225 (1970).

11. R. F. Muirhead, Proc. Edinburgh Math. Soc. 19, 36 (1901); 21, 144 (1903); 24, 45 (1906).

12. E. T. Jaynes, Phys. Rev. 106, 620 (1957).

13. H. Mori, J. Phys. Soc. Japan 11, 1029 (1956).

14. S. C. Park, Thesis, University of Minnesota, 1978.

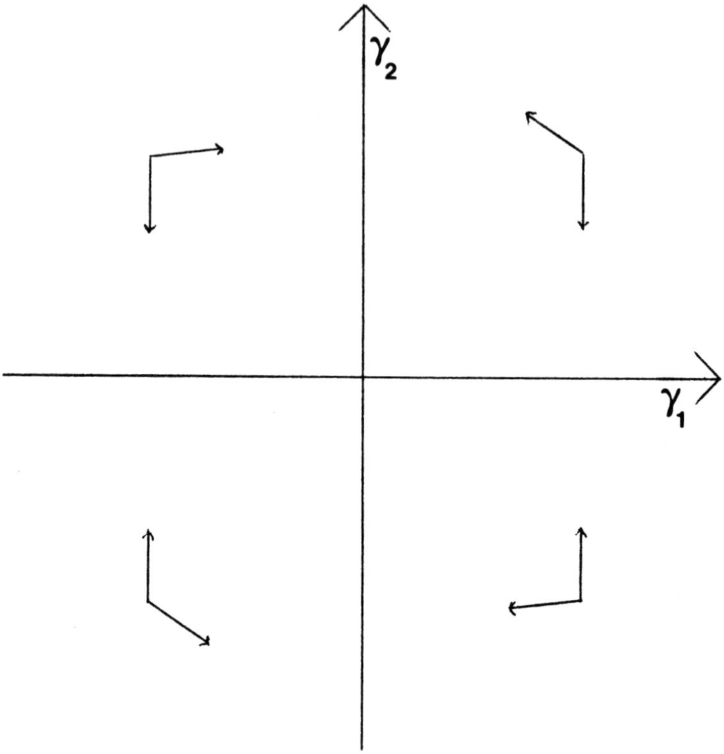

Figure 1: Restrictions imposed on heat and matter exchange by requirement of decreasing mixing distance. In each region, motion is restricted to directions within the angle formed by the arrows.

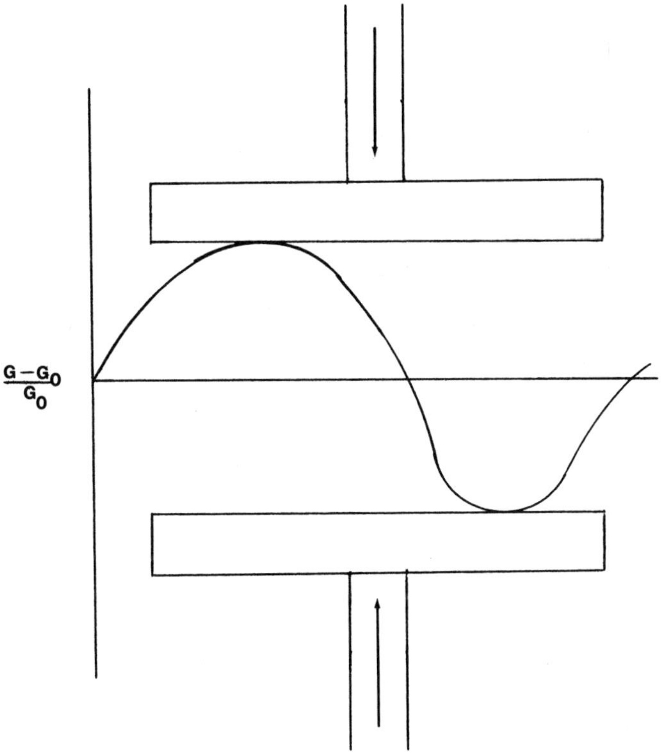

Figure 2: Time development of correlation function G(k) as determined by requirement of decreasing mixing distance. The curve is squeezed toward its equilibrium form from above and below, as if by a vise.

APPLICATION OF MAXIMUM ENTROPY TO NONEQUILIBRIUM STATISTICAL MECHANICS

Baldwin Robertson

This paper reviews the use of the maximum entropy formalism (MEF) on an isolated system that may be arbitrarily far from thermodynamic equilibrium. Information for one instant of time is collected because it gives the entire equilibrium thermodynamic formalism for a system far from equilibrium. In particular, the MEF entropy is then a suitable expression for the thermodynamic entropy for a nonequilibrium system. The MEF yields a general expression for a projection operator. This is used with the Liouville equation to derive an exact expression for the nonequilibrium statistical density. This immediately gives exact closed equations of motion for the time dependence of the expectations that appear as constraints in the MEF. The equations are integrodifferential equations whose kernels are generalizations of rate-rate time-correlation functions. The equations are also written using flux operators satisfying conservation equations, and the memory-retaining nonlocal generalizations of the equations of nonequilibrium thermodynamics are obtained. The kernels in these equations are flux-flux time-correlation functions. The appearance of the projection operator in the correlation functions has important consequences for their asymptotic time dependence. References to specific applications of the formalism and to related work are given.

1. Introduction

The equations of motion to be derived are for the expectations of a set of dynamical variables of interest. In Sec. 2 these expectations are defined in terms of the statistical density that satisfies the Liouville equation. The latter equation is written for both classical[1] and quantum[2] statistical mechanics in terms of the (classical[3] or quantum[4,5]) Liouville operator. This permits the theory to have the same form for classical and quantum statistical mechanics.

The MEF[6-9] is set up in Sec. 3 for a nonequilibrium system using the values of the expectations at one instant of time as constraints. This use of the MEF on systems possibly far from equilibrium[10,11] is implicit in Jaynes's publications,[6-8] which motivated the work to be reviewed. Jaynes's ideas have also been expanded in this direction independently by Schwegler,[12] and, earlier, applied to a nonequilibrium quantum gas by Landsberg.[13] The same use of equilibrium-like formulas was made much earlier by Elsasser[14] and by Bergman and Thompson[15] and also by Richardson[16] although they started from a different point of view than Jaynes.

The equilibrium-like thermodynamic formulas resulting from the MEF are not reproduced here because the present audience may be assumed to be familiar with their derivation.[6-8] The only difference in using the MEF is the present emphasis that the dynamical variables appearing as constraints are not all constants of the motion and that the system is not in equilibrium.[11]

The MEF is much more general than the local equilibrium formalism used e.g. by Mori[17] and McLennan.[18] They make the assumption that the exact nonequilibrium statistical density is in some sense approximately equal to the local equilibrium one.

The formalism in the present paper does not make this assumption; the dynamical variables used in the constraints for the MEF need not be approximate local constants of the motion as assumed for local equilibrium. No assumption is involved here since the equations of motion to be obtained follow exactly from the Liouville equation.

Zubarev and Kalashnikov[19] also use the MEF on a nonequilibrium system with information collected at one instant. Unfortunately, they do not distinguish this from the less general local equilibrium. The book by Zubarev[20] summarizes this work, and the book by Hobson[21] also discusses the MEF for nonequilibrium systems. Books on the MEF and equilibrium systems are by Tribus,[22,23] Baierlein,[24] and Katz.[25]

Also in Sec. 3 the MEF using information collected at one instant is compared with the MEF using information collected over a prior interval of time. The latter theory for a nonequilibrium system was initiated by Jaynes and Scalapino[26] and developed independently by Zubarev and Kalashnikov.[19,28]

In Sec. 4, the MEF density is shown to be a homogeneous functional of the expectations used as constraints.[29]

In Sec. 5, the MEF is used to obtain an explicit expression for a generalization of the Nakajima-Zwanzig[5,30] projection operator. This operator is given for an arbitrary time-dependent MEF density and is exact for systems arbitrarily far from equilibrium. It is used to derive an exact expression for the nonequilibrium statistical density that satisfies the Liouville equation. This statistical density, unlike that of Zubarev[31] and that of Zubarev and Kalashnikov,[32] does not require introducing a convergence factor to select a retarded part and does not require the volume to be infinite. The relation between all these formalisms[11,31,32] and also the time-interval MEF[19,28] is discussed in detail by Zubarev and Kalashnikov.[33]

In Sec. 6 the nonequilibrium statistical density is used to obtain closed equations of motion for the expectations. In Sec. 7, the equations are rewritten using flux operators satisfying conservation equations, and the memory retaining nonlocal generalization of the equations of nonequilibrium thermodynamics are obtained.[34] These equations contain generalizations of the Green-Kubo[35,4] correlation function expressions for transport coefficients. Expressions for these (generalized) correlation functions are given in terms of the MEF density.

In Secs. 8 and 9 formulas are derived that are useful in simplifying the expressions for the correlation functions, and a time-dependent generalization[36] of Mori's projection operator[37] is obtained from the operator of Sec. 5. This projection operator and the equations of motion of Sec. 6 reduce to Mori's in a near-total-equilibrium approximation.[11] However, the present formalism is more useful and relevant than Mori's approximate one, not only because it is exact, but also because it contains the physically important thermodynamic conjugate variables (the MEF multipliers), which do not appear in Mori's formalism. Also, all calculations use the MEF density appropriate for a system that may be far from equilibrium, rather than using an equilibrium statistical density as in Mori's formalism. This is physically desirable, and it permits including all nonlinearities exactly.

In Sec. 10, the connection with the Zwanzig projection operator[30] is established, and the projection operators used by Kawasaki and Gunton[38,39] and by Grabert[40] are shown to be identical to those of Secs. 5 and 9.

The projection operator is an important feature of the formalism in that it repeatedly projects out the expected motion of the dynamical variables. This has the following three consequences. First, it automatically accomplishes a subtraction of fluxes.[41,42] Second, the order in which space and time integrals of the correlation functions are carried out is irrelevant, and the space integral need not be over an infinite volume; neither property is true for the Green-Kubo expressions.[43] Third, the asymptotic time dependence of the correlation functions can not be predicted using a Markovian approximation for the equations of motion containing these correlation functions, as has been asserted for the Green-Kubo correlation functions.[44] These consequences are discussed in Sec. 11.

The formalism reviewed in this paper has been applied to the derivation of an energy transport equation.[45] The MEF used in this derivation provides a natural definition of temperature for a nonequilibrium system, and the usual[35] expression for the heat conductivity comes out without, however, requiring any of the usual assumptions described e.g. in the reviews by Chester[46] and Zwanzig.[27]

The formalism has also been applied by a number of authors to the derivation of equations of motion for a variety of physical systems. Piccirelli[47] has applied it to derive generalized hydrodynamic equations from classical statistical mechanics. A lucid treatment of similar work appears in a paper given at this conference by Oppenheim in collaboration with Levine, et al.[48] They, as well as Grabert,[40] use the same projection operator to derive a generalized Langevin equation and use it to obtain the same equations of motion. The formalism has also been applied to derive generalized hydrodynamic equations from quantum statistical mechanics.[49]

It has been applied to nonlinear shear and normal stress effects by Kawasaki and Gunton,[43] and to obtain the acoustic propagation constant at high frequency by Piccirelli and Im.[50] Taylor and Evans have applied it to derive two fluid equations for superfluid helium,[51] and Noethe and Taylor have applied it to derive hydrodynamic equations for ^{3}He - He II mixtures.[52] Badiali and Gabrielli have applied it to obtain equations for an electrochemical system.[53]

It has been applied to derive a kinetic equation by Ochiai[54] and to open systems using Mitchell's concept of thermal driving.[55]

It has been applied to derive magnetic relaxation equations for nuclear magnetic resonance,[56] for magnetoacoustic resonance by Khenner,[57] for linear response of a system in a strongly driven state and double resonance by Rebsch,[58] and for longitudinal relaxation in an isolated nuclear spin system by Kadyrov and Shaposhnikov.[59]

Closely related developments have also been given by a number of authors.[60-66]

2. Statement Of Problem

Let $F_1(\vec{r}), \cdots, F_m(\vec{r})$ be the time independent dynamical variables whose expectations correspond to the variables that are observed or controlled in the experiment considered. Here $\vec{r}$ denotes position in space. In classical statistical mechanics the dynamical variables are functions of phase, i.e., the positions and momenta of all of the molecules. In quantum statistical mechanics they are linear Hermitian operators that in general do not commute with each other. If there is no spatial dependence in the problem, then the dynamical variables are denoted by $F_1, \cdots, F_m$. In order to have a compact notation, let F stand for a column vector whose elements are $F_1(\vec{r}), \cdots, F_m(\vec{r})$ or in the spatially independent case $F_1, \cdots, F_m$.

The expectation of F is

$$\langle F\rangle_t = \mathrm{Tr}[F\rho(t)], \tag{2.1}$$

where $\rho(t)$ is the statistical density. In classical statistical mechanics, Tr denotes an integration over all phase space, and in quantum statistical mechanics, it denotes a sum of the diagonal elements of a matrix representation of the operators to its right. The normalization of $\rho(t)$ is

$$\mathrm{Tr}[\rho(t)] = 1. \tag{2.2}$$

Although $\langle F\rangle_t$ is defined above in terms of $\rho(t)$, it is ultimately to be calculated by solving the equations of motion to be derived.

The time dependence arises from the Liouville equation[1,2]

$$\partial\rho(t)/\partial t = -iL\rho(t), \tag{2.3}$$

where L is the Liouville operator. In classical mechanics,[3]

$$L = i\sum_{s=1}^{3N}\left(\frac{\partial H}{\partial q_s}\frac{\partial}{\partial p_s} - \frac{\partial H}{\partial p_s}\frac{\partial}{\partial q_s}\right), \tag{2.4}$$

so that for any function of phase A,

$$LA = i\{H,A\}, \tag{2.5}$$

where H is the Hamiltonian and { , } is a Poisson bracket. In quantum mechanics,[4,5] for any operator A,

$$LA = \hbar^{-1}[H,A], \tag{2.6}$$

where H is the Hamiltonian operator and [,] is a commutator bracket.

A formal solution to (2.3) is

$$\rho(t) = \exp(-iLt)\rho(0). \tag{2.7}$$

This is of limited usefulness since it does not directly reveal the form the equations for $\langle F \rangle_t$ must take.

Since $\rho(t)$ satisfies the Liouville equation (2.3), the quantity

$$-k\mathrm{Tr}[\rho(t)\ \ell n\ \rho(t)] \tag{2.8}$$

is independent of time t. In the classical case this follows from (2.4) by integration by parts. In the quantum case the noncommutivity of $\rho(t)$ and $\partial\rho(t)/\partial t$ can be circumvented [in proving (2.8) is time independent] by taking the trace of the identity

$$\delta \exp[xA] = \int_0^x dx' \exp[x'A](\delta A)\exp[(x - x')A] \tag{2.9}$$

with $x = 1$ and $A = \ell n\rho(t)$ and by using the cyclic invariance of the trace and then (2.2). Since (2.8) is time independent, it cannot be the thermodynamic entropy of a nonequilibrium system.

A suitable expression for the latter is defined in the next section.

3. MEF for A Nonequilibrium System

The maximum entropy formalism (MEF) used in the derivation of the equations of motion is based on information obtained at one instant only. Thus the constraints are

$$\langle F\rangle_t = \mathrm{Tr}[F\sigma(t)], \tag{3.1}$$

where

$$\sigma(t) = \exp[-\lambda(t)*F] \tag{3.2}$$

is the MEF density. Here the * denotes a sum and integral

$$\lambda(t)*F = \sum_n \int d^3r\lambda_n(\vec{r},t)F_n(\vec{r}), \tag{3.3}$$

and $\lambda(t)$ is a row vector whose elements $\lambda_1(\vec{r},t),\cdots,\lambda_m(\vec{r},t)$ are the multipliers. The latter are the thermodynamic conjugates of the expectations $\langle F_1(\vec{r})\rangle_t$, $\cdots$, $\langle F_n(\vec{r})\rangle_t$. If there is no spatial dependence, the $\vec{r}$ dependence and the integral can be deleted.

The normalization of (3.2) is obtained as follows. In many applications, it turns out that a linear combination of the $F_n(\vec{r})$ equals 1. By letting ζ represent a column vector whose elements are the coefficients of this linear combination, we may express this equality as

$$\zeta * F = 1 \tag{3.4}$$

in the notation of (3.3). There is no loss of generality in requiring this since, if (3.4) is not already true, an additional variable $F_0 \equiv 1$ can always be included in the set $\{F_n(\vec{r})\}$ so that a ζ satisfying (3.4) will exist. Equations (3.4), (2.1), and (2.2) give

$$\langle 1\rangle_t = 1 \tag{3.5}$$

The normalization of (3.2) follows from (3.1), (3.4), and (3.5).

The above formalism defines $\lambda(t)$ as a functional of $\langle F\rangle_t$. This can be seen by inserting (3.2) into (3.1) to get

$$\langle F\rangle_t = \mathrm{Tr}\{F\exp[-\lambda(t)*F]\}. \tag{3.6}$$

The expectation $\langle F\rangle_t$ is defined by (2.1) and hence is considered as a known quantity in (3.6). The thermodynamic conjugate $\lambda(t)$ is obtained by solving the nonlinear equation (3.6) for it in terms of $\langle F\rangle_t$.

As is well known the above MEF leads to the entire formalism of equilibrium thermodynamics when F consists of just the constants of the motion. For the following, however, F may contain any dynamical variables whatsoever, and the above MEF is to be used no matter how far from equilibrium the system is driven. Thus the entire equilibrium thermodynamic formalism is extended for use on a nonequilibrium system.

In order to ensure that the above MEF includes equilibrium thermodynamics as a special case, a linear combination of the $F_n(\vec{r})$ must equal the Hamiltonian H. In the notation of (3.3), this requires that there exist an η such that

$$\eta * F = H. \tag{3.7}$$

[If other macroscopic constants of the motion exist, there should be additional equations like (3.7) but with the constant of the motion to the right, and the following should be modified accordingly.] Again there is no loss of generality in requiring this since, if (3.2) is not already true, an additional variable $F_1 \equiv H$ can always be included in the set $\{F_n(\vec{r})\}$ so that an η satisfying (3.7) will exist. Thus when $\lambda(t)$ reduces to a linear combination of ζ and η, the MEF density (3.2) will reduce to the Gibbs canonical density, and equilibrium thermodynamics will result.

In order to simplify the equations of motion to be derived, let the initial condition $\rho(0)$ be the Gibbs canonical density. Then $\langle F \rangle_0$, calculated from (2.1) and inserted into (3.6), will give $\lambda(0)$ so that

$$\sigma(0) = \rho(0). \tag{3.8}$$

This is not an unduly restrictive initial condition since most systems on which nonequilibrium experiments are performed are in equilibrium at one time or another, and one such time is denoted here as $t = 0$.

Also, if the system is disturbed from equilibrium for $t>0$ and then settles down to equilibrium again, $\sigma(t)$ will initially differ from a canonical density for $t>0$ and then settle down to a new canonical density different from (3.8). The latter can be proved with an argument similar to the one for $t = 0$. The statement about $\sigma(t)$ of course is not true for $\rho(t)$, which unlike $\sigma(t)$ must satisfy the Liouville equation.

The maximum entropy for the constraint (3.1) is

$$S(t) = -k \, \mathrm{Tr}[\sigma(t) \ln \sigma(t)]. \tag{3.9}$$

If the (isolated) system is suddenly driven from equilibrium at $t = 0$ and then relaxes toward equilibrium again with H independent of time for $t > 0$, then $S(t)$ will approach the Gibbs entropy for the new equilibrium as $t \to \infty$. This occurs because $\sigma(t)$ becomes canonical for $t \to \infty$. The approach of $S(t)$ to its new equilibrium value $S(\infty)$ may not be monotonic, but $S(t)$ will satisfy the inequalities

$$S(0) \leq S(t) \leq S(\infty). \tag{3.10}$$

The first inequality follows from $-x \ln x \leq 1-x$ for $x > 0$ and[10] from the time independence of (2.8); the second is true because $S(\infty)$ is the maximum entropy with just constants of the motion as constraints while $S(t)$ is the maximum entropy with the more restrictive constraints (3.1).

The above MEF is to be contrasted with the MEF with information collected over a time interval.[26,19,28] For the latter, the MEF density equals $\rho(t)$. Hence the maximum entropy is given by (2.8), which is constant and so is not suitable for a nonequilibrium system. Thus the MEF with information obtained at one instant has the following advantages: It gives a simple equilibrium-like formalism for a nonequilibrium system; all expressions depend only on present values of the thermodynamic coordinates $\langle F \rangle_t$ and $\lambda(t)$. Because of this the thermodynamic coordinates reduce to the correct equilibrium values whenever the system settles down to equilibrium. Also, the one-instant

MEF leads to an explicit general expression for the projection operator, and a generalization of nonequilibrium thermodynamics results, as will be seen.

4. $\sigma(t)$ A Homogeneous Function Of $\langle F\rangle_t$.

In (3.2) the MEF density $\sigma(t)$ is expressed as a functional of $\lambda(t)$, which is defined in (3.6) as a functional of $\langle F\rangle_t$. Hence $\sigma(t)$ is a functional of $\langle F\rangle_t$ denoted here by $\sigma[\langle F\rangle_t]$. This functional is homogeneous of degree 1 as will be shown in this section.

If there exists a ξ such that

$$\xi * F = 0, \tag{4.1}$$

then the solution to (3.6) for $\lambda(t)$ in terms of $\langle F\rangle_t$ will not be unique. This is true since adding a constant times ξ to any solution for $\lambda(t)$ yields a different solution. Hence it is necessary to restrict F such that there does not exist a ξ satisfying (4.1). Then $\lambda(t)$ will be a unique functional of $\langle F\rangle_t$ denoted here by $\lambda[\langle F\rangle_t]$.[55]

Multiply (3.6) by a positive real number a, take the factor a into the exponent as an added term ℓn a, and use (3.4) to get

$$\lambda[a\langle F\rangle_t] = \lambda[\langle F\rangle_t] - \zeta \ \ell n \ a. \tag{4.2}$$

Insert (4.2) into (3.2) and use (3.4) to get

$$\sigma[a\langle F\rangle_t] = a \ \sigma[\langle F\rangle_t]. \tag{4.3}$$

This states that $\sigma[\langle F\rangle_t]$ is a homogeneous functional of $\langle F\rangle_t$ of degree 1. This result is a direct consequence of the restriction (3.4).

The homogeneity of $\sigma[\langle F\rangle_t]$ can be used to obtain an expression involving its functional derivative. Differentiate (4.3) with respect to a and set a = 1 to get

$$\frac{\delta\sigma(t)}{\delta\langle F\rangle_t} * \langle F\rangle_t = \sigma(t). \tag{4.4}$$

This result will be used in Sec. 10.

5. Nonequilibrium Statistical Density

An expression for the nonequilibrium statistical density $\rho(t)$ in terms of the MEF density $\sigma(t)$ can be obtained by starting with the time derivative of $\sigma[\langle F\rangle_t]$. This can be calculated with the chain rule for a total derivative and expressed concisely as

$$\frac{\partial\sigma(t)}{\partial t} = P(t)\frac{\partial\rho(t)}{\partial t}, \tag{5.1}$$

where P(t) is a linear operator defined by

$$P(t)A = \frac{\delta\sigma(t)}{\delta\langle F\rangle_t} * \mathrm{Tr}(FA). \tag{5.2}$$

Here A is an arbitrary function of phase (for classical statistical mechanics) or an arbitrary operator (for quantum statistical mechanics).

The expression for $\rho(t)$ is obtained by use of an integrating factor T(t,t') defined to be the solution to

$$\partial T(t,t')/\partial t' = T(t,t')[1-P(t')]iL, \tag{5.3}$$

with the initial condition

$$T(t,t) = 1. \tag{5.4}$$

Since both P(t) and L are linear operators, so is T(t,t'), and since P(t) is a functional of $\langle F \rangle_t$, so is T(t,t'). Equations (5.3), (5.1), and (2.3) give

$$\partial\{T(t,t')[\rho(t')-\sigma(t')]\}/\partial t' = -T(t,t')[1-P(t')]iL\sigma(t'). \tag{5.5}$$

Integrate this over t' from 0 to t and use the initial conditions (5.4) and (3.8) to get

$$\rho(t) = \sigma(t) - \int_0^t dt' T(t,t')[1-P(t')]iL\sigma(t'). \tag{5.6}$$

This is the desired expression for the nonequilibrium statistical density $\rho(t)$ as a functional of $\langle F \rangle_t$.

This exact result is superior to Zubarev's nonequilibrium statistical operator[19,20,28,31-33] because the latter requires an ad hoc convergence factor $\exp(\varepsilon t)$ to select the retarded part of an integrand and because his expression is not a solution to the Liouville equation until the $\varepsilon \to 0$ limit is taken, which cannot be done until the calculation of the integral is completed and the volume of the system made infinite.

The result (5.6) is also more useful than the solution (2.7) because it has the structure of a reversible term plus an irreversible term on the right and because irrelevant ancient history drops out of the memory integral on the right, while $\rho(t)$ in (2.7) has no apparent structure and obviously always depends explicity on $\rho(0)$. The structure in (5.6) appears also in the equations of motion derived in the next section.

6. Equations Of Motion For $\langle F\rangle_t$

The closed equations for $\langle F\rangle_t$ can be derived by substituting (5.6) for $\rho(t)$ on the right of the Liouville equation (2.3), multiplying the result on the left by F, applying Tr, and using (2.1) to get

$$\partial\langle F\rangle_t/\partial t = -\mathrm{Tr}[FiL\sigma(t)] + \int_0^t dt'\mathrm{Tr}\{FiLT(t,t')[1-P(t')]iL\sigma(t')\}. \tag{6.1}$$

This can be written in a more suggestive way as follows. For arbitrary A and B, the identity

$$\mathrm{Tr}(ALB) = -\mathrm{Tr}[(LA)B] \tag{6.2}$$

follows from (2.4) by an integration by parts or from (2.6) by using the cyclic invariance of the trace. As a result the first term on the right of (6.1) becomes just $\langle\dot{F}\rangle_t$, where the dot is defined by

$$\dot{A} = iLA, \tag{6.3}$$

and the angular brackets are defined by

$$\langle A\rangle_t = \mathrm{Tr}\{A\sigma(t)\}, \tag{6.4}$$

where A is an arbitrary function of phase (for classical statistical mechanics) or an arbitrary operator (for quantum statistical mechanics).

The last term in (6.1) can be transformed using

$$iL\sigma(t) = -\lambda(t) * \bar{\dot{F}}\sigma(t), \tag{6.5}$$

where the dot is defined by (6.3), and the overbar notation is as follows. For classical statistical mechanics, the overbar may be completely ignored, and (6.5) follows from (3.2) because (2.4) is a first derivative operator. For quantum statistical mechanics, the overbar is defined for an arbitrary operator A by

$$\bar{A} = \int_0^1 dx\ \sigma(t)^x A\sigma(t)^{-x}, \tag{6.6}$$

and (6.5) follows by integrating the derivative

$$i\hbar^{-1}d(\sigma^x H\sigma^{1-x})/dx = -\lambda\sigma^x\dot{F}\sigma^{1-x}$$

over x from 0 to 1. Note that if $\sigma(t)$ commutes with A, the integration gives $\bar{A} = A$, which is the classical result.

When (6.2)-(6.6) are applied to the last term of (6.1), that equation becomes

$$\frac{\partial \langle F\rangle_t}{\partial t} = \langle\dot{F}\rangle_t + \int_0^t dt'\ K(t,t')*\lambda(t'), \tag{6.7}$$

with

$$K(t,t') = \langle\dot{F}T(t,t')[1-P(t')]\bar{\dot{F}}'\rangle_{t'}, \tag{6.8}$$

where the prime on $\bar{\dot{F}}'$ indicates that the t in (6.6) for the overbar is primed. Note that because P and T in (6.8) operate to the right, $\sigma(t)$ must be on the right in the Tr in the definition (6.4) of the angular bracket.

The kernel (6.8) is a generalization of an autocorrelation of $\dot{F}$. Both the kernel and $\langle \dot{F} \rangle_t$ are functionals of just $\lambda(t)$ or equivalently $\langle F \rangle_t$. Hence the equations are closed. The first term on the right of (6.7) is a functional of just the present value of $\lambda(t)$ or $\langle F \rangle_t$ and does not by itself change the entropy and hence is called the reversible term. The last term in (6.7) depends on $\lambda(t')$ or $\langle F \rangle_{t'}$, over the interval $0 \leq t' \leq t$ and causes relaxation changing the entropy and hence is called the irreversible term.

Although the equations (6.7), (6.8), and (3.6) are closed in the sense that $\lambda(t)$ and $\langle F \rangle_t$ are the only unknowns, calculation of the kernel (6.8) often is difficult. Application of the formalism requires approximation in this calculation. One method valid for linear deviations from equilibrium, involves a continued fraction expansion.[37] A simple method of truncating this kind of expansion has been used to predict a nuclear magnetic resonance line shape in excellent agreement with experiment.[67]

7. Derivation of Nonequilibrium Thermodynamics

Often the $\dot{F}_n(\vec{r})$ can be written as the divergence of a flux $\vec{J}_n(\vec{r})$,[34, 49]

$$\nabla \cdot \vec{J}_n(\vec{r}) = -\dot{F}_n(\vec{r}), \qquad \vec{r} \text{ inside } R, \tag{7.1}$$

with

$$\vec{J}_n(\vec{r}) = 0, \qquad \vec{r} \text{ outside } R, \tag{7.2}$$

where R is the region of space occupied by the system. Integration of (7.1) using (7.2) and (6.3) shows that this can be done only when

$$L\int F_n(\vec{r})d^3r = 0 \tag{7.3}$$

i.e. when $F_n(\vec{r})$ is the density of a conserved quantity. When the above is true, the equations (6.7) can be rewritten as generalizations of the equations of nonequilibrium thermodynamics.

Insert (7.1) into (6.7) and integrate by parts using (7.2) to get

$$\partial\langle F_n(\vec{r})\rangle_t/\partial t = -\nabla\cdot\langle \vec{J}_n(\vec{r})\rangle_t$$
$$-\nabla\cdot\sum_{n'}\int_0^t dt'\int d^3r'\overleftrightarrow{K}_{nn'}(\vec{r},t,\vec{r}',t')\cdot\nabla'\lambda_{n'}(\vec{r}',t'),$$
$$n = 1, 2, \cdots, m, \tag{7.4}$$

where

$$\overleftrightarrow{K}_{nn'}(\vec{r},t,\vec{r}',t') = \langle \vec{J}_n(\vec{r})T(t,t')[1-P(t')]\overline{\vec{J}_{n'}(\vec{r}')}'\rangle_{t'}. \tag{7.5}$$

Equations (7.4) are the memory-retaining nonlocal generalizations of the equations of nonequilibrium thermodynamics. Here the $\langle\vec{J}_n(\vec{r})\rangle_t$ are the reversible fluxes, the integrals on the right of (7.4) are the irreversible fluxes, the $\nabla\lambda_n(\vec{r},t)$ are the thermodynamic forces, and in a linear approximation the $\overleftrightarrow{K}_{nn'}(\vec{r},t,\vec{r}',t')$ satisfy reciprocity relations.[11]

Provided $\{F_n(\vec{r})\}$ consists of densities of all the conserved quantities, the $\langle F_n(\vec{r})\rangle_t$, and hence their thermodynamic conjugates the $\lambda_n(\vec{r},t)$, may be expected to vary slowly compared with the $\overleftrightarrow{K}_{nn'}(\vec{r},t,\vec{r}',t')$, which will decay to zero as $t-t'$ becomes large. Also if $\nabla\lambda_n(\vec{r},t)$ varies slowly compared with $\overleftrightarrow{K}_{nn'}(\vec{r},t,\vec{r}',t')$ as $|\vec{r}-\vec{r}'|$ becomes large, the $\nabla'\lambda_{n'}(\vec{r}',t')$ in the integral in (7.4) may be replaced with $\nabla\lambda_{n'}(\vec{r},t)$, and (7.4) becomes

$$\partial\langle F_n(\vec{r})\rangle_t/\partial t = -\nabla\cdot\langle\vec{J}_n(\vec{r})\rangle_t$$

$$-\nabla\cdot\sum_{n'}\overset{\Rightarrow}{L}_{nn'}\cdot\nabla\lambda_{n'}(\vec{r},t), \qquad n = 1, 2, \cdots, m, \tag{7.6}$$

where

$$\overset{\Rightarrow}{L}_{nn'} = \int_0^t dt'\int d^3r'\overset{\Rightarrow}{K}_{nn'}(\vec{r},t,\vec{r}',t'). \tag{7.7}$$

Equation (7.6) is the only approximate result in this paper; all other equations are exact. For t much larger than the time required for (7.5) to decay to zero, (7.7) will be constant. Thus the equations (7.6) are the equations of nonequilibrium thermodynamics, and for $t \to \infty$ (7.7) is the Green-Kubo[35,4] expression for the transport coefficients, except that the correlation function contains the operator P(t). This operator yields a number of advantages as will be discussed in the following.

8. Properties of P(t)

The operator P(t) satisfies several relations that are useful in simplifying the equations of motion. These relations are obtained as follows.

Let $A = -\lambda * F$ in the identity (2.9) and apply (3.2) to get

$$\delta\sigma(t)/\delta\lambda(t) = -\bar{F}\sigma(t), \tag{8.1}$$

where the overbar is defined by (6.6). Left multiply this by F, apply Tr, and use (6.4) to get

$$\delta\langle F\rangle_t/\delta\lambda(t) = -\langle F\bar{F}\rangle_t\,. \tag{8.2}$$

Now the identity

$$\langle A\bar{B}\rangle_t = \langle B\bar{A}\rangle_t \tag{8.3}$$

may be verified directly from the definitions (6.4) and (6.6), where A and B are arbitrary (except that they are not column vectors, just their elements). Thus $\langle F\bar{F}\rangle_t$ is a symmetric matrix, and by (8.2) so is $\delta\langle F\rangle_t/\delta\lambda(t)$.

Equations (5.2), (8.1) and (8.2) and the definition of a total derivative give

$$P(t)A = \bar{F}\sigma(t) * \langle F\bar{F}\rangle_t^{-1} * \mathrm{Tr}(FA) \tag{8.4}$$

for arbitrary A. This and (6.4) give

$$P(t)\bar{A}\sigma(t) = \bar{F}\sigma(t) * \langle F\bar{F}\rangle_t^{-1} * \langle F\bar{A}\rangle_t, \tag{8.5}$$

which with (8.1) - (8.3) becomes

$$P(t)\bar{A}\sigma(t) = \bar{F}\sigma(t) * \delta\langle A\rangle_t / \delta\langle F\rangle_t \tag{8.6}$$

provided $\delta A/\delta\langle F\rangle_t = 0$. If F is substituted for A, (8.6) reduces to

$$P(t)\bar{F}\sigma(t) = \bar{F}\sigma(t). \tag{8.7}$$

This may be used to show that the kernel (7.5) is zero whenever the flux $\vec{J}_{n'}(\vec{r}\,')$ on the right of T(t,t') is a linear combination of the dynamical variables $F_n(\vec{r})$. A similar result holds for the kernel (6.8).

Further relations can be obtained as follows. Equations (5.2), and (3.1) and the definitions of the functional derivative give

$$\mathrm{Tr}[FP(t)A] = \mathrm{Tr}(FA), \tag{8.8}$$

which immediately gives

$$P(t)P(t')A = P(t)A, \tag{8.9}$$

where A is arbitrary and t' need not equal t. When this is applied on the left to a series expansion of T(t,t') obtained from (5.3) and (5.4), only the first term survives, giving $P(t)T(t,t') = P(t)$ so that

$$P(t)T(t,t')[1-P(t')] = 0. \tag{8.10}$$

Equations (8.10) and (8.8) may be used to show that the kernel (7.5) is zero whenever the flux $\vec{J}_n(\vec{r})$ on the left of T(t,t') is a linear combination of the dynamical variables $F_n(\vec{r})$. A similar result holds for the kernel (6.8).

9. The Projection Operator $\mathcal{P}(t)$

Because the kernels (6.8) and (7.5) contain the operator P(t), the rates $\dot{F}$ in (6.8) and the fluxes $\dot{\vec{J}}_n(\vec{r})$ in (7.5) appear in subtracted form. This can be shown using a projection operator $\mathcal{P}(t)$ related to the operator P(t) as follows.

The Hermitian conjugate of the operator P(t) is an operator $\mathcal{P}(t)$ defined by

$$\mathrm{Tr}\{[P(t)A]B\} = \mathrm{Tr}[AP(t)B], \tag{9.1}$$

where A and B are arbitrary. Thus the Hermitian conjugate of (5.2) is

$$P(t)A = F * \delta\langle A\rangle_t / \delta\langle F\rangle_t \tag{9.2}$$

as follows from (6.4), while the Hermitian conjugate of (8.4) is

$$P(t)A = F * \langle F\bar{F}\rangle_t^{-1} * \langle F\bar{A}\rangle_t \tag{9.3}$$

as follows from (6.4) and (8.3).

The operator $P(t)$ can be seen to be a projection operator as follows. Either (9.2) or (9.3) immediately give

$$P(t)F = F \tag{9.4}$$

and hence

$$P(t')P(t) = P(t). \tag{9.5}$$

Since $\langle F\bar{F}\rangle_t^{-1}$ is a symmetric matrix, (9.3) and (8.3) give

$$\langle [P(t)A]\bar{B}\rangle_t = \langle A\overline{[P(t)B]}\rangle_t \tag{9.6}$$

Now it follows from (6.6), (6.4), and (3.2) that $\langle A\bar{A}\rangle_t$ is positive for all nontrivial Hermitian A. Hence $\langle A\bar{B}\rangle_t$ in (8.3) satisfies the conditions of an inner product between A and B with respect to the weight $\sigma(t)$. Equation (9.6) states that $P(t)$ is Hermitian with respect to the weight $\sigma(t)$. Thus $P(t)$ is a projection operator with respect to the weight $\sigma(t)$.

This projection operator efficiently accomplishes the subtraction of the rates $\dot{F}$ in the kernel (6.8) and of the fluxes $\vec{J}_n(\vec{r})$ in the kernel (7.5). The subtraction on the left of the $T(t,t')$ in these kernels follows from (8.10) and (9.3). The subtraction on the right of $T(t,t')$ follows from

$$P(t)\bar{A}\sigma(t) = [\overline{\mathcal{P}(t)A}]\sigma(t), \tag{9.7}$$

which can be proved using (8.5) and (9.2). The resulting rate-rate correlation function is

$$K(t,t') = \left\langle \Delta\dot{F}\; T(t,t')\overline{\Delta\dot{F}}' \right\rangle_{t'}, \tag{9.8}$$

and the resulting flux-flux correlation function is

$$\overleftrightarrow{K}_{nn'}(\vec{r},t,\vec{r}',t') = \left\langle \Delta\vec{J}_n(\vec{r})T(t,t')\overline{\Delta\vec{J}_{n'}(\vec{r}')}' \right\rangle_{t'}, \tag{9.9}$$

where the Δ is defined by

$$\Delta A = [1-\mathcal{P}(t)]A \tag{9.10}$$

for arbitrary A. This subtraction is essential for the correlation functions to approach zero for large $t-t'$.[41,42]

10. Comparison With Other Formalisms

The functional equation (4,4) may be combined with the definition (5.2) of $P(t)$ and the definition (2.1) of $\langle F\rangle_t$ to obtain

$$P(t)\rho(t) = \sigma(t). \tag{10.1}$$

Both (5.1) and (10.1) are satisfied, even though P(t) is non-trivially time dependent, because $\sigma(t)$ is homogeneous of degree 1 in $\langle F \rangle_t$, which occurs because a linear combination of the $F_n(\vec{r})$ equals 1.

The result (10.1) shows that the present formalism parallels Zwanzig's,[30] which has a time-independent operator P operating on $\rho(t)$ and projecting out its 'relevant' part. However, Zwanzig does not give a general expression for P or for the relevant part of $\rho(t)$. In the present formalism the relevant part of $\rho(t)$ is the MEF density $\sigma(t)$, and an explicit expression for P(t) is given by (5.2) or (8.4).

Mori[37] does give an explicit expression for a projection operator, but his theory involves an expansion about equilibrium and also does not contain the physically relevant thermodynamic conjugates $\lambda(t)$. The operator (9.3) is a generalization of his in that it is a projection operator with respect to the weight $\sigma(t)$ rather than with respect to an equilibrium density.

The present formalism with F_o chosen to be equal to 1[29] is exactly the same as the formalism of Kawasaki and Gunton.[38,39] The first term that appears in their expression for P(t) (their Eq. 1 or 2.5) is just the term corresponding to $F_o = 1$ in the sum in (5.2). To see this expand the sum explicitly by separating the column vector F into two parts,

$$F = \begin{pmatrix} 1 \\ A \end{pmatrix}, \tag{10.2}$$

where, following their notation, A is a column vector whose elements are the dynamical variables. Insert this into (5.2) to get

$$P(t)X = \frac{\delta\sigma(t)}{\delta\langle 1\rangle_t} \, Tr(X) + \frac{\delta\sigma(t)}{\delta\langle A\rangle_t} * Tr(AX), \tag{10.3}$$

where X is an arbitrary function of phase or quantum mechanical operator. Also, insert (10.2) into (4.4) to get

$$\frac{\delta\sigma(t)}{\delta\langle 1\rangle_t} \langle 1\rangle_t + \frac{\delta\sigma(t)}{\delta\langle A\rangle_t} * \langle A\rangle_t = \sigma(t). \tag{10.4}$$

Eliminate $\delta\sigma(t)/\delta\langle 1\rangle_t$ between the last two equations and use $\langle 1\rangle_t = 1$ to get

$$P(t)X = \left[\sigma(t) - \frac{\delta\sigma(t)}{\delta\langle A\rangle_t} * \langle A\rangle_t\right] Tr(X) + \frac{\delta\sigma(t)}{\delta\langle A\rangle_t} * Tr(AX). \tag{10.5}$$

This is the Kawasaki-Gunton expression, which has been shown to be identical to (5.2). Their formalism is therefore identical to the present one.

The Hermitian conjugate of this is

$$\mathcal{P}(t)X = \langle X\rangle_t - \langle A\rangle_t * \frac{\delta\langle X\rangle_t}{\delta\langle A\rangle_t} + A * \frac{\delta\langle X\rangle_t}{\delta\langle A\rangle_t}, \tag{10.6}$$

which follows from (9.1). This is Grabert's projection operator (his Eqs. 2.8 and 2.9), which has been shown here to be identical to (9.2) or (9.3). The resulting equations of motion (his Eq. 4.11) are identical to (6.1) with the order of the operators in the trace in the integrand reversed.

11. Conclusion

Exact closed equations of motion have been derived for the expectations that appear as constraints in the MEF. These equations are integrodifferential equations whose kernels are generalizations of rate-rate or flux-flux time-correlation functions. The derivation uses a projection operator, which is defined in terms of the MEF density. The presence of this projection operator in the expression for the correlation functions has the following three consequences.

First, it automatically accomplishes a subtraction of the dynamical variables F from the rates or fluxes in the correlation functions (9.8) - (9.10). Without this substraction, the correlation functions would not vanish for large t - t'.[41.42]

Second, the order in which the space and time integrals are carried out in the expression (7.7) for the transport coefficients is irrelevant, and the space integral does not have to be over an infinite volume. This is not true for the Green-Kubo expressions, which do not involve a projection operator and for which an infinite spatial integral must be carried out before the upper time limit can become infinite. That is, the Fourier and Laplace transform of the Green-Kubo correlation function has a singularity at $k = 0$ and $s = 0$, and the limit $k \to 0$ must be carried out before the limit $s \to 0$ in order to obtain a sensible result.[43] No such singularity occurs in the foregoing formalism.

Third, the asymptotic time dependence of the correlation functions (9.8) - (9.9) is not given by the phenomenological equations (7.6) as has been asserted for the Green-Kubo correlation functions.[44] On the contrary, it is given by the difference between the time dependence of the expectations $\langle F\rangle_t$ and that time dependence predicted by the phenomenological equations (7.6). This can be seen by comparing the Markovian equations (7.6) with the integrodifferential equations (7.4). Equations (7.6) can be obtained by substituting the Dirac delta functions

$$\overset{\Rightarrow}{K}_{nn'}(\vec{r},t,\vec{r}',t') = \overset{\Rightarrow}{L}_{nn'}\delta(\vec{r}-\vec{r}')\delta(t-t'+0) \qquad (11.1)$$

for the kernels of (7.4) and carrying out the integrals. This correlation function is zero for $t > t'$ and hence has no long time tail. The absence of that tail is what causes the equations (7.6) to be Markovian. To the extent that the solutions of the exact equations (7.4) agree with the solutions of the Markovian local equations (7.6), the kernel of the former will approximimately equal the delta functions (11.1) and have no long time tail. It is only the difference between these solutions that is associated with the asymptotic time dependence of the correlation functions.

The above mentioned differences between the Green-Kubo correlation functions and the correlation functions (9.8) - (9.9) can arise only because of the presence of the projection operator in the latter. Thus the projection operator not only facilitates the derivation of the exact equations of motion with a desirable form, but it also causes the correlation functions in these equations to have desirable properties.

References

1. J. Liouville, Journ. de. Math 3, 349 (1838).

2. P. A. M. Dirac, Proc. Cambridge Phil. Soc. 25, 62 (1928).

3. B. O. Koopman, Proc. Nat'l Acad. Sci., U.S. 17, 315 (1931).

4. R. Kubo, J. Phys. Soc. Japan 12, 570 (1957).

5. S. Nakajima, Progr, Theoret. Phys. (Kyoto) 20, 948 (1958).

6. E. T. Jaynes, Phys. Rev. 106, 620 (1957).

7. E. T. Jaynes, Phys. Rev. 108, 171 (1957).

8. E. T. Jaynes, in Statistical Physics (1962 Brandeis Lectures, Vol. 3), edited by K. W. Ford (W. A. Benjamin Inc., 1963).

9. E. T. Jaynes, Am. J. Phys. 33, 391 (1965).

10. B. Robertson, dissertation, Stanford University, 1964. Also Bull. Am. Phys. Soc. 9, 733 (1964).

11. B. Robertson, Phys. Rev. 144, 151 (1966).

12. H. Schwegler, Z. Naturforsch, 20a, 1543 (1965).

13. P. T. Landsberg, Proc. Roy. Soc. 74, 486 (1959).

14. W. M. Elsasser, Phys. Rev. 52, 987 (1937).

15. P. G. Bergmann and A. C. Thompson, Phys. Rev. 91, 180 (1953).

16. J. M. Richardson, J. Math. Anal. Appl. 1, 12 (1960).

17. H. Mori, J. Phys. Soc. Japan 11, 1029 (1956); Phys. Rev. 112, 1829 (1958); 115, 298 (1959); H. Mori, I. Oppenheim, and J. Ross, in Studies in Statistical Mechanics (North Holland Publishing Co., 1962), Vol I, p. 271.

18. J. A. McLennan, Phys. Fluids 4, 1319 (1961); Adv. Chem. Phys. 5, 261 (1963).

19. D. N. Zubarev and V. P. Kalashnikov, Teor. Mat. Fys. 1, 137 (1969) [Theor. Math. Phys. 1, 108 (1969)].

20. D. N. Zubarev, Nonequilibrium Statistical Thermodynamics (Consultants Bureau, 1974) translation of 1971 Russian original, p. 266.

21. A. Hobson, Concepts in Statistical Mechanics (Gordon and Breach, 1971).

22. M. Tribus, Rational Descriptions, Decisions and Designs (Pergamon Press, 1969).

23. M. Tribus, Thermostatics and Thermodynamics (D. Van Nostrand Co., 1967).

24. R. Baierlein, Atoms and Information Theory (W. H. Freeman and Co., 1971).

25. A. Katz, Principles of Statistical Mechanics (W. H. Freeman and Co., 1967).

26. E. T. Jaynes, private communication. The only published discussion of this is in Ref. 27, page 88.

27. R. Zwanzig, Ann. Rev. Phys. Chem. 16, 67 (1965).

28. D. N. Zubarev and V. P. Kalashnikov, Physica 46, 550 (1970). Also Ref. 20, page 436.

29. B. Robertson, Phys. Rev. 160, 175 (1968), Appendix C.

30. R. Zwanzig, J. Chem. Phys. 33, 1338 (1960).

31. D. N. Zubarev, Dokl. Acad. Nauk SSSR 140, 92 (1961); 164, 537 (1965) [Sov. Phys. Dokl. 6, 776 (1962); 10, 850 (1966)].

32. D. N. Zubarev and V. P. Kalashnikov, Teor. Mat. Fiz. 3, 126 (1970); 5, 406 (1970) [Theor. Math. Phys. 3, 395 (1970); 5, 1242 (1971)].

33. D. N. Zubarev and V. P. Kalashnikov, Teor. Mat. Fiz. 7, 372 (1971) [Theor. Math. Phys. 7, 600 (1971)].

34. Ref. 29, Sec II.

35. M. S. Green, J. Chem. Phys. 20, 1281 (1952); 22, 398 (1954).

36. Ref. 29, Appendices A and B.

37. H. Mori, Progr. Theoret. Phys. (Kyoto) 33, 423 (1965); 34, 399 (1965).

38. K. Kawasaki and J. D. Gunton, Phys. Lett. 40A, 35 (1972).

39. K. Kawasaki and J. D. Gunton, Phys. Rev. A8, 2048 (1973).

40. H. Grabert, Z. Physik B27, 95 (1977).

41. M. S. Green, Phys. Rev. 119, 829 (1960).

42. J. A. McLennan, Progr. Theoret. Phys. (Kyoto) 30, 408 (1963).

43. R. W. Zwanzig, J. Chem. Phys. 40, 2527 (1964).

44. M. H. Ernst, E. H. Hauge, and J. M. J. van Leeuwen, Phys. Rev. Lett. 25, 1254 (1970).

45. Ref. 29, Sec. III.

46. G. V. Chester, Rept. Progr. Phys. 26, 411 (1963).

47. R. A. Piccirelli, Phys. Rev. 175, 77 (1968).

48. I. Oppenheim, R. Levine, et. al., paper at this conference.

49. B. Robertson, J. Math. Phys. 11, 2482 (1970). For an introduction to the field operators used in this paper see B. Robertson, Am. J. Phys. 41, 678 (1973).

50. R. A. Piccirelli and K. H. Im, J. Non-Equilib. Thermo. 1, 43 (1976).

51. A. W. B. Taylor and J. W. Evans, Collective Phenomena 1, 37 (1972).

52. L. Noethe and A. W. B. Taylor, Z. Physik B26, 217 (1977).

53. J.-P. Badiali and C. Gabrielli, J. Chim. Phys. 69, 1725 (1972).

54. M. Ochiai, Phys. Lett. 44A, 145 (1973).

55. B. Robertson and W. C. Mitchell, J. Math. Phys. 12, 563 (1971).

56. B. Robertson, Phys. Rev. 153, 391 (1967).

57. E. K. Khenner, Zh. Eksp. Teor. Fiz. 63, 261 (1972) [Soviet Phys. JETP 36, 137 (1973)].

58. J.-T. Rebsch, Physica 84A, 143 (1976).

59. D. I. Kadyrov and I. G. Shaposhnikov, Zh. Eksp. Teor. Fiz. 71, 2330 (1976) [Soviet Phys. JETP 44, 1229 (1976)].

60. R. Haberlandt and G. Vojta, J. Phys. Soc. Japan 26 Supplement, 222 (1969).

61. C. R. Willis and R. H. Piccard, Phys. Rev. A9, 1343 (1974).

62. V. P. Kalashnikov and M. I. Auslender, Dokl. Akad Nauk SSSR 215, 810 (1974) [Sov. Phys. Dokl. 19, 193 (1974)].

63. T. Shimizu, Physica 83A, 486 (1976).

64. M. H. Ernst, E. H. Hauge, and J. M. J. van Leeuwen, J. Stat. Phys. 15, 23 (1975).

65. V. P. Kalashnikov and M. I. Auslender, Dokl. Akad. Nauk SSSR 215, 810 (1974) [Sov. Phys. Dokl. 19, 143 (1974)]; Physica 85A, 71 (1976).

66. H. Grabert, Phys. Lett, 57A, 105 (1976).

67. B. Robertson, Bull. Am. Phys. Soc. 12, 1141 (1967).

RELATIVE STABILITY IN THE DISSIPATIVE STEADY STATE

Rolf Landauer

Introduction

The statistical mechanics of open dissipative systems has been a subject of increasing interest as a result of the emergence of a number of new viewpoints. The maximum entropy principle of this conference is only one of these. A review of other approaches, more concerned with the detailed kinetics of the active system, and less with the attempt to extrapolate general principles from equilibrium, will be found in the works of Haken.[1] My own work has included some attempts to generalize $TdS=dQ$, *with an equality sign*, to the slowly changing dissipative state. It turns out that, in some cases, this is possible with surprisingly simple generalizations of both T and dQ. This work has been described in considerable detail in a recent paper,[2] and will not be included in the present discussion.

Dissipative steady state systems can be multistable, i.e. have more than one state which is stable with respect to small disturbances. It has become fashionable to emphasize this in connection with chemical reactions, which are continually supplied with new reactants, and where the reaction products get carried away.[3] Multistability is, however, a common and old phenomena. It is illustrated, for example, by gas jets and arc lights, which can be either "lit" or quiescent under the same externally imposed conditions. A pendulum clock or a gasoline engine are somewhat more complex. In their interesting state they are oscillating or rotating, and this active state is itself highly degenerate, permitting a free choice of phase. Our discussion will emphasize the simpler systems, exemplified by the gas jet and the arc light. Information storage in computers utilizes bistable systems. These can be nondissipative, e.g. using magnetic media, or else can be active circuits requiring a continuous energy supply.

Let us first consider thermal equilibrium. If a potential U has competing states of local stability, i.e. several minima, then relative stability is easily discussed via the Boltzmann distribution, $\exp(-\beta U)$. The lowest minimum corresponds to the highest probability density. If we allow for the extension, in phase space, of the minima, then we can compare the total probability in the vicinity of each minimum, rather than the probability density. In that case we must compare free energies, instead of energies. The coexistence of two states of relative stability, for the same externally imposed parameters, is characteristic of a first-order transition. In the neighborhood of thermal equilibrium between two phases there will be a range where both phases are locally stable. Except *at* phase equilibrium, one of the two will be metastable. As already emphasized, open dissipative systems far from equilibrium, can also have a multiplicity of locally stable states, toward which the system will be driven, if fluctuations are neglected.

The literature contains two viewpoints toward relative stability in open systems. One view was stated[4] in 1962, and elaborated later.[5-7] These discussions stress the role of noise in the determination of relative stability, and emphasize that not only the fluctuations near the states being compared are relevant, but that the fluctuations at the improbable intervening states can be equally critical. Other authors have expressed very similar viewpoints.[1,8-10] Most recently this viewpoint has been confirmed by Nicolis and coauthors.[11] The alternative view stresses an analogy to the van der Waals diagram in the liquid-gas transition. Under this viewpoint phase equilibrium, even in open dissipative systems, can be characterized by a generalization of Maxwell's equal area rule, i.e. through a macroscopic deterministic relationship, without a detailed concern with fluctuations. We cite only a sampling of the papers advocating this viewpoint.[12-16]

Our resolution: The viewpoint of Ref. 4 is valid for small systems in which the phase transition is made in a spatially uniform way. The alternative viewpoint can be a reasonable approximation for very large systems in which the phase transition is made through a nucleation event, followed by spatial expansion of the favored phase. This resolution is implicit in an earlier discussion,[17] but has not been generally appreciated. We will analyze a simple example, intended to supplement that of Ref. 17, and will not settle the question with guaranteed generality.

One Degree of Freedom

Let us first consider systems with only one interesting degree of freedom, q. As explained by Haken[1] this one degree of freedom can be one which in turn controls a number of others that adjust very rapidly to the momentary value of the interesting variable. A distribution function $\rho(q)$ gives rise to a flux, j, of probability, often given by the Smoluchowski equation $j = \rho v(q) - D\partial\rho/\partial q$. Here v describes the restoration velocity, along q, to points of local stability, as given by the macroscopic equations of motion. The r.h.s. diffusion term describes the tendency of ensemble members, initially at the same value of q, to separate as a result of fluctuations. The conditions under which a broadly valid master equation can be replaced by the Smoluchowski equation we have just invoked, have received a great deal of discussion. A few key citations are given in Ref. 2. It suffices here to state that the simple equation we have written, is often valid. In many cases where it is not valid, the resulting more complex equations still permit the points to be made that we shall try to establish.

In the steady state ρ typically (but not necessarily) vanishes at $q = \pm\infty$, i.e. there is no injection of new systems into the space under consideration. In that case $j = 0$, and the Smoluchowski equation can be integrated to give $\rho = C \exp[\int (v/D) dq]$. This shows that the relative probability of two states of local stability is determined by an integral over the states between them. We can play tricks with the behavior of (v/D), in the intermediate parts of the range of integration, without affecting the behavior near the points of local stability, and thus control relative stability, without varying the behavior of the system near the points of stability. Note that, in general, D is a function of q, not a constant. That point was already emphasized in Ref. 4, and in much of the later literature. If we ignore that, as is done in some places, then relative probabilities depend on $D^{-1}\int v\, dq$. When this integral vanishes the two states to be compared become equally probable. This simplified "equilibrium" condition invokes only the macroscopic laws of motion, reflected by v, and not the fluctuations represented by D. Unfortunately, this generalization of Maxwell's equal area rule requires the generally invalid assumption that D does not depend on q.

A number of "tricks" have been described[5,7,18] in which the behavior of (v/D) in the range between two points of local stability is modified, without changing the behavior near the points of local stability. We will review one of these[5] here.

Consider the overdamped bistable potential of Fig. 1 and a particle (or set of independent particles) in it. If the system is in equilibrium, at a uniform temperature, the deeper right hand well is preferred. Now take the shaded portion of the right hand well and bring it to a much higher temperature. Escape out of the right hand well is made much easier, but escape out of the left hand well is left unmodified. In the new steady state, therefore, the number of particles in the right hand well will be greatly reduced, and the magnitudes involved can be adjusted so that the right hand well becomes the less likely state.

Let us establish the result we have just quoted a little more carefully. Fig. 2(a) shows a bistable potential, with C as the favored state in thermal equilibrium. In thermal equilibrium: $\rho \sim \exp[-U/kT]$. Thus the ordinate U in Fig. 2(a) can also be considered to be a plot of $-kT \log \rho$. Now let us ask how Fig. 2(a) changes if section BD is brought to a much higher temperature. To make this more physical we can imagine the potential U to represent a force field acting on an electrostatically charged particle, in a thin and long insulating tube. The force field can be set up by depositing charges on the exterior of the tube. The moving particle makes frequent collisions with the tube walls. These collisions determine both its temperature and viscosity, but do not change its charge.

The heavy damping will insure that a particle will come to its new temperature very quickly after crossing the temperature discontinuities at B and at D. Thus, in the steady state, in the interior of each of the three regions of constant temperature, we can expect a Boltzmann distribution characteristic of the region's temperature. Let us then consider $-kT \log \rho$ for this inhomogeneous temperature case, but with the coefficient T multiplying $\log \rho$ held at its original temperature, i.e. the temperature still prevailing to the left of B and to the right of C. This is plotted in Fig. 2(b). The shape of the curve to the left of B and to the right of C is unchanged. In the range BD the distribution function is flattened out, corresponding to the much lesser variation of $\exp(-U/kT_H)$, at the higher temperature T_H. The matching problem at the temperature discontinuities B and D remains to be discussed.

If the damping is strong enough, then we can view this matching problem as one determined by the temperature discontinuity and unrelated to the much slower scale of the potential variation. Fast high temperature particles arrive at B from the right, while slow low temperature particles arrive at B from the left. In the steady state the net flux through B vanishes and therefore the distribution function ρ must be higher on the low temperature side. In Fig. 2(b) where $-kT \log \rho$ is plotted this lowering of density, as we move thorugh B, to the right, shows up as an increase in $-\log \rho$. Our most important point, however: The *ratio* by which ρ changes in going through the temperature discontinuity is a function only of the temperatures involved, and not of the absolute magnitude of ρ. Thus the two discontinuities at B and D, plotted on a logarithmatic scale as in Fig. 2(b), must have the same magnitude. The size of this temperature discontinuity thus has no effect on the relative vertical positions of A and C in Fig. 2(b). We see that the elevated temperature in BD has two effects: 1. The levelling effect on the distribution function in BD can cause C to become the less likely state. 2. The density ρ, in BD, has been depressed below that of its surroundings. The particles get shaken out of the hot zone. This is a point which we will utilize in a later section.

Our example shows that relative stability can be controlled by the particle behavior in states where the particle is unlikely to be found. *Thus any criterion which only invokes entropy, entropy production, excess entropy production, etc., heavily weighting the likely states, cannot help us in the evaluation of relative stability.* This statement is not intended as an unqualified assertion that the maximum entropy principle is useless. It is only an assertion that the constraints applied to the system, before maximizing entropy, must do justice to the kinetics in the unlikely states.

Spatially Extended Systems

Let a series of identical double well potentials, of the sort shown in Fig. 1, be placed next to each other, along the x-axis, while the particle displacement is maintained in the z-direction. Let adjacent particles be coupled, exerting a force on each particle proportional to the separation in z from the neighboring particles, and tending to bring adjacent particles to the same value of z. We will concentrate on the continuum limit of this model. The wells supply damping forces which are large compared to the inertial forces and also act as sources of random thermal agitation for the particles. Our model is related to the one-dimensional dislocation models proposed decades ago,[19,20] and also used to treat open systems.[21] In recent years treatment of these models has been extended to include thermal fluctuations.[22] It is the interaction with an external thermal reservoir, introducing noise and heavy damping, which distinguishes our current model from the conservative Hamiltonian systems of Refs. 19 and 20. An overdamped noisy systems has recently been analyzed,[23,24] differing from ours only in the exact form of the particle potential. Ours is bistable, while that in Refs. 23 and 24 is periodic. In an Appendix we discuss the relationship of our model to the currently fashionable soliton theory in more detail. For a short enough system the lateral coupling between wells will lock all particles to the same z position, and the discussion associated with Fig.1 applies. Now consider more extended systems. The equation of motion of our system is[23]

$$\gamma\dot{z} = -\frac{\partial V}{\partial z} + \beta\frac{\partial^2 z}{\partial x^2} + L. \qquad (1)$$

The r.h.s. of Eq. (1) represents the force on a unit length of the chain, whose damping constant is γ. V is the potential, per unit length of chain, of the form shown in Fig. 1. β is the strength of lateral coupling. L is the random force due to thermal agitation. Consider first the behavior of the chain in the absence of L, and for uniformly moving transitions, $z(x-ut)$, in which the particles are shifted from one well to the other. Eq. (1) then becomes

$$\beta \frac{\partial^2 z}{\partial x^2} + u\gamma \frac{\partial z}{\partial x} - \frac{\partial V}{\partial z} = 0. \tag{2}$$

This is equivalent to an equation of motion

$$m \frac{\partial^2 z}{\partial t^2} + \alpha \frac{\partial z}{\partial t} + \frac{\partial U}{\partial z} , \tag{3}$$

for a particle of mass $m = \beta$, damping constant $\alpha = u\gamma$, and in a potential $U = -V$. The potential in Eq.(3) is that of Eq. (2) turned upside down, with the well bottoms becoming maxima, as in Fig. 3. Consider the solid curve of Fig. 2, where the two maxima of U lie at the same value of U. A particle can start from rest at $z = a$ and make a transition, coming to rest again at $z - b$, if and only if the damping term vanishes, i.e. if $u = 0$. This stationary transition, therefore, exists only if the potential minima of V are equal. For the dotted line in Fig. 3 a particle can make the transition from a to b, coming to rest at b, only if the damping constant α has just the right value, corresponding to a positive value for u. The dotted line, as shown in Fig. 3, corresponds to a *higher* value for the potential V(b). If the transition from a to b, in Eq.(2) is made with advancing x, then a positive value of u implies a transition from b to a in time, i.e. toward the lower energy valley. The propagation velocity u, for the transition toward the favored valley, can be shown to be proportional to the energy difference between valleys, for small departure from the static case. We are dealing with an overdamped solitary wave in which this energy difference determines a velocity, in contrast to the more conventionally fashionable undamped soliton. In the latter case a driving force produces an acceleration, rather than a velocity. Our case is equivalent to a phase boundary velocity in a first order phase transition, proportional to the degree of superheating or supercooling. The transition from $z = a$ to $z = b$ described in Fig.3 is clearly the only solution of Eq. (2) allowing that transition, except of course that the location of the transition in time in Eq. (3), or in position in Eq. (2), can still be freely chosen.

The uniformly moving transitions between potential wells that we have just described permit an energetically favored valley to propagate, once some portion of the chain of particles has been deposited in it. Our basic point: Fluctuations, as described by L in Eq. (1), affect this propagation process only in a minor way. In the spatially extended system the fluctuations have a major role in first getting a portion of the chain from one valley, over the intervening barrier, into the other valley, but not thereafter. Fluctuations, therefore, do not control relative stability in the extended system in the same drastic way as in the one-dimensional or short system. The limited role of fluctuations in Eq. (1) can be seen most clearly at low temperatures. We cannot expect weak random and oscillatory forces to appreciably alter the behavior described in connection with Fig. 3.

Now consider noise in Eq. (1) more explicitly. In the simplest case, where both γ and the statistical behavior of L are independent of x and z, where the fluctuating forces on different particles are uncorrelated, and L is uncorrelated over times which matter, Eq. (1) leads to a many dimensional version of our earlier equation for the probability flux j, in which curl $\vec{v} = 0$. The resulting steady state distribution function is exp $(-E/kT')$. Here E is the total energy of the chain, including the coupling energy. The effective temperature T' is proportional to the ratio of $\langle L^2 \rangle$ to γ. This Boltzmann distribution applies if we allow all possible chain configurations, and also if we limit ourselves to a subspace in which $z = a$ at $x = -\infty$ and $z = b$ at $x = +\infty$. We are, however, interested in more general situations, as illustrated in Fig. 1, corresponding to a dependence of L on z. Furthermore, the solution just described (just like the distribution $\exp[Fx/kT]$ for a simple particle subject only to a force F acting in the x-direction) does not correspond to net motion and is not the only solution.

Let us continue considering the extended chain, but as in Fig. 1, create a hot zone on one side of the potential barriers, near their peak. If we assume that motion is constrained so that z is independent of x, then we immediately return to the argument of Fig. 1 and Ref. 5. In the more general case, if we start from a chain lying along a well bottom, and consider nucleation of a segment into the other well, then the hot zone will aid the transition out of the heated side, and thus make this nucleus more likely in the case of equal well depths, or else will overcome some difference in well depths, in the case of unequal well depths. What happens once a domain of the alternative phase and the associated domain wall or walls has been formed? The domain wall is subject to two influences: The driving forces due to the coherent forces of the first two r.h.s. terms of Eq. (1) and also the fluctuations. The fluctuations do not exert a net force on the particles toward one well or the other, and therefore will not drive the wall in either direction, but only cause it to diffuse. The greater fluctuations in the hot zone will cause particles to be shaken out of that hot zone, and will give the particles a tendency to "condense" toward the colder zones, as already discussed in connection with region BD of Fig. 2(b). This causes the domain wall to become narrower, in the hot zone, above and beyond the effect due to the elevated potential in that zone. Thus the hot zone will make complex and esoteric domain structures, even *less* likely than in the absence of noise terms. Instead of invoking an intuitive physical picture which refers to "particles shaken out of a hot zone," we can be more analytical. The many dimensional Smoluchowski equation derivable from Eq. (1) generates matching or boundary condition problems, as particles enter or leave the hot zone, just as in the one-dimensional case. A straightforward generalization of the reasoning applied to that case leads to the conclusion that the steady state distribution function drops by the same factor, for each additional particle that crosses into the hot zone.

The coherent forces in Eq. (1) will have two effects. First of all, they determine the domain thickness in the usual way, balancing the interparticle coupling energy which causes wide walls, against the potential energy V, which causes narrow walls. Additionally, the potential V will drive the chain toward the lower potential valley, and there is nothing in the fluctuations that can offset that. The direction of motion, predicted from the arguments associated with Fig. 3 remains unaffected by fluctuations. Fluctuations can affect nucleation, and the detailed width of the wall. If the fluctuations are large enough to interact nonlinearly with the wall motion, they may also act as an additional source of damping. But fluctuations cannot change the tendency of a domain to expand or contract, once it is formed. The direction of domain motion (and its velocity, if the fluctuations are sufficiently modest) is correctly given by the deterministic, noiseless, equations.

As the chain under consideration becomes longer, the importance of the nucleation process, relative to the subsequent wall motion, becomes less significant. Thus the effect of a hot zone on the determination of the steady state becomes less important. Note, however, that this dimunition of the effect of a hot zone does not increase indefinitely with the length of the chain. A really long chain will make a transition by independent nucleation in different places. The easier nucleation event, starting from the hotter valley, can offset a small retarding force seen in the subsequent domain expansion. Thus the hot zone does have an effect, but one that will be very small compared to that for the short chain making a spatially uniform transition. As was already emphasized in Ref. 17, the various paths between a chain lying completely in one well, and one completely in the other do not give zero transition rate, under the same conditions. This distinguishes our open system from ordinary thermodynamic equilibrium. In the latter all paths give zero flux under phase equilibrium conditions.[1]

In the case of the single degree of freedom of Fig. 1 the hot zone aids escape from the heated side. It does so much less effectively in the extended chain. It still does have an effect, however, and we cannot expect to calculate relative stability *precisely*, without full consideration of the fluctuations along the path between the two valleys. It is likely, however, that a good approximation can be obtained without this precaution.

In conclusion we stress that in biological applications we are likely to be dealing with *small* systems. As a consequence the fluctuations along the transition paths between one state of relative stability, and another one, can be significant. Thus, as argued in Ref. 6, in the explanation of evolutionary sequences and the origin of life, *we cannot expect thermodyna-*

mic or information-theoretic shortcuts, which only compare initial and final states. If, however, we accept the very appealing arguments of Kuhn[25] that extremely modest departures from thermal equilibrium are needed for the origin of life, then our considerations, here, become irrelevant.

Appendix

While it is not needed for our discussion of relative stability, we will here make a more detailed connection between systems of the sort we have discussed in this paper and earlier notes,[17,26] and the currently more fashionable soliton theory.[27] The systems I have discussed are dissipative open systems. These exhibit energy dissipation even in the steady state. Furthermore, the energy dissipation is not just the sort found for particles moving in a viscous medium under an applied field, where a continuing influx of new ensemble members is needed at one end of the medium, and an outflow at the other. Our "open" systems exhibit energy dissipation even for a fixed ensemble. In the example of the present paper this energy dissipation is provided by thermal transport, with the particles carrying heat from the hot zone into the adjacent colder areas. The transitions, domain walls, or "kinks" that we have considered in this paper involve only two favored states, thus they are like kinks in a ϕ^4 potential. While not utilized in this paper, open systems can also manifest kinks in systems which have an unlimited number of locally stable states, similar to kinks characterized by the Sine-Gordon equation and the examples of Refs. 19, 20, 22-24. Stratonovich[28] has discussed oscillators synchronized by an external signal at a frequency differing from the natural frequency of the oscillator, and in the presence of noise. He has shown that the phase drift of such an oscillator is described by the equation describing the Brownian motion of a particle in a heavily damped sinusoidal potential, with an additional superimposed spatially uniform bias force. A chain of such oscillators, coupled to each other in a way which tends to keep adjacent oscillators in phase, is an open system which has a periodic sequence of points of local stability for each of the individual members of the chain. These points of local stability correspond to an oscillator which is in phase with its synchronization signal. The oscillator's tendency to run at its natural frequency, if it is different from that of the injected signal, will manifest itself through occasional phase jumps. The size of the injected signal determines the amplitude of oscillation of the "effective" sinusoidal potential, while the frequency difference determines the "tilt" of the potential, i.e. the force which induces phase jumps.

There are a number of obvious variations on this basic oscillator chain scheme. Ref. 21, for example, discussed an oscillator chain synchronized by an injected signal near a frequency twice that of the oscillators. In that case the points of stability of the oscillator phases are only π apart, instead of the more common 2π. That's an unimportant change of scale, until one starts to play tricks with the boundary conditions applied to a chain of limited length, e.g. invert the phase instead of just applying periodic boundary conditions.

References

1. H. Haken, Rev. Mod. Phys. 47, 67 (1975); *Synergetics, An Introduction* (Springer, Heidelberg, 1977).

2. R. Landauer, Phys. Rev. A, to be published.

3. *Physical Chemistry of Oscillatory Phenomena*, Faraday Society Conf. Proceedings, University Press, Aberdeen (1975).

4. R. Landauer, J. Appl. Phys. 33, 2209 (1962).

5. R. Landauer, Phys. Rev. A 12, 636 (1975).

6. R. Landauer, Ber. Bunsenges physik chem. 80, 1048 (1976).

7. R. Landauer, in *Bifurcation Theory and its Applications in Scientific Disciplines*, O. E. Rössler and O. Gurel, eds. (Ann. N. Y. Acad. Sci.) to be published.

8. R. L. Stratonovich, *Topics in the Theory of Random Noise*, Vols. I and II (Gordon and Breach, New York, 1963 and 1967).

9. I. Matheson, D. F. Walls, and C. W. Cardiner, J. Stat. Phys. 12, 21 (1975).

10. J. Keizer, *Maxwell-Type Constructions for Multiple Nonequilibrium Steady States*, preprint.

11. G. Nicolis, and R. Lefever, Phys. Lett. 62A, 469 (1977); G. Nicolis, J. W. Turner, in *Bifurcation Theory and its Applications in Scientific Disciplines*, O. E. Rössler and O. Gurel, eds. (Ann. N. Y. Acad. Sci.) to be published.

12. I. Prigogine, R. Lefever, J. S. Turner, and W. Turner, Phys. Lett. 51A, 317 (1975).

13. B. Ross, and J. D. Litster, Phys. Rev. A, 15, 1246 (1977).

14. K. Takeyama, and K. Kitahara, J. Phys. Soc. Japan 39, 125 (1975).

15. D. Bedeaux, P. Mazur, and R. Pasmanter, Physica 86A, 355 (1977).

16. I. Procaccia, and J. Ross, J. Chem. Phys. 67, 5565 (1977).

17. R. Landauer, Phys. Rev. A 15, 2117 (1977).

18. R. Landauer, J. Stat. Phys. 13, 1 (1975).

19. J. Frenkel and T. Kontorova, Phys. Zeit. Sowjetunion, 13, 1 (1938); J. Phys. USSR, 1, 137 (1939).

20. F. C. Frank and J. H. van der Merwe, Proc. Roy. Soc. (London), Sect. A 198, 205 (1948); 198, 216 (1948); 200, 125 (1949); 201, 261 (1950).

21. R. Landauer, "Moebius Strip Coupling of Bistable Elements," unpublished laboratory report of 1955, reprinted in 1976 as IBM Research Report RC 6093, available from the author or his laboratory.

22. J. A. Krumhansl, and J. R. Schrieffer, Phys. Rev. B 11, 3535 (1975); C. M. Varma, Phys. Rev. B 14, 244 (1976); A. R. Bishop, and J. A. Krumhansl, Phys. Rev. B 12, 2824 (1975).

23. S. E. Trullinger, M. D. Miller, R. A. Guyer, A. R. Bishop, F. Palmer, and J. A. Krumhansl, Phys. Rev. Lett. 40, 206 (1978).

24. T. Schneider, E. P. Stoll and R. Morf, to be published.

25. H. Kuhn, in *Biophysik*, W. Hoppe, W. Lohmann, H. Markl, and H. Ziegler, eds. (Springer, Heidelberg, 1977).

26. R. Landauer, J. Phys. Soc. Japan 41, 653 (1976).

27. See for example the proceedings of the Symposium on Nonlinear Structure and Dynamics in Condensed Matter, Oxford, June 27-29, 1978.

28. Ref. 8, Vol. II, Chapt. 9, p. 222.

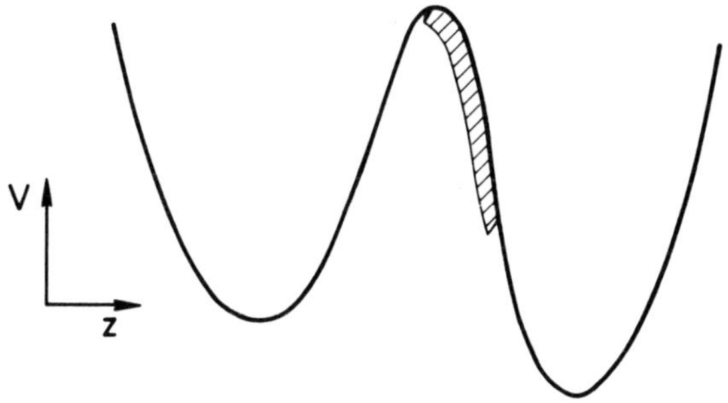

Fig. 1. Bistable well. Shaded portion on right hand side of barrier can be elevated in its temperature.

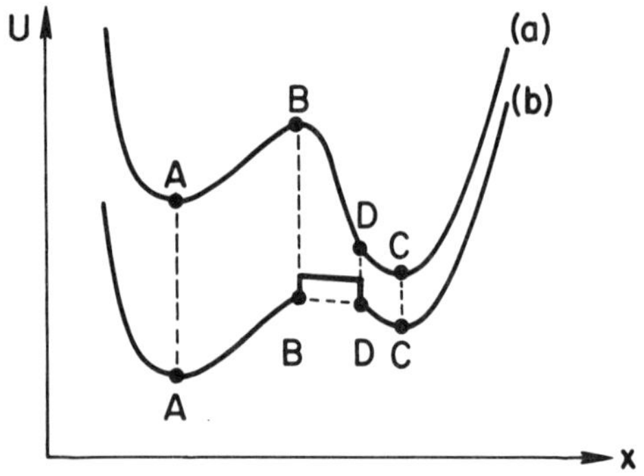

Fig. 2 (a) Bistable potential well with A metastable relative to C. (b) D has been raised almost to level of B to illustrate effect of heating BD.

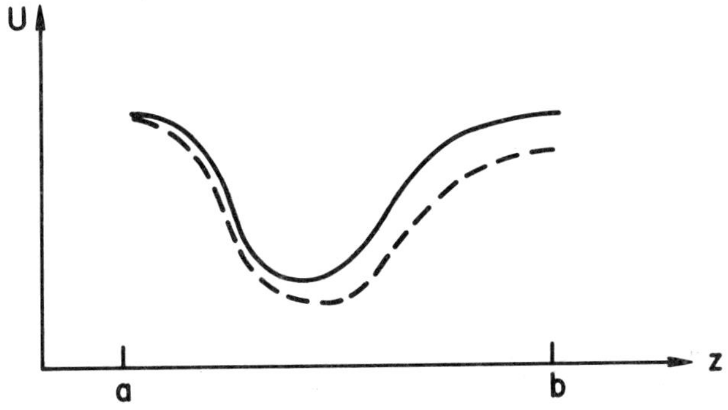

Fig. 3. Effective potential in Eq.(3). Solid line is the case of equally deep wells, dashed line has shallower well in Eq. (2) at $z = b$.

A NEW LOOK AT THE RELATION BETWEEN INFORMATION THEORY AND SEARCH THEORY

John G. Pierce

Abstract. Some apparent contradictions in the operations research literature of search theory and information theory are reviewed, and the source of the conflict is isolated. The information processes connected with a search operation are examined by studying the geometric properties of information surfaces in search-allocation space. Some new theorems are developed which show that the connection between search and information theory is much stronger than previously believed.

It is demonstrated that the amount of search effort, C, is a critical parameter in the relation between search and information gain. For very small values of C, there is no unique relation between detection probability and information gain; for very large values of C, the optimal whereabouts search policy produces the greatest information gain; for a broad intermediate range of C (including many cases of practical interest) the optimal detection search policy produces the greatest information gain.

1. Introduction

Background. The relation between search theory and information theory has had a thorny history. A 1967 view was summarized concisely by Koopman (reference 1): "Ever since the mid-nineteen-forties when the theories of information and of search became subjects of general interest, attempts have been made to apply the theory of information to problems of search. These have proved disappointing; neither the formulas nor the concepts of the former theory have found a place in clarifying the problems of the latter." These views doubtless reflected much of Koopman's own research on the subject.

An independent contribution to the generally pessimistic view of the matter was the 1961 paper by Mela (reference 2). He presented numerical examples for simple search models, demonstrating that a search policy that is designed to maximize detection probability does not necessarily maximize either the gain in expected information or the probability of correctly committing forces on the basis of search outcomes. Mela concluded: "...it does not seem likely that there is any intimate connection between search theory and information theory." And: "...search theory should be considered in connection with the general theory of statistical decisions rather than with information theory."

This negative view was reinforced by a later paper by Pollock (reference 3) in 1971. He presented yet another numerical example, in which different search policies were necessary to maximize either: the probability of detection as a result of the search, or the probability of guessing the position of the target after the search, or the information gained during the search.

These negative findings had a clearly inhibiting effect on research, and relatively little effort has been devoted to the connections between information and search for the past fifteen years. Nonetheless, the intuitive appeal of information theory remains strong, and more recently the tide of pessimism has been stemmed somewhat. In 1973, for instance, Richardson (reference 4) used Monte Carlo simulation to explore alternative surveillance policies in a false-target environment. Target motion based on a Markov process was also included. Richardson's policy options included: (a) an optimal single-stage look-ahead policy (about the same as maximizing the probability of a right guess after every search state); (b) a policy to maximize information gain; (c) a policy to search the cell with the highest a priori probability; (d) a uniform surveillance policy (cycling systematically through all cells).

Richardson's results showed that the maximum-information-gain policy was generally (but not uniformly) best. In his summary, he said: "The principal conclusion...is that the maximum information gain policy...appears to have desirable characteristics in the idealized surveillance scenario considered. Among these characteristics...are good initial behavior in the early stages, and good asymptotic behavior in the later stages. The initial behavior is measured principally by comparison with the optimal single-stage look-ahead policy ...which is designed to be good in the early stages. The asymptotic behavior is measured principally by comparison with the uniform surveillance policy, which, for a stationary target, is guaranteed to converge to 1...."

Finally, inspired by Richardson's simulation findings, Barker (reference 5) proved these two theorems analytically:

"I: Suppose that the detection function is given by $b(u) = 1 - e^{-u}$ for $u \geq 0$. Then, subject to a constraint on total search effort, the allocation of search effort which maximizes the probability of detection also maximizes the entropy of the posterior distribution."

And the converse:

"II: Let b be a regular detection function. Suppose that for every prior target location distribution and for every constraint on total search effort, any allocation of search effort which maximizes the probability of detection also maximizes the entropy of the posterior distribution. Then, $b(u) = 1 - e^{-au}$ for some positive constant a ."

These findings, taken as a whole, appear to be inconsistent. In particular, Mela's numerical example seems, at least superficially, to be in clear contradiction to Barker's theorems.

If science teaches us anything, it should be that where contradictions abound, there lies a fertile field for research.

Accordingly, this paper reexamines the relation of search theory to information theory and attempts to reconcile past results, to clarify the current state of knowledge, and to remove some impediments to further work on the subject.

Plan of the Paper. Section 2 establishes the mathematical framework of the problem. A model for the search process is constructed. A description of the problem in terms of information theory is provided, and distinctions are drawn among several possible defintions of information. These are then related to quantities discussed in the earlier literature.

Section 3 reexamines the examples of Mela and Pollock and extends their results numerically to the case of a large amount of search effort. Section 4 treats analytically the case when a large search effort is applied; section 5 treats analytically the case in which the search effort is infinitesimal.

Section 6 summarizes the findings and places the results of this paper in a framework with past work.

Notation. This paper frequently uses a summation over a set J . Unless otherwise noted, the symbol $\sum$ implies summation from 1 to J over the index in the summand.

2. Problem Statement

The Search Model. We assume a simple model for the search process. A single target is in one of J cells. Our prior knowledge of its position is described by a set of probabilities, p_j , $j \in J$, $\sum p_j = 1$. We conduct a search operation in which various amounts of search effort, z_j , are applied to each of the J cells. The total amount of search effort, C , is fixed, so that the z_j amounts of search are subject to a constraint:

$$C = \sum_{j=1}^{J} z_j \ . \qquad (2.1)$$

For our general treatment, the z_j need not be quantized. Special cases, for which the search effort is applied in multiples of a fixed unit (the "look"), appear in the examples of Mela and Pollock.

We assume that the detection process is governed by an exponential detection function. This means that the conditional detection probability for cell j, given that the target is in cell j and that an amount of search effort, z_j, is applied there, is:

$$\Pr(\text{det } j|j,z_j) = 1 - \exp(-\alpha_j z_j) \ . \qquad (2.2)$$

α_j is a characteristic parameter of the detection process in cell j.

This formulation subsumes the discrete-look cases of Mela and Pollock. If q_j is the single-look conditional detection probability, then, for n_j looks in cell j:

$$\Pr(\text{det } j|j,n_j) = 1 - (1-q_j)^{n_j} \ , \qquad (2.3)$$

which is equivalent to eq. 2.2 when n_j is identified with z_j, and α_j is identified with $-\ln(1-q_j)$.

Finally, we assume that no false alarms occur.

Within the framework of this basic model, a specific allocation of search effort may lead to $J+1$ possible outcomes: either a single detection in any of the J cells or no detections at all. Because we have assumed no false alarms, a detection in cell k produces a posterior probability distribution:

$$p'_k = 1; \ p'_j = 0, \ j \neq k \in J \ . \qquad (2.4)$$

Failure to achieve detection in any cell produces a posterior distribution:

$$p'_j = p_j \exp(-\alpha_j z_j) / \left[\sum_{k=1}^{J} p_k \exp(-\alpha_k z_k) \right] . \tag{2.5}$$

For convenience of notation, we define

$$\varphi_j = \varphi(z_j, \alpha_j, p_j) \equiv p_j \exp(-\alpha_j z_j) . \tag{2.6}$$

(These φ_j functions correspond to the "failure densities," $\beta(j)$, used by Barker.)
For nondetection, the posterior distribution is:

$$p'_j = \varphi_j / \sum_{k=1}^{J} \varphi_k . \tag{2.7}$$

The unconditional probability of detection in the jth cell is, in this notation:

$$PD_j = p_j (1 - \exp(-\alpha_j z_j)) = p_j - \varphi_j , \tag{2.8}$$

and the overall detection probability is:

$$PD = \sum_{j=1}^{J} PD_j = 1 - \sum_{j=1}^{J} \varphi_j . \tag{2.9}$$

The Information Description. It is possible to define many "information" quantities related to search operations. Values for self information and mutual information, for conditional information and unconditional information, for information as a random variable and for ensemble average information -- any of these can be computed at each step of a search. To avoid ambiguity, we proceed with caution in defining the specific "information" we wish to discuss in this paper and in relating that definition to the quantities discussed in the references.

First, we define two probability spaces, X and Y , that are concerned, respectively, with the true position of the target and the observed response of the search sensor. Let the event x_j represent the presence of the target in cell j . The space X has dimension J , and the discrete event probabilities are given by $p_X(x_j) = p_j$, $j \in J$. Let the event y_0 be the null response to search, and let the event

y_k represent a detection in cell k. The event probabilities in the Y space can be constructed from the appropriate probabilities in the X space, together with the conditional detection probabilities: $p_Y(y_k) = \sum p_{Y|X}(y_k|x_j)p_X(x_j)$. In our present example, the dimension of the Y space is $J + 1$; it could be larger if multiple detections or false alarms were considered.

Our principal concern in this paper is the ensemble average mutual information between the X and Y ensembles. By definition (reference 6), the mutual information between the events x_j and y_k is:

$$I_{jk} = \ln [(p_{X|Y}(x_j|y_k))/p_X(x_j)] . \tag{2.10}$$

This quantity is a random variable in the product space $X \otimes Y$. Its ensemble average is

$$\begin{aligned} I_E &= \sum_{j,k} p_{X,Y}(x_j,y_k)\, I_{jk} , \\ &= \sum_{j=1}^{J} \sum_{k=0}^{J} p_{X|Y}(x_j|y_k)p_Y(y_k) \ln (p_{X|Y}(x_j|y_k))/p_X(x_j) . \end{aligned} \tag{2.11}$$

To compute this ensemble average, we use the following relationships. Implicit in all of these is an allocation of search effort, $\{z_j\}$:

$$p_X(x_j) = p_j \text{, the prior probabilities;} \tag{2.12}$$

$$p_Y(y_0) = \sum_{j=1}^{J} \varphi_j \text{, from (2.9);} \tag{2.13}$$

$$p_Y(y_k) = p_k - \varphi_k \text{, from (2.8);} \qquad k = 1,\ldots,J \tag{2.14}$$

$$p_{X|Y}(x_j|y_0) = \varphi_j / \sum_{k=1}^{J} \varphi_k ; \tag{2.15}$$

$$p_{X|Y}(x_j|y_k) = \delta_{jk} . \qquad k = 1,\ldots,J \tag{2.16}$$

The latter two equations are derived by appropriate applications of Bayes's rule. When these are substituted into eq.

2.11, we obtain:

$$I_E = -\sum_{j=1}^{J} p_j \ln p_j - \left[-\sum_{j=1}^{J} \varphi_j \ln \varphi_j + \left(\sum_{k=1}^{J} \varphi_k \right) \ln \left(\sum_{\ell=1}^{J} \varphi_\ell \right) \right] \quad (2.17)$$

The first term is recognized as the entropy of the prior ensemble, X , which we will designate H_0 . By a general result of information theory, the mutual information between two ensembles, X and Y , can be written:

$$I_E = H(X) - H(X|Y) . \quad (2.18)$$

Thus, we can identify the bracketed term in eq. 2.17 as the conditional entropy of X , given Y . For reasons discussed below, we designate this as H_E . Thus:

$$H_E = -\sum_{j=1}^{J} \varphi_j \ln \varphi_j + \left(\sum_{k=1}^{J} \varphi_k \right) \ln \left(\sum_{\ell=1}^{J} \varphi_\ell \right) . \quad (2.19)$$

The mutual information is then:

$$I_E = H_0 - H_E . \quad (2.20)$$

For a given prior distribution, H_0 is a constant; H_E , and consequently I_E , are functionals of the allocation of search effort, $\{z_j\}$, as well as functions of the prior probabilities, p_j , and the set of detection parameters, α_j .

We will eventually wish to investigate the extrema of I_E , but H_E is more convenient to work with. We note, therefore, that because of the minus sign in eq. 2.20, a maximum of H_E will mean a minimum of I_E , and vice versa.

Throughout this paper, we will discuss the ensemble average mutual information, I_E and the conditional entropy of X given Y --- $H_E \equiv H(X|Y)$ --- almost interchangeably. The reader should therefore bear their relationship (eq. 2.20) in mind as he proceeds.

An alternative approach is to consider the self-entropy of the posterior ensemble. If detection does not occur, posterior probabilities p'_j are as given by eq. 2.5 or 2.7; we can, therefore, define a posterior self-entropy, conditioned on non-detection of the target, as:

$$H_{ND} \equiv -\sum_{j=1}^{J} p'_j \ln p'_j \tag{2.21}$$

$$= -\sum_{j=1}^{J} \left(\varphi_j / \sum_{k=1}^{J} \varphi_k\right) \ln \left(\varphi_j / \sum_{\ell=1}^{J} \varphi_\ell\right)$$

$$= \left[-\sum_{j} \varphi_j \ln \varphi_j + \left(\sum_{k} \varphi_k\right) \ln \sum_{\ell} \varphi_\ell\right] / \sum_{m=1}^{J} \varphi_m \tag{2.22}$$

Conversely, if a detection occurs (and by our assumption there are no false detections) eq. 2.4 holds for the posterior distribution, and the posterior entropy, conditioned on detection in any $k \in J$, is:

$$H_D = -p'_k \ln p'_k - \sum_{j \neq k} p'_j \ln p'_j \tag{2.23}$$

$$= -1 \ln 1 - (J-1)\ 0 \ln 0 \equiv 0 \quad . \tag{2.24}$$

By computing these two conditional entropies with the appropriate probabilities of occurrence, we can obtain the expected entropy of the posterior ensemble: unconditional

$$H_E = P_D H_D + P_{ND} H_{ND}$$

$$= P_D H_D + (1-P_D) H_{ND} \quad . \tag{2.25}$$

Using equations 2.9, 2.22, and 2.24, we find an expression for H_E that is identical to eq. 2.19. The two approaches are thus equivalent, and the expected entropy, H_E, defined in terms of the expectation over the posterior distribution, is the same as that arising from computing the average mutual information between the two ensembles. The latter is the more fundamental approach, however; it should be followed in more complex cases such as those involving false alarms.

Having defined our basic terms, we can now return to the references to see what, in fact, the previous authors have discussed. None of those authors recognized explicitly the possibilities for confusion arising from the existence of various information and entropy values. Each, however, was consistent in doing calculations with the specific value he felt to be appropriate. Both Mela and Pollock used I_E. Richardson did his calculation with H_E and inferred max I_E based on min H_E; despite the consistency of his calculations, he did not draw a careful enough distinction between H_E and H_{ND} in presenting his conclusions. Barker's theorems refer to H_{ND}.

Because Mela and Barker were discussing different quantities, no inference of direct contradiction can immediately be drawn. The ramifications of their findings will now be discussed.

3. The Examples Cited by Mela and Pollock

The examples devised by Mela and Pollock have two features in common: Both consider discrete looks only, and both choose the total number of looks to be small--no more than the number of search cells, J. They differ in that Mela considers situations in which both the prior probabilities and the conditional detection probabilities are identical in all search cells; Pollock, on the other hand, treats the case in which both prior probabilities and conditional detection probabilities vary. Tables 1-3 list the particulars of their examples, expressed in the notation of this paper.

The starred entries in the tables indicate the respective maxima of P_D and I_E for the various search policies. It is obvious that the same policy leads to maxima of P_D and I_E in Mela's second example, and that different policies are required to maximize P_D and I_E in Mela's first example and in Pollock's example. From this it is self-evident that a given search policy does not always maximize both detection probability and information gain at the same time.

What is not self-evident -- yet is often asserted -- is that this fact further implies that there is no necessary connection between the extrema of P_D and I_E. As we show below, this assertion is, in fact, false. The existence of counterexamples, based on a small number of policy options and a small amount of search effort, shows that the connection between the extrema of P_D and I_E is not universal. In a

TABLE 1

MELA'S FIRST EXAMPLE

$J = 2;$

$p_1 = p_2 = .5;$

$\alpha_1 = \alpha_2 = .693$

SEARCH POLICY	P_D	$I_E = H_O - H_E$
(1,1)	.5*	.347
(2,0)	.375	.380*

TABLE 2

MELA'S SECOND EXAMPLE

$J = 3;$

$p_1 = p_2 = p_3 = 1/3;$

$\alpha_1 = \alpha_2 = \alpha_3 = .693$

SEARCH POLICY	P_D	$I_E = H_0 - H_E$
(1,1,1)	.5*	.549*
(2,1,0)	.417	.541
(3,0,0)	.292	.478

TABLE 3

POLLOCK'S EXAMPLE

$J = 3;$

$p_1 = .1,\ p_2 = .3;\ p_3 = .6;$

$\alpha_1 = +\infty,\ \alpha_2 = .511,\ \alpha_3 = .357$

SEARCH POLICY	P_D	$I_E = H_0 - H_E$
(1,0,0)	.1	.14*
(0,1,0)	.12	.07
(0,0,1)	.18*	.04

wide range of circumstances, however, the extrema of P_D and I_E do indeed depend on identical search policies.

We examine these first by relaxing the limitation of a small amount of search effort and by examining P_D and I_E as surfaces in a (J-1)-dimensional search allocation space. Although Mela and Pollock do not deal with H_{ND}, we shall examine that quantity numerically also, as a basis for analysis relating to Barker's work.

For Mela's first example, we consider a large number of discrete looks, C, allocated between the two cells; m looks are applied in cell 1, and C-m in cell 2. We then calculate P_D, H_{ND}, and H_E as functions of m for $0 \leq m \leq C$. Analytically:

$$P_D = 1 - [2^{-m} + 2^{-(C-m)}]/2 \tag{3.1}$$

$$H_{ND} = \ln \{[2^{-m} + 2^{-(C-m)}]/2\} + \{[2^{-m} + 2^{-(C-m)}]/2\}^{-1} \{(\ln 2)/2\} \cdot \{(m+1)2^{-m} + (C-m+1)2^{-(C-m)}\} \tag{3.2}$$

$$H_E = \{[2^{-m} + 2^{-(C-m)}]/2\} \ln [2^{-m} + 2^{-(C-m)}] + [(\ln 2)/2] \cdot [m2^{-m} + (C-m)2^{-(C-m)}] \tag{3.3}$$

These quantities are plotted in figs. 1-3 for the case C=30. H_{ND} is plotted on a linear scale (fig. 2); P_D and H_E are plotted logarithmically, as $-\log\ (1-P_D)$ and $\log\ H_E$, respectively (figs. 1 and 3).

The obvious content of figs. 1 - 3 and their supporting computations is that each of the three quantities -- P_D, H_{ND}, and H_E -- has a maximum at the same position, m=C/2=15. Equal allocation of search effort to the two cells produces the maximum detection probability, in keeping with Mela's finding for C=2, and the maximum value of H_{ND} as predicted by Barker's first theorem. The surprising finding is that the expected information gain, I_E , shows a minimum ($\max H_E$) for equal allocation of search effort. Though this is consistent with Mela's calculation, it could not be anticipated on the basis of previously published results.

We use a similar approach to Mela's second example, with ℓ looks allocated to the first cell, m looks to the second, and $C-\ell-m$ to the third. The expressions for P_D (ℓ,m) , H_{ND} (ℓ,m) , and H_E (ℓ,m) are completely analogous to those of equations 3.1 - 3.3 (but too unwieldly to reproduce here). Figs. 4 - 6 show the three-dimensional plots of P_D , H_{ND} , and H_E , for $C=30$, $0\leq m$, $\ell\leq 30$, $0\leq m+\ell\leq 30$. As before, P_D and H_E are plotted logarithmically; the H_{ND} scale is linear.

Figs. 4 - 6 give a general idea of the shape of the surfaces. The perspective of the figures is so oblique, however, that precise conclusions must rest on the back-up computations. These say that the uniform search allocation (10,10,10) produces the maximum of P_D and H_{ND} , and the minimum of H_E (maximum I_E) . The maxima of P_D and H_{ND} are both local and global maxima. But the minimum of H_E is a local minimum only. Smaller values occur on the boundaries, at (15,15,0), (15,0,15), and (0,15,15). Again, these findings are consistent with Mela's $C=3$ case and with Barker's theorem. But they are significantly different from Mela's first example, in that the uniform allocation now produces a maximum information gain rather than a minimum and that this maximum is relative, not absolute.

Pollock's example is modified slightly here for graphical presentation. The essence of Pollock's basic example is the variation of both α_i and p_i from cell to cell. To achieve extreme variation from cell to cell, Pollock chooses a value of 1 for q_1 , the conditional detection probability in the first cell. The assumption $q_1 = 1$ is equivalent to $\alpha_1 \rightarrow +\infty$ in our notation. This has the unfortunate effect of introducing a singularity into our formalism, because, in computing φ_1 , we are faced with the fact that:

$$\lim_{\alpha_1 \to \infty} \lim_{z_1 \to 0} p_1 e^{-\alpha_1 z_1} \neq \lim_{z_1 \to 0} \lim_{\alpha_1 \to \infty} p_1 e^{-\alpha_1 z_1} .$$

To avoid this singularity, we take $q_1 = .99$, which implies $\alpha_1 = 4.605$. This constitutes "Pollock's Example-Modified," which is presented in figs. 7 - 9, in a format comparable to that of Mela's second example. The effect of the cell-to-cell variation is preserved by the modified example, while the added complications of the singularity are circumvented.

As we shall see in a later section, the cell-to-cell variation of either α_i or p_i will destroy the symmetry of the surfaces in (ℓ,m) space. Variation of α_i alone will displace the relative locations of the stationary points of P_D, H_{ND}, and H_E, and can in fact change the whole character of the surface of H_E.

These points are not all obvious from figs. 7 - 9. It is clear that the symmetry has been destroyed in all three figures. H_E no longer has a local extremum within the surface. A saddle point appears at (2,10,18), but the extreme values of H_E occur at edge points. P_D and H_{ND} retain local maxima, at (2,11,17) and (1,11,18) respectively.

The different positions for the maxima of P_D and H_{ND} could be inferred from Barker's second theorem; the different positions for the maximum of P_D and the minimum of H_E are consistent with Pollock's findings. But the totally different character of the H_E surface (a saddle point in place of a local extremum) could not have been anticipated from published work.

The numerical experiments presented in this section were designed to be suggestive, not definitive, and to point the way to fruitful lines of analysis. What the experiments have suggested can be summarized concisely:

a. When p_i and α_i are the same for all search cells:

- P_D, H_{ND}, and H_E (or I_E) all have local extrema at a common point.
- The local extrema of P_D and H_{ND} are maxima.
- The local extremum of H_E may be either a maximum or a minimum.

- The local maxima of P_D and H_{ND} are also global maxima, but the local extremum of H_E is not necessarily a global extremum.

b. When the p_i and α_i differ from cell to cell:

- H_E does not necessarily have a local extremum.
- P_D and H_{ND} have local maxima. The positions of their respective maxima are close, but not identical.

The next section of this paper consists of analysis based on these observations.

4. Analysis -- Large Search Effort

In this section we pursue analytically some of the ideas suggested in the previous section by the numerical extension of the Mela and Pollock examples to the limit of large search effort. The mathematical framework is that established in section 2. We take the three functionals of the search allocation $\{z_j\}$: P_D , defined in eq. 2.9; H_{ND} , defined in eq. 2.22; and H_E , defined in eq. 2.19. We seek to determine the extrema of these functionals, subject to the constraint on total search effort (eq. 2.1).

We defer temporarily the precise specification of a "large amount of search effort." We assert that for $C > C^*$, for some C^* , the conventional calculus of variation technique using Lagrange multipliers to incorporate constraints is applicable and will lead us to the desired extrema. The determination of C^* in terms of the p_j and α_j will follow as part of the analysis.

First, it is important to note that the three quantities P_D , H_{ND} , and H_E depend on only two independent functionals of the $\{z_j\}$:

$$A \equiv \sum_{j=1}^{J} \varphi_j(z_j) \ln \varphi_j(z_j) \tag{4.1}$$

and

$$B \equiv \sum_{j=1}^{J} \varphi_j(z_j) \quad . \tag{4.2}$$

Specifically:

$$P_D = 1 - B \tag{4.3}$$

$$H_{ND} = \ln B - A/B \tag{4.4}$$

$$H_E = B \ln B - A \quad . \tag{4.5}$$

We can obtain some general insights using this simplified representation. Let the operator δ signify constrained variation; that is, for a function f, δf refers to the set of J quantities $\partial f/\partial z_j$, subject to $C = \sum z_j$. With this notation:

$$\delta P_D = -\delta B \tag{4.6}$$

$$\delta H_{ND} = \delta A(-(1/B)) + \delta B\left[(1/B) + (A/B^2)\right] \tag{4.7}$$

$$\delta H_E = -\delta A + \delta B(1 + \ln B) \quad . \tag{4.8}$$

From this set of equations we see that if there is an allocation of search effort $\{z_j^*\}$, such that δA and δB are both zero, then all three functionals P_D, H_{ND}, and H_E have extrema for that particular allocation $\{z_j^*\}$. Conversely, if δA and δB are not simultaneously equal to zero, then the extrema of the three functionals will occur for different search allocations, given respectively by:

$$\delta B = 0 \tag{4.9}$$

$$\delta A = \delta B(1 + (A/B)) : \quad (\delta A \neq \delta B \neq 0) \tag{4.10}$$

$$\delta A = \delta B(1 + \ln B) : \quad (\delta A \neq \delta B \neq 0) \tag{4.11}$$

Returning to the specific functional form of A and B, we find that:

$$\begin{aligned} \delta A &= (\partial\varphi_k/\partial z_k)\,(1 + \ln \varphi_k) - \lambda \\ &= -\alpha_k\varphi_k(1 + \ln \varphi_k) - \lambda \end{aligned} \tag{4.12}$$

for all $k \in J$. λ is a Lagrange multiplier. Similarly:

$$\delta B = -\alpha_k \varphi_k - \lambda \quad . \tag{4.13}$$

If δA and δB are to be simultaneously zero, we must have:

$$\alpha_k \varphi_k (1 + \ln \varphi_k) = \alpha_j \varphi_j (1 + \ln \varphi_j) \tag{4.14}$$

$$\alpha_k \varphi_k = \alpha_j \varphi_j \tag{4.15}$$

for all $j,k \in J$. Eliminating φ_j and φ_k between these equations, we obtain:

$$\alpha_j = \alpha_k \ , \quad \text{for all} \quad j,k \in J \quad . \tag{4.16}$$

Thus, δA and δB are simultaneously zero only if α is the same for each of the j search cells in J .

When all the α_j are equal, eqs. 4.14 and 4.15 reduce to:

$$\varphi_k (1 + \ln \varphi_k) = \varphi_j (1 + \ln \varphi_j) \tag{4.17}$$

and

$$\varphi_k = \varphi_j \ , \quad \text{for all} \quad j,k \in J \quad . \tag{4.18}$$

The only solution for these sets of equations is:

$$\varphi_k = \text{const} = K \ , \qquad \text{for all} \quad k \in J \quad . \tag{4.19}$$

This equation cannot be satisfied for small values of the total search effort, C . By requiring it to be satisfied, we can determine C^* , the lower bound on the search effort for which this method of analysis is valid. Using the definition of φ_k , (eq. 2.6), we infer from eq. 4.19:

$$p_1 e^{-\alpha_1 z_1} = p_2 e^{-\alpha_2 z_2} = \ . \ . \ . = p_J \quad . \tag{4.20}$$

In this formulation we have assumed that the p_j values are arranged in nonincreasing order, so that $p_J = \underset{J}{\text{Min}} \{p_j\}$.

From eq. 4.20, with $\alpha_j = \alpha_k$:

$$z_j = (1/\alpha)\ \ln\ (p_j/p_J)\ ; \qquad j = 1,2\ \text{--}\ J-1 \tag{4.21}$$

or:

$$C^* = (1/\alpha)\left(\sum_{j=1}^{J} \ln\ p_j - J\ \ln\ p_{min}\right) . \tag{4.22}$$

For $C > C^*$, eq. 4.19, leads to

$$z_k = (1/\alpha)\ \ln\ (p_k/K) \quad . \tag{4.23}$$

Further, because of the constraint on the z_k , this constant K can be evaluated in terms of the total search effort; thus:

$$z_k = (C/J) + (1/\alpha)\left[\ln\ p_k - (1/J) \sum_{j=1}^{J} \ln\ p_j\right] . \tag{4.24}$$

This is a well known result, closely related to that originally derived by Koopman, (reference 8) and expounded in Stone (reference 9). The optimum allocations for Mela's two examples follow from this trivially.

We may now summarize these results in two theorems:

Theorem I: For the assumptions of our model (section 2.1) with exponential detection functions and with $C > C^*$, the allocation of search effort $\{z_k^*\}$ that leads to an extremum of P_D also leads to an extremum of H_{ND} only if the α_k values are the same for all $k \in J$. (This is the alternative statement of one of Barker's results.)

Theorem II: For the assumptions of our model (section 2.1) with exponential detection functions and with $C > C^*$, the allocation of search effort $\{z_k^*\}$ that leads to an extremum of P_D also leads to an extremum of H_E only if the α_k values are the same for all $k \in J$. (New result.)

The allocation $\{z_k^*\}$ that leads to simultaneous extrema of all three functionals P_D , H_{ND} , and H_E is given by eq. 4.24.

We must now find out whether the extrema thus determined are maxima or minima. To do this, we examine the second derivatives. Let the second constrained variation, denoted by $\delta^2 f$, be defined by the quadratic form

$$\sum_{i,j} \zeta_i \zeta_j \, (\partial^2 f/\partial z_i \, \partial z_j) \text{ , subject to } C = \sum z_j \quad .$$

To compute $\delta^2 f$, it is more convenient to include the constraint explicitly, rather than use the Lagrange multiplier technique. This is done by noting that only J-1 of the z_j are independent. The first J-1 of the φ_j are taken as explicit functions of their respective z_j , and the remaining φ_J is taken as a function of

$$C - \sum_{j=1}^{J-1} z_j \quad .$$

In computing derivatives with respect to a specific z_k , therefore, both φ_k and φ_J must be considered.

With this convention, the various second partial derivatives can be written, after some manipulation, as:

$$\partial^2 P_D/\partial z_k^2 = - \left[\alpha_k^2 \varphi_k + \alpha_J^2 \varphi_J\right] \tag{4.25}$$

$$\partial^2 P_D/\partial z_k \partial z_\ell = - \alpha_J^2 \varphi_J \quad .$$

$$\partial^2 H_{ND}/\partial z_k^2 = (1/\sum_j \varphi_j)^2 \; (\alpha_k \varphi_k - \alpha_J \varphi_J)^2 \tag{4.26}$$

$$- \; (1/\sum_j \varphi_j) \; (\alpha_k^2 \varphi_k + \alpha_J^2 \varphi_J)$$

$$\partial^2 H_{ND}/\partial z_k \partial z_\ell = - \alpha_J^2 \varphi_J \left[1 + \ln \varphi_J - (\sum_{j=1}^{J} \varphi_j \ln \varphi_j)/(\sum_{j=1}^{J} \varphi_j)\right] /$$

$$(\sum_{j=1}^{J} \varphi_j) + (- \alpha_k \varphi_k + \alpha_J \varphi_J)(- \alpha_\ell \varphi_\ell + \alpha_J \varphi_J)$$

$$\partial^2 H_E/\partial z_k^2 = -(\alpha_k^2\varphi_k + \alpha_J^2\varphi_J) + (1/\sum_j \varphi_j)(\alpha_k\varphi_k - \alpha_J\varphi_J)^2 \quad (4.27)$$

$$-\alpha_k^2\varphi_k(\ln \varphi_k - \ln \sum_j \varphi_j)$$

$$-\alpha_J^2\varphi_J(\ln \varphi_J - \ln \sum_j \varphi_j)$$

$$\partial^2 H_E/\partial z_k \partial z_\ell = -\alpha_J^2\varphi_J \left[1 + \ln \varphi_J - \ln (\sum_{j=1}^{J} \varphi_j)\right]$$

$$+ (-\alpha_k\varphi_k + \alpha_J\varphi_J)(-\alpha_\ell\varphi_\ell + \alpha_J\varphi_J)$$

These are the second variations evaluated at the extrema, where the relationships obtained by equating the first variation to zero have been used to simplify the expressions, especially in the case of H_{ND} .

For the case discussed above, $\alpha_j = \alpha$ and $\varphi_j = K$ for all j , the second variations simplify to:

$$\delta^2 P_D = -\alpha^2 KD \quad (4.28)$$

$$\delta^2 H_{ND} = -((\alpha^2)/J)D \quad (4.29)$$

$$\delta^2 H_E = -\alpha^2 K(1 - \ln J)D \quad . \quad (4.30)$$

D is the residual quadratic form:

$$D = \sum_{j=1}^{J} \zeta_j^2 + \left(\sum_{j=1}^{J} \zeta_j\right)^2 \geq 0 \quad .$$

Noting that K , by definition, is a positive constant between zero and one, we have:

Theorem III: For the assumptions of our model (section (2.1)) with exponential detection functions, with $C > C^*$ and all α_j equal, the extrema of P_D and H_{ND} are always maxima. The extremum of H_E is a maximum for $J=2$ and a minimum for all $J > 2$.

This result clarifies a great deal of the mystery and confusion resulting from Mela's paper. His two cases, J=2 and J=3 , gave contradictory indications about H_E and its relation to P_D , and produced the false impression that there was no connection between detection probability and information gain I_E (Mela's I_E is equal to $(\ln J - H_E)$ in our terminology.) What we have shown here is that for the important case in which the conditional detection probability is uniform throughout the search area, optimizing the detection probability is identical with optimizing the change in the amount of information. Moreover, in all instances except the anomalous case J=2 , maximizing the detection probability is identical with maximizing the information gain.

The procedures leading to Theorems I-III guarantee only that we have found local extrema of the expected entropy surface. To complete the analysis, we must further investigate:

- Whether there are additional stationary points on the H_E surface.
- Whether there are in general points on the boundaries of the surface for which the value of H_E exceeds the value at the local extremum, as suggested by the numerical results in section 3.

We note first the values of the pertinent quantities at their local stationary points. Using the solution (eq. 4.24), we find that:

$$\mathrm{Max}\,\{P_D\} = 1 - \exp\left[\ln J + (1/J) \sum_{i=1}^{J} \ln p_i\right] \exp(-\alpha C/J) \quad (4.31)$$

$$\mathrm{Max}\,\{H_{ND}\} = \ln J$$

$$\mathrm{Extr}\,\{H_E\} = \ln J \left[\exp(\ln J + (1/J) \sum_{i=1}^{J} \ln p_i)\right] \exp(-\alpha C/J). \quad (4.32)$$

For equal prior probabilities, $p_i = 1/J$, these reduce to:

$$\mathrm{Max}\,\{P_D\} = 1 - \exp(-\alpha C/J) \quad (4.33)$$

$$\mathrm{Max}\,\{H_{ND}\} = \ln J \quad (4.34)$$

$$\text{Extr}\,\{I_E\} = H_0 - \text{Extr}\,\{H_E\} = \ell n\, J\,(1 - \exp(-\alpha C/J)) \qquad (4.35)$$

To check for the existence of other stationary points on the H_E surface, we return to eq. 4.11. That equation can be put in the form:

$$\alpha\varphi_k\,(\ln \varphi_k - \ln \sum_{i=1}^{J} \varphi_i) = \lambda \qquad (4.36)$$

or, by further manipulation:

$$-\gamma_k \ln\gamma_k = G \;, \qquad (4.37)$$

where

$$\gamma_k = \varphi_k / \sum_{j=1}^{J} \varphi_j \;;\; G = \lambda/\alpha \sum_{j=1}^{J} \varphi_j \;. \qquad (4.38)$$

The γ_k must satisfy $\sum \gamma_k = 1$, and, for a solution to exist:

$$0 \leq G \leq 1/e \;, \qquad (4.39)$$

as can be seen from figure 10.

The nature of the solutions can be argued from the shape of the curve in figure 10. Assume, first, that $G = 1/e$. Then, $\gamma_k = 1/e$, and $\sum \gamma_k = J/e$, which is less than 1 for $J=2$, and greater than 1 for J greater than 2. To drive $\sum \gamma_k$ toward 1, we decrease G . For $J=2$, the points must move down the right branch of the curve, increasing γ_k , until $\gamma_k = \frac{1}{2}$, $G = \frac{1}{2} \ln 2$. For all $J > 2$, the points must move down the left branch of the curve, decreasing γ_k , until $\gamma_k = 1/J$, $G = (\ln J)/J$. These represent the solutions $\varphi_k = K$, already discussed in equation 4.19 and those which follow.

If, however, we continue to decrease G , with all γ_k on the left branch of the curve, $\sum \gamma_k$ approaches zero in such a way that for some G^* , we can again make $\sum \gamma_k = 1$ by switching any one of the γ_k to the right branch of the curve. The condition for this is $\varphi_k = K1$; $\varphi_j = K2$; $j \neq k \in J$, for any specific k. Because of the shape of the curve, $\gamma_k \sim 1$, $\gamma_j \sim 0$, and thus $K1 >> K2$.

Let $K1 = \mu K2$, and $n = J-1$. Then, from eq. 4.37:

$$[\mu/(\mu+n)] \ln [\mu/(\mu+n)] = [1/(\mu+n)] \ln [1/(\mu+n)] \quad (4.40)$$

The second equation for determining the two free parameters, K1 and K2, is the constraint on total search effort. Using the definitions of the φ_k :

$$\alpha C/J = (1/J)\left[\sum_{k=1}^{J} \ln p_k - \ln \mu\right] - \ln K2 \quad . \quad (4.41)$$

Eq. 4.40 cannot be solved for $\mu(n)$, but it can be solved for $n(\mu)$:

$$n(\mu) = -\mu + \exp[(\mu \ln \mu)/(\mu-1)] \quad . \quad (4.42)$$

This equation is plotted in figure 11, with $\ln \mu$ as a function of n. For large μ :

$$n \sim \ln \mu + (1/\mu)[(\ln \mu)^2/2 + \ln \mu] + 0(\mu^{-2}) \quad . \quad (4.43)$$

For $n \geq 8$ $(J \geq 9)$, $n \sim \ln \mu$, $\mu \sim e^n$ within a few percent of accuracy. It is important to note from figure 11 that this type of solution exists only for $n > e-1$ $(J > e)$. This means that for $J = 2$, the only solution is the one with both γ points on the right branch of the curve 10 ; for $J \geq 3$, however, the two classes of solutions exist: a single solution (Type 1) with all γ points on the left branch of figure 10, and J solutions (Type 2) with any one of the J γ points on the right branch of the curve and the other $J-1$ on the left branch.

There is an additional condition on the amount of search effort, C, however. Just as there was a C^* threshold for the existence of the Type 1 solution, there are J additional thresholds for the appearance of each Type 2 solution. This is best seen when the search allocation solutions in the asymptotic limit are calculated:

$$z_k \cong (C/J) + (1/\alpha)\,\ell n\, p_k - (1/\alpha J) \sum_{\ell=1}^{J} \ell n\, p_\ell - (J-1)^2/\alpha J \quad (4.44)$$

$$z_j \cong (C/J) + (1/\alpha)\,\ell n\, p_j - (1/\alpha J) \sum_{\ell=1}^{J} \ell n\, p_\ell + (J-1)/\alpha J \quad (4.45)$$

Since the $z_k \geq 0$, eq. 4.44 provides the threshold criteria for C :

$$\alpha C_k^* \cong (J-1)^2 + \sum_{j=1}^{J} \ln p_j - J \ln p_k \quad (4.46)$$

$k = 1 \ldots J$

This, together with eq. 4.22, provides the values of C for which each of the $J+1$ stationary points of the H_E surface first appears. In the case of equal prior probabilities, $C^*=0$ from (4.22), and $\alpha\ C_k^*$ equals $(J-1)^2$ for all k .

Having found the approximate locations of the J additional stationary points of the H_E surface, we now find out what type of stationary points they are. Using eq. 4.27, we find:

$$\partial^2 H_E/\partial z_k^2 = -\alpha^2\ K2\ (1 - \ln J) \quad (4.47)$$

for those coordinates such that $\varphi_k = \varphi_J = K2$. Thus, for $J > 2$, we have a minimum of H_E in $J-1$ of the coordinates. For the remaining case, $\varphi_k = K1$, $\varphi_j = K2$.

$$\partial^2 H_E/\partial z_k^2 = -\alpha^2 K2\ [(\mu+1) - 2(\mu\,\ell n\,\mu)/(\mu-1) \quad (4.48)$$

$$- (\mu-1)^2 \exp(-(\mu\,\ell n\,\mu)/(\mu-1))]$$

where eq. 4.42 has been used to eliminate n . Direct calculation for $1 \leq \mu \leq \infty$ shows that $\delta^2 H_E \geq 0$ for all μ . This result seems to suggest, paradoxically, that <u>all</u> $J+1$ stationary points are minima.

To resolve this apparent paradox, we must be considerably more careful with our coordinates. We seek, in particular, at least one linear combination of the z_i, such that its second derivative has the opposite sign at the stationary point.

Let us introduce a variable x, $0 \le x \le 1$, such that:

$$z_k = (1-x)\left[(C/J) + (1/\alpha)\,\ell n\, p_k - (1/\alpha J)\sum_{\ell=1}^{J} \ell n\, p_\ell\right] \tag{4.49}$$

$$z_j = \left[(C/J) + (1/\alpha)\,\ell n\, p_j - (1/\alpha J)\sum_{\ell=1}^{J} \ell n\, p_\ell\right] \tag{4.50}$$

$$+ (x/(J-1))\left[(C/J) + (1/\alpha)\,\ell n\, p_k - (1/\alpha J)\sum_{\ell=1}^{J} \ell n\, p_\ell\right]$$

$j \neq k \in J$.

For ease of notation, define:

$$\beta_k/\alpha = (C/J) + (1/\alpha)\,\ell n\, p_k - (1/\alpha J)\sum_{\ell=1}^{J} \ell n\, p_\ell \quad . \tag{4.51}$$

Then equations 4.49 and 4.50 become:

$$z_k = (1-x)(\beta_k/\alpha) \tag{4.52}$$

$$z_j = (\beta_j/\alpha) + (x/(J-1))(\beta_k/\alpha) \quad . \tag{4.53}$$

By introducing this parameter, we achieve several results. Note that when $x=0$:

$$z_k = \beta_k/\alpha \; ; \; z_j = \beta_j/\alpha \tag{4.54}$$

These are the solutions for optimal detection search from eq. 4.24. When $x=1$, $z_k=0$, and:

$$z_j = C/(J-1) + (1/\alpha)\,\ell n\, p_j - (1/\alpha(J-1))\sum_{\ell\neq k}^{J-1} \ell n\, p_\ell \tag{4.55}$$

The latter are recognized as the optimal detection search allocations in J-1 cells, when cell k is not searched. These represent the class of whereabout searches, whose aim is to locate an object by searching all cells but one, then guessing the position on the basis of the search outcome. Of this class, the one for which $p_k = \operatorname{Max}_J(p_j)$ in the unsearched cell is called the "optimal whereabouts search." This search maximizes the probability of correctly guessing the object's location (reference 10).

Finally, for $x = (J-1)^2/J\beta_k$ we find:

$$z_k = \beta_k/\alpha - (J-1)^2/\alpha J \tag{4.56}$$

$$z_j = \beta_j/\alpha + (J-1)/\alpha J \ . \tag{4.57}$$

These are identical to the asymptotic forms of the positions of the stationary points of the H_E surface, determined in equations 4.44 and 4.45.

Thus, by use of the parameter x, we have introduced a set of J hyperplanes. Each of these hyperplanes connects (a) the Type 1 stationary point corresponding to the allocation for optimal detection search; (b) one of the J Type 2 stationary points; and (c) one of the J boundary points corresponding to a whereabouts search. We now investigate the curve formed by intersection of these J hyperplanes with the H_E surface.

Using equations 4.52 and 4.53, we can put H_E in this form:

$$H_E = \left(\exp - [(C/J)-(1/\alpha J \sum_{\ell=1}^{J} \ell n\, p_\ell]\right) \cdot \left\{- \beta_k x\, [\exp(\beta_k x)-\exp(-\beta_k x/n)] + [\exp(\beta_k x) + n \exp(-\beta_k x/n)]\, \ell n\, [\exp(\beta_k x) + n \exp(-\beta_k x/n)]\right\} \tag{4.58}$$

By computing $dH_E/dx = 0$, it can be shown that an equation identical to eq. 4.42 is obtained, with μ now identified as:

$$\exp\,[(\beta_k x)(n+1)/n] \quad . \tag{4.59}$$

Its solution is again the curve shown in figure 11, and the asymptotic form leads to:

$$x \simeq n^2/(n+1)(\beta_k) \quad . \tag{4.60}$$

This is indeed the position of the Type 2 stationary point as determined earlier.

The second derivative of H_E at the stationary point is:

$$d^2H_E/dx^2 \sim \Big[[(\mu\,\ell n\,\mu)/(\mu-1) + \mu^2\,\exp[-(\mu\,\ell n\,\mu)/(\mu-1)] - (\mu+1)]\Big] \quad . \tag{4.61}$$

In contrast to eq. 4.48, this is negative for all $1 \leq \mu < \infty$. Thus, along the curves generated by the J hyperplanes, H_E has a max at the stationary point, and we may safely describe the J Type 2 stationary points as saddle points.

Since the entropy has a max at the Type 2 stationary point along the x path, and then turns downward again, we may ask whether it later reaches a value below the minimum at the optimal detection search allocation. In fact, it does, if C is large enough. To find this threshold of C, we simply equate $H_E(0) = H_E(1)$, yielding:

$$J\,\ell n\,J = \Big[[\exp(\beta_k) + n\,\exp(-\beta_k/n)]\,\ell n\,[\exp(\beta_k) + n\,\exp(-\beta_k/n)]\Big] - \beta_k[\exp(\beta_k) - \exp(-\beta_k/n)] \quad . \tag{4.62}$$

In the dual limit $J \gg 1$, $\beta_k \gg \ln\,(J-1)$, this equation can be solved approximately:

$$\alpha C_{EQ}^{\;k} \cong \sum_{\ell=1}^{J} \ln p_\ell - J\,\ln p_k + J(J-1)\,\ln J \quad . \tag{4.63}$$

The minimum of this over the set J is:

$$\alpha C_{EQ}^{MIN} \cong \sum_{\ell=1}^{J} \ln p_\ell - J\,\ln p_{MIN} + J(J-1)\,\ln J \quad . \tag{4.64}$$

For any $C > C_{EQ}^{MIN}$, there is a whereabouts search that yields more expected information than does the optimal detection search.

We can now give a complete description of the expected information surface in allocation space, with the total search effort C as a parameter:

1. For $C < C^*$ (eq. 4.22), I_E has no stationary points. The extreme values of I_E occur on the boundaries.

2. For $C^* < C < C_k^*$ (eq. 4.46), and for $J > 2$, I_E has only one stationary point. That point is both relative and absolute maximum, and it occurs at the allocation of search effort corresponding to optimal detection search.

3. As C increases through the series C_k^* (k = 1, 2, -- J), J new stationary points appear on the I_E surface. All J are saddle points, whose coordinates are given approximately by equations 4.44 and 4.45. The one maximum point remains both relative and absolute maximum.

4. When C reaches C_{EQ}^{MIN} (eq. 4.64), the relative maximum is no longer the absolute maximum. The new absolute maximum of I_E occurs for the search allocation corresponding to the optimal whereabouts search. As C increases still further, new border regions are added to the surface, for which I_E exceeds the local maximum. The optimal whereabouts search allocation remains the absolute maximum of the surface, however.

Finally, it is interesting to compute the values of P_D and of the expected information at C_{EQ}^{MIN}, where the crossover occurs.

These are, for optimal detection search allocation:

$$P_D = 1 - p_{MAX}\, J^2 \exp(-J \ln J) \; ; \tag{4.65}$$

for optimal whereabouts search allocation:

$$P_D = (1 - p_{MAX}) - p_{MAX}\, (J-1) \exp(-J \ln J) \; ; \tag{4.66}$$

and, for the information in either case:

$$(I_{MAX} - I_E)/I_{MAX} = H_E/H_0 = (p_{MAX}/H_0) J^2 \ln J \exp\,(-J \ln J). \tag{4.67}$$

From these it can be seen that for even modest J ($J = 5$, say) P_D is extremely close to 1, and I_E is extremely close to I_{MAX}, by the time the crossover occurs.

We return now to the case in which the α_j are not equal. This corresponds to Pollock's example. In this case, the allocations $\{z_j\}$ that determine the extrema of P_D, H_{ND}, and H_E are given by the solutions of equations 4.9 - 4.11. They are, of course, not identical.

Using the Lagrange multiplier notation, we obtain the following equations. For $\delta P_D = 0$:

$$\alpha_k \varphi_k = \lambda_P , \qquad \text{all } k \in J ; \tag{4.68}$$

for $\delta H_{ND} = 0$:

$$\alpha_k \varphi_k \left\{ \left[(\ln \varphi_k) / \sum_j \varphi_j \right] - \left(\sum_j \varphi_j \ln \varphi_j \right) \Big/ \left(\sum_j \varphi_j \right)^2 \right\} = \lambda_{ND} ; \tag{4.69}$$

for $\delta H_E = 0$:

$$\alpha_k \varphi_k \left[\ln \varphi_k - \ln \sum_j \varphi_j \right] = \lambda_E . \tag{4.70}$$

The solution to eq. 4.68 is straightforward and well known:

$$z_k^{*P} = (\ell n\, \alpha_k p_k)/\alpha_k - \left[\left(\sum_{j=1}^{J} (\ell n\, \alpha_j p_j)/\alpha_j \right) \Big/ \alpha_k \sum_{j=1}^{J} (1/\alpha_j) \right] \tag{4.71}$$

$$+ C/\alpha_k \sum_{j=1}^{J} (1/\alpha_j)$$

Equations 4.69 and 4.70 are transcendental equations and appear more formidable, but, in point of fact, only eq. 4.70 is so formidable. If we divide eq. 4.69 by α_k and then sum on the index k, we find that:

$$\lambda_{ND} \sum_{k=1}^{J} (1/\alpha_k) = \left[\left(\sum_{k=1}^{J} \varphi_k \, \ell n \, \varphi_k\right) / \sum_{j=1}^{J} \varphi_j\right] \tag{4.72}$$

$$- \left[\left(\sum_{k=1}^{J} \varphi_k \sum_{j=1}^{J} \varphi_j \, \ell n \, \varphi_j\right) / \left(\sum_{j=1}^{J} \varphi_j\right)^2\right] = 0 \quad .$$

Since the $1/\alpha_k$ are all nonzero quantities of the same sign, we must have:

$$\lambda_{ND} = 0 \quad . \tag{4.73}$$

Putting this back into eq. 4.69, we get a much simplified equation. Since $\alpha_k \varphi_k$ is not zero, we must have:

$$\ln \varphi_k = \left(\sum_{j=1}^{J} \varphi_j \ln \varphi_j\right) / \sum_{j=1}^{J} \varphi_j \quad . \tag{4.74}$$

Because the right side is independent of k, the only solution to this set of equations is:

$$\varphi_k = \text{const} = K \, , \qquad \text{for all} \quad k \in J \quad . \tag{4.75}$$

This is identical in form to the result found in the case of equal α_k. But the solution for z_k^* is somewhat different:

$$z_k^{*ND} = (\ell n \, p_k)/\alpha_k - \left[\left(\sum_{j=1}^{J} (\ell n \, p_j)/\alpha_j\right) / \alpha_k \sum_{j=1}^{J} (1/\alpha_j)\right] \tag{4.76}$$

$$+ \, C/\alpha_k \sum_{j=1}^{J} (1/\alpha_j) \quad .$$

The difference between the solutions (equations 4.71 and 4.76) is a functional of the α_k only, and does not depend on the prior probabilities p_k. Specifically:

$$\Delta_k \equiv z_k^{*P} - z_k^{*ND} \tag{4.77}$$

$$= (1/\alpha_k)\left[\ell n\, \alpha_k - \left(\sum_{j=1}^{J} (\ell n\, \alpha_j)/\alpha_j\right) / \sum_{j=1}^{J} (1/\alpha_j)\right] .$$

Note that when all α_k are equal, both z_k^{*P} and z_k^{*ND} reduce to z_k (eq. 4.24), and the $\Delta_k \equiv 0$ for all k .

Equations 4.71 and 4.76 can be used to calculate the optimum allocation of search effort in Pollock's modified example. The analytic solutions are, for P_D :

$z_1 = 1.497$

$z_2 = 11.334$

$z_3 = 17.169$

and, for H_{ND} :

$z_1 = .996$

$z_2 = 11.126$

$z_3 = 17.878$.

These values should be compared with those found in section 3 to be the maximum points on the integer grids: (2,11,17) and (1,11,18), respectively.

It is significant that the analytically determined maxima of P_D and H_{ND} are fairly close. With α_1 large, Pollock's example represents a major departure from a uniform probability of conditional detection. Nevertheless, the maxima show only a small displacement from each other, although they show a large displacement from their locations when all the α_j values are equal. (A representative example, using Pollock's p_j , but taking $\alpha_j = \alpha = .693$ for all j , yields $z_1 = 8.61$, $z_2 = 10.19$, $z_3 = 11.20$). Thus, in practical situations, optimizing H_{ND} will lead to search allocation nearly equal to that required for optimum detection probability.

These solutions, too, are valid for large search efforts only. The thresholds for their validity are determined from equations 4.68 and 4.75, using arguments analogous to those previously given. The results are, for P_D :

$$C^* = \sum_{j=1}^{J} (\ln \alpha_j p_j)/\alpha_j - \left(\sum_{j=1}^{J} (1/\alpha_j) \right) \ln [\alpha p]_{min} \tag{4.78}$$

and, for H_{ND} :

$$C^* = \sum_{j=1}^{J} (\ln p_j)/\alpha_j - \left(\sum_{j=1}^{J} (1/\alpha_j) \right) \ln p_{min} \tag{4.79}$$

Of greater theoretical significance than the z_k are the relative fractions of the total search effort assigned to each cell:

$$\zeta_k = z_k/C \quad .$$

These are trivally determined from equations 4.71 and 4.76:

$$\zeta_k^{*P} = \left(1/\alpha_k \sum_{j=1}^{J} (1/\alpha_j) \right) + (1/C) \left[(\ln \alpha_k p_k)/\alpha_k - \left(\sum_{j=1}^{J} (\ln \alpha_j p_j)/\alpha_j \right) /\alpha_k \sum_{j=1}^{J} (1/\alpha_j) \right] \tag{4.80}$$

$$\zeta_k^{*ND} = \left(1/\alpha_k \sum_{j=1}^{J} (1/\alpha_j) \right) + (1/C) \left[(\ln p_k)/\alpha_k - \left(\sum_{j=1}^{J} (\ln p_j)/\alpha_j \right) /\alpha_k \sum_{j=1}^{J} (1/\alpha_j) \right] \tag{4.81}$$

These clearly converge to the same asymptotic limit as $C \to \infty$:

$$\zeta_{k\infty}^{*P} = \zeta_{k\infty}^{*ND} = 1/\alpha_k \sum_{j=1}^{J} (1/\alpha_j) \tag{4.82}$$

No simple strategem is available for dealing with equation (4.70), and closed-form solutions for the optimum search allocation do not appear possible in the case of H_E. Some limited statements can be made about the nature of the solutions, however. If we eliminate λ_E from any two equations in eq. 4.70, and use ζ_i rather than z_i, we get:

$$\begin{aligned} &\left[\alpha_i p_i \exp(-\alpha_i \zeta_i C)\right] \cdot \\ &\quad \ln\left[\left(p_i \exp(-\alpha_i\zeta_i C)\right) / \sum_{j=1}^{J} p_j \exp(-\alpha_j\zeta_j C)\right] \\ &\quad = \left[\alpha_k p_k \exp(-\alpha_k\zeta_k C)\right] \cdot \\ &\quad \ln\left[\left(p_k \exp(-\alpha_k\zeta_k C)\right) / \sum_{j=1}^{J} p_j \exp(-\alpha_j\zeta_j C)\right] . \end{aligned} \tag{4.83}$$

This can be further manipulated into the form:

$$\begin{aligned} (\alpha_i/\alpha_k) = &\left\{\left[(p_i/p_k) \exp\left(-C(\alpha_k\zeta_k - \alpha_i\zeta_i)\right)\right] \cdot \right. \\ &\quad \ln\left[\sum_{j=1}^{J} (p_j/p_k)\exp\left(-C(\alpha_j\zeta_j - \alpha_k\zeta_k)\right)\right] \\ &\quad \left. / \ln\left[\sum_{j=1}^{J} (p_j/p_i) \exp\left(-C(\alpha_j\zeta_j - \alpha_i\zeta_i)\right)\right]\right\} . \end{aligned} \tag{4.84}$$

Since the left side of this equation is independent of C, the right side must be, as well, if the equation is to remain valid in the asymptotic limit $C \to \infty$.

Because the α_k values are given and independent of C and because, by definition, $0 \le \zeta_i \le 1$, then $\alpha_i\zeta_i$ must be $O(C^0)$ in its leading time. The most general form that satisfies these requirements is:

$$\zeta_i = \left[1/\alpha_i \sum_{j=1}^{J} (1/\alpha_j)\right] + (f_i/\alpha_i\ C) \tag{4.85}$$

This renders eq. 4.47 independent of C, to all orders, but still makes ζ of order C^0. The functions f_i are functions of the α_j, and p_j; they must satisfy:

$$\left[\alpha_i p_i \exp(-f_i)\right] \ell n \left[\left(p_i \exp(-f_i)\right) / \sum_{j=1}^{J} p_j \exp(-f_j)\right] = \lambda \tag{4.86}$$

and

$$\sum_{j=1}^{J} (f_j/\alpha_j) = 0 \tag{4.87}$$

An explicit solution for the f_i of eq. 4.86 is no more feasible than an explicit solution for the φ_i of eq. 4.70. But, by the artifice of eq. 4.85, we have separated out the asymptotic part of the solution. This leads to the very important Theorem IV:

Theorem IV: Under the assumptions of section 2.1, the allocations of the fraction of search effort among the J search cells that produce stationary points of P_D, H_{ND}, and H_E have a common asymptotic limit as the amount of search effort C approaches infinity. That limit is:

$$\zeta_i^\infty = \left[1/\alpha_i \sum_{j=1}^{J} (1/\alpha_j)\right] \tag{4.88}$$

The final topic to address in this section is whether the extrema are maxima or minima in the case of differing α_j. This is easily answered in the case of P_D and H_{ND}. Combining equations 4.25 and 4.68, we get:

$$\delta^2 P_D \sim -\lambda_P(\alpha_k + \alpha_J) \qquad \text{for all } k \in J \ . \tag{4.89}$$

λ_P , α_k , and α_J are all positive, and P_D , therefore, always has a maximum.

Similarly, when we combine equations 4.26 and 4.75:

$$\delta^2 H_{ND} \sim -2\alpha_k \alpha_J / J^2 - (\alpha_k^2 + \alpha_J^2)\,[(1/J) - (1/J^2)] \qquad (4.90)$$

This is always negative for positive integer J ; therefore, H_{ND} , too, always has a maximum.

No general rule appears appropriate for application to $\delta^2 H_E$. Though analytical results are not forthcoming, we have already seen in the numerical examples that maxima, minima, and saddle points are all possible as the stationary points of the H_E surface.

5. Analysis -- Infinitesimal Search Effort

We turn now to the opposite limit, in which the total amount of search effort approaches zero. In particular, we consider the case in which an infinitesimal amount of search effort Δz_k is applied in only one cell, $k \in J$. We then apply the usual limiting processes of calculus to compute the rates of change of P_D , H_{ND} , and H_E as a result of searching in cell k .

The quantities A and B , defined in equations 4.1 and 4.2 are, for search in a single cell:

$$A = -H_0 - (p_k \ln p_k)[1 - \exp(-\alpha_k \Delta z_k)] - \alpha_k \Delta z_k p_k \exp(-\alpha_k \Delta z_k) \; ; \qquad (5.1)$$

$$B = 1 - p_k\,[1 - \exp(-\alpha_k \Delta z_k)] \qquad (5.2)$$

where H_0 has been defined in eq. 2.17. When A and B are expanded to first order in Δz_k and substituted in equations 4.3, 4.4, and 4.5, we obtain:

$$P_D = (\Delta z_k)\alpha_k p_k + O(\Delta z_k^2) \qquad (5.3)$$

$$H_{ND} = H_0 + (\Delta z_k)\alpha_k p_k (H_0 + \ln p_k) \qquad (5.4)$$

$$+ O(\Delta z_k^2)$$

$$H_E = H_0 + \Delta z_k \alpha_k p_k \ln p_k \quad . \tag{5.5}$$

Thence:

$$\lim_{\Delta z_k \to 0} (P_D - P_0)/\Delta z_k \equiv dP_D/dz_k = \alpha_k p_k \tag{5.6}$$

$$\lim_{\Delta z_k \to 0} (H_{ND} - H_0)/\Delta z_k \equiv dH_{ND}/dz_k = \alpha_k p_k (H_0 + \ln p_k) \tag{5.7}$$

$$\lim_{\Delta z_k \to 0} (H_E - H_0)/\Delta_k \equiv dH_E/dz_k = \alpha_k p_k \ln p_k \quad . \tag{5.8}$$

Our basic concerns will be with the magnitude of these rates of change, and with the extreme values of the magnitudes over the set J ; we shall determine these extreme values. First, however, some observations about the signs are significant. Recalling that $0 \le \alpha_k < \infty$, and $0 \le p_k \le 1$ for all $k \in J$, we note immediately that:

$$dP_D/dz_k \ge 0 \; ; \qquad \text{all } k \in J \tag{5.9}$$

$$dH_E/dz_k \le 0 \; ; \qquad \text{all } k \in J \quad . \tag{5.10}$$

The corresponding result for dH_{ND}/dz_k is less obvious. However, the reader may persuade himself by selected numerical examples that dH_{ND}/dz_k can be either positive or negative, depending on the set of the prior probabilities p_j and on the searched cell, k . (Alternatively, we may note that $\sum (1/\alpha_k) dH_{ND}/dz_k = 0$, implying that the signs of dH_{ND}/dz_k are not the same for all k .

The observation concerning dP_D/dz_k is scarcely surprising. But those related to the entropies are significant:

Theorem V. For any search operation conducted according to the assumptions of section 2.1, the expected entropy never increases (and the expected information never decreases), regardless of the cell chosen for search. The entropy that is conditioned on non-detection of the target, however, may either increase or decrease, depending on local conditions.

It is worth pointing out that the rates of change of P_D and H_E depend entirely on local conditions, i.e., on the prior probability p_k in the cell being searched; the rate of change of H_{ND}, on the other hand, depends globally on the entire prior ensemble, through H_0. It is therefore conceivable that for two distinct prior ensembles, the rates of change of H_{ND} could vary in both magnitude and sign, even though the individual cells chosen for search in the two cases had identical prior probabilities.

We now determine whether the results obtained in section 4 still apply in the limit $\Delta z_k \to 0$. Specifically, does the allocation that maximizes P_D also maximize H_{ND} and produce an extremum (either maximum or minimum) in H_E ?

We have already noted that in the case in which the α_k are different in different cells, no results of useful generality were obtained. We therefore restrict ourselves to the case $\alpha_k = \alpha$, all $k \in J$, and -- for simplicity in this section -- let $\alpha = 1$. Then, the rates of change for P_D, H_{ND}, and H_E are:

$$p_k, p_k(H_0 + \ln p_k), \text{ and } p_k \ln p_k ,$$

respectively.

Let us assume that the cells are numbered in order of non-increasing magnitude of p_k : $p_1 \geq p_2 \geq --- \geq p_J \geq 0$. By this device, we make sure that infinitesimal search in cell 1 will produce the highest rate of increase in P_D, and hence, maximize P_D in the limit $\Delta z \to 0$. We then investigate the extrema over the set J for the rates of change of H_{ND} and H_E.

$\underline{H_{ND}}$. We noted earlier that the sign of dH_{ND}/dz_k might be either positive or negative, depending on the prior ensemble and on the searched cell, k. This general observation can be made somewhat more precise. Some preliminary observations:

1. For a given prior ensemble, H_0 is a fixed positive constant.

2. The factor p_k is monotonically nonincreasing with the index k because of the assumed ordering. It is always ≥ 0 .

3. The factor $H_0 + \ln p_k$ is monotonically nonincreasing with the index k from some initial value, and may approach $-\infty$ for a p_k that is small enough. Whether the maximum value of $H_0 + \ln p_k$ is positive or negative remains to be determined.

We shall later prove that the max of $H_0 + \ln p_k$ is in fact nonnegative for any prior ensemble. For the moment, we assume that this is true and derive the consequences.

By the assumption, there is a range of k , $1 \leq k \leq K < J$, for which $H_0 + \ln p_k \geq 0$. Then, in this range, $p_k(H_0 + \ln p_k)$ is a positive, monotonically nonincreasing function of k , because it is a product of two positive, monotonically nonincreasing functions of k . In this range:

$$p_1(H_0 + \ln p_1) \geq p_k(H_0 + \ln p_k) \tag{5.11}$$

for $1 < k \leq K$. Outside this range $H_0 + \ln p_k < 0$, and the monotonicity of the product is no longer assured. Since the product is then negative, however, it is obviously less than $p_1(H_0 + \ln p_1)$. Thus, if $p_1 \geq p_k$ for any $1 < k \leq J$, then $p_1(H_0 + \ln p_1) \geq p_k(H_0 + \ln p_k)$ for any $1 < k \leq J$.

Now we prove the assumption. We need to show that for any prior distribution, there is at least one p_k such that $H_0 + \ln p_k$ is nonnegative. This is easily done by noting that

$$H_0 + \ln p_k = \sum_{j=1}^{J} p_j(-\ln p_j + \ln p_k) = \sum_{j=1}^{J} p_j \, (\ln(p_k/p_j)) \, . \tag{5.12}$$

If p_k is $\underset{J}{\text{Max}}\, p_j$, then each term in the sum is nonnegative, and

$$H_0 + \ln p_{max} \geq 0 \, . \tag{5.13}$$

One final comment about dH_{ND}/dz_k . We have seen that, for the largest p_k in J , $dH_{ND}/dz_k \geq 0$; in addition, for sufficiently small p_k values, $dH_{ND}/dz_k \to 0^-$. This sort of behavior suggests that for some intermediate value of p_k , dH_{ND}/dz_k has a relative minimum < 0 (which, in fact, is an absolute minimum, too). This is best illustrated by a specific model.

Consider a countably infinite set of search cells, in which the (ordered) prior probabilities are:

$$p_k = (1 - \gamma)\,\gamma^{k-1} \tag{5.14}$$

Clearly

$$\sum_{k=1}^{\infty} p_k = 1$$

when the sum converges $(0 \leq \gamma < 1)$.

This model is a useful tool for investigating some situations involving finite J . General analytical results can be obtained with the infinite model, yet that model can be made to approximate a finite model reasonably well by truncation of the series after J-1 terms and lumping all the remaining probability

$$\sum_{k=J}^{\infty} p_k$$

together as the J th term. For $\gamma < \frac{1}{2}$, the total residual probability in (J,∞) is less than the J-1 term, and the decreasing ordering is preserved.

For this model, with $\alpha_k = \alpha = 1$, we can show that:

$$dH_{ND}/dz_k = (1/\gamma)\,|\ell n\,\gamma|\;\exp(-k|\ell n\,\gamma|)[1 - k(1-\gamma)] \tag{5.15}$$

This expression is positive for small k and negative for large k . The crossover point is given by:

$$k^+(\gamma) = 1/(1-\gamma) \tag{5.16}$$

which is monotonically increasing from 0 to ∞ as γ goes from 0 to 1^-. Thus, by choice of the parameter γ, we can make an arbitrarily large number of cells have positive values of dH_{ND}/dz_k.

Treating k as a continuous variable and differentiating, we find that:

$$\partial(dH_{ND}/dz_k)/\partial k = 0 \qquad \text{for } k = k^* , \tag{5.17}$$

where:

$$k^*(\gamma) = 1/(1-\gamma) + 1/(|\ln \gamma|) \tag{5.18}$$

This can be shown to be a minimum of dH_{ND}/dz_k. Like $k^+(\gamma)$, $k^*(\gamma)$ is also a monotonically increasing function of γ, going from 1 to $+\infty$ as γ goes from 0 to 1^-. The position of the minimum can also be situated at arbitrarily large values of k, depending on the choice of γ, although clearly the choices of $k^+(\gamma)$ and $k^*(\gamma)$ are not independent of each other.

$\underline{H_E}$. An analysis of the relative magnitudes over $\{k\}$ of $p_k \ln p_k$ is closely related to a problem already solved by Browning (reference 11). Browning treats a slightly more complicated case, in which the amounts of search effort are finite ("looks") rather than infinitesimal. That makes his proofs more complex than the present case requires, but it does not alter their validity or their relevance to the infinitesimal limit.

We shall simply state the necessary results with qualitative justifications. The reader may consult Browning for analytical proofs.

We shall consider the rate of change of the expected information, rather than the expected entropy: $dI_E/dz_k = -p_k \ln p_k \geq 0$. Figure 12 shows the function $y(p) = -p \ln p$ in the domain $0 \leq p \leq 1$. This curve is asymmetrical, having a maximum at $(1/e, 1/e)$. We must visualize along this curve the set of points $y_k = y(p_k)$, subject to $\sum p_k = 1$, and to the ordering $p_1 \geq p_2 \geq p_3 \geq - - - \geq p_j \geq 0$.

Browning has shown that the following cases should be distinguished:

1. $1/e \geq p_1$. In this case all the y_k are on the increasing part of the curve $y(p)$; therefore, the ordering of the p_k guarantees the ordering of the y_k :

$y_1 \geq y_2 \geq y_3 \geq - - - \geq y_k \geq 0$.

2. $1/e < p_2$. In this case, which can still satisfy the constraint

$$\sum_{k=1}^{J} p_k = 1 ,$$

y_1 is on the decreasing segment of $y(p)$; therefore, $y_2 \geq y_1$.

3. $p_2 \leq 1/e \leq p_1$. This case is problematical and is best resolved by numerical calculation.

Before turning to relevant calculations, we must resolve the question whether y_3 could be the largest of the y_k under any conditions. Reference to the figure shows this is obviously not possible if $p_2 \leq 1/e$, since then $y_3 \leq y_2$. On the other hand, if both p_1 and p_2 are $\geq 1/e$, then the maximum value possible for p_3 is $(1 - 2/e) < 1/e$, attained when $p_1 = p_2 = 1/e$. There again $y_3 < y_1 , y_2$. Any increase in p_1 and p_2 above $1/e$ can occur only at the expense of p_3 , which further decreases y_3 at a rate faster than either y_1 or y_2 because of the steeper slope for $p < 1/e$. Thus, there are no conditions for which $y_3 = \underset{J}{\mathrm{Max}}\, y_k$; we need therefore consider only cells 1 and 2 when allocating search effort to produce the maximum rate of information increase.

Figure 13 shows the relevant portion of the p_1,p_2 plane. The boundary curves are $p_1 + p_2 = 1$ (the limiting case of $\sum p_k = 1$) , $p_1 = p_2$ (the limiting case of the assumed ordering $p_1 \geq p_2$) , and $y(p_1) = y(p_2)$. In region I, $y_1 > y_2$; in region II, $y_2 > y_1$.

It is important to note that the line $p_1 + p_2 = 1$ is a boundary of region II only. This implies that for the case

$J=2$, $y_2 > y_1$, except for the endpoint $y_2(0) = y_1(1) = 0$. When there are only two search cells, the maximum rate of increase of information is always attained by searching the cell with the lower prior probability, and the minimum rate of increase of information is always attained by searching the cell with the higher prior probability. This in agreement with the results obtained for the asymptotic case $C \to \infty$.

When $J > 2$, however, a new feature enters. There is a significant region in the p_1 , p_2 plane, where the maximum rate of information increase is attained by searching in the cell with the second highest prior probability. The cell with the highest prior probability, which is searched to maximize the growth of P_D , does not represent either a maximum or a minimum in the rate of information growth. The results of the asymptotic analysis, therefore, do not carry over into the infinitesimal limit, and we cannot assert universally that the allocation that maximizes P_D also produces an extremum in H_E (or I_E) .

Reference to our infinite-number-of-cells model will illustrate some of these points. For that model:

$$dI_E/dz_k = -(1/\gamma)(1-\gamma)\exp(-k|\ell n\ \gamma|)[\ell n(1-\gamma) + (k-1)\ \ell n\ \gamma] \quad ,(5.19)$$

which is, of course, always ≥ 0 for any k . Again, treating k as a continuous variable, we find that dI_E/dz_k has a maximum at:

$$k_E^*(\gamma) = 1 - [\ell n(1-\gamma)]/\ \ell n\ \gamma - 1/\ \ell n\ \gamma \quad . \qquad (5.20)$$

The behavior of $k_E^*(\gamma)$ differs significantly from that of $k_{ND}^*(\gamma)$. As γ goes from 0 to 1^- , $k_E^*(\gamma)$ starts at 1 , increases slightly to a maximum of ~ 1.5422 at $\gamma \cong .351575$, then decreases monotonically toward $-\infty$. Of the positive integers, only $k = 1$ or 2 are closest integers to this curve for any γ , as suggested by our previous discussion. Thus, only cells 1 or 2 need be considered candidates for $\underset{J}{\mathrm{Max}}\{dI_E/dz_k\}$.

The results of this section are summarized as follows. For ordered probabilities $p_1 \geq p_2 \geq p_3 \geq - - - \geq p_J \geq 0$:

a. The rate of change of P_D is always positive. It has

a maximum for search in cell 1 and a minimum for search in cell J .

b. The rate of change of H_{ND} may be either positive or negative. It has a nonnegative maximum for search in cell 1, and a negative minimum for search in some intermediate cell m : $1 < m < J$.

c. The rate of change of I_E is always positive. If J=2 , it has a maximum for search in cell 2 and a minimum for search in cell 1. If $J > 2$, then, either (1) it has a maximum for search in cell 1 and a minimum for search in cell J , or (2) it has a maximum for search in cell 2 and a minimum for search in either cell J or cell 1.

6. Summary

The relation between search theory and information theory remains complex.

Previous attempts to attack the problem have focused on only a narrow part of the relationship. Their narrow focus and the lack of clear definition of "information" or "entropy" created an unwarranted impression that the body of work was contradictory, if not incorrect. It is not.

The work reported in the earlier references is both correct and, when viewed from a broader perspective, consistent. The individual results reported earlier have been useful in establishing this broader perspective.

In the present paper, some of these earlier results have been rederived from a different viewpoint. Other new results have been added. Taken together, these findings are enough to allow us to sketch a coherent, though still incomplete, picture of the relation between search theory and information theory.

Equally important, we can now identify the gaps somewhat more clearly and can direct future research toward those gaps with the confidence that there is an underlying body of theory to be completed, rather than a mere collection of isolated observations.

Before summarizing the findings, it is important for us to recapitulate the assumptions. These are the most critical assumptions of our model, which are used consistently throughout this paper: There is a single target; it is stationary; no false detections occur; and the detection process is governed by an exponential detection function. Relaxation of any of these assumptions would add considerable complexity to the theory and might change the conclusions significantly. Within the framework of these general assumptions, this paper has examined subcases in which (a) the amount of search effort is either very large or very small, and (b) the conditional detection probability is either uniform or variable throughout the search cells.

Summary of Findings: Uniform Conditional Detection Probability.

a. The allocation of search effort that maximizes the posterior entropy conditioned on nondetection H_{ND} also maximizes the probability of detection P_D . This result was first established by Barker and was confirmed by the present author in both the large and small search effort limits. Since Barker's proof is not restricted to these limiting cases, the result is valid in the intermediate range as well.

b. In the large search effort limit, the allocation of search effort that maximizes the probability of detection, P_D also produces a local extremum in the expected information I_E . For two search cells, this extremum is a local minimum of I_E . For any number of search cells greater than two, this extremum is a local maximum of I_E . This is consistent with Mela's calculations. (Although the number of looks is only equal to J in Mela's examples, this still corresponds to the large search effort limit, since $C^* = 0$ by eq. 4.22.)

c. The local extremum of I_E is not necessarily the global extremum. For an intermediate range of C , $C^* \leq C \leq C_{EQ}^{MIN}$, the local extremum is also the global extremum. For $C_{EQ}^{MIN} < C$, there is always at least one whereabouts search (and there may be as many as J), corresponding to an edge point of the I_E surface, which produces more expected information than the optimal detection search. In practical cases the distinction appears unimportant, since $P_D \simeq 1$ and $I_E \simeq I_{MAX}$, when C reaches C_{EQ}^{MIN} .

d. In the small search effort limit, the exact correspondence between the maximum of P_D and the extremum of I_E breaks down. We achieve the maximum rate of increase of I_E by searching in the cell that has either the largest or second largest rate of increase of detection probability. In the two-cell case, this is at least consistent with the large search effort limit. For more than two cells, however, this is a distinct phenomenon, constituting another demonstration that the connection between optimum search and optimum information gain is not universal. Working with P_D and I_E directly, rather than with rates of change, Browning has obtained comparable results in the intermediate search effort region for cases $J=2$ and $J=3$. There remains an open question: precisely how the transition between the two limits takes place and whether the C^* determined in (4.22) is in fact the minimum value for which the large search effort limit is valid.

e. The rate of change of I_E is always positive for any search allocation. The rate of change of H_{ND} can be either positive or negative, depending both on the search allocation and on the ensemble of prior probabilities in the search cells.

Summary of Findings: Variable Conditional Detection Probability.

a. In the large search effort limit, both P_D and H_{ND} have local maxima. The search allocations leading to those maxima are not the same, as is demonstrated by eq. 4.77. This finding is consistent with Barker's second theorem.

b. In the large search effort limit, no closed-form solution for the stationary points of I_E can be determined. Even the nature of the stationary points cannot be established analytically, but the numerical examples have shown that maximum, minimum, and saddle point are all possibilities. In the asymptotic limit $(C \to \infty)$, the positioning of the stationary points of P_D , H_{ND} , and I_E all converge to a common limit:

$$(C/\alpha_i)\ (1/\sum_{j=1}^{J} (1/\alpha_j))$$

c. In the small search effort limit, no systematic relationships have been established among the maxima of P_D , H_{ND} , and I_E .

d. That the rate of change of I_E is positive for any search allocation, even when the conditional detection probabilities vary, remains valid.

Interpretation. Only in the case of uniform conditional detection probability do we have enough results to attempt some interpretation. There, however, some simple statements can do much to dispel the legacy of past confusion.

First, using Barker's theorems and our present results on the rates of change of P_D and H_{ND}, we have noted that identical search policies produce the maximum increases of both detection probability and posterior conditional entropy. Like all entropies, H_{ND} is a negative information. Thus, maximizing its rate of increase is equivalent to maximizing the rate of decrease of an information. The information that is being decreased is that contained in the ensemble up to the time of detection: the information about the target location that is expressed by the prior probabilities and that changes as the search progresses. With this view, we can state one general conclusion, already correctly anticipated by Richardson:

- The search policy that maximizes the probability of detection is the one that uses up the information contained in the prior ensemble at the maximum rate.

Second, we have shown in this paper that in most cases the optimum policies for P_D and I_E are equivalent. Here, however, information is being created rather than used up. Because this is expected information, visualizing the overall process of information creation may be harder. Consider a repeated search in one of the J cells. If that cell has a prior probability p_k of containing the object, its initial self information is $\ln p_k$. At each stage of of the search, if detection does not occur, p_k decreases; consequently, the actual self information of the cell also decreases. The probabilities of the other J-1 cells, however, increase simultaneously (because of the constraint $\sum_{j=1}^{J} p_j = 1$), as do their values of self information. If detection does occur in the kth cell after some number of unsuccessful searches, its self information undergoes a sudden increase; the self information of the other cells then drops.

When we speak of ensemble averages, as we have throughout this paper, the competing effects of these information-creating and -destroying processes are combined, with a probabilistic weighting. Until the time of detection, the loss of information caused by failure to detect in the cell being searched is weighed against the inferred gain in information in the other cells. The net change of information may be either positive or negative, as we noted in our discussion of the rate of change of H_{ND} . If detection occurs, the positive and negative information contributions from that process must also be added in, with the proper weighting. Then, as indicated by the rate of change of I_E , the overall information gain is net positive.

This interpretation of the creation and destruction of information leads to our second general conclusion:

• In a wide range of practical cases, the search policy that maximizes the detection probability is the one that creates expected information at the maximum rate. We achieve that maximum rate of creation by using the existing information of the prior ensemble as fast as possible, thereby gaining an early detection, with its concomitantly large increase of expected information.

Except for the anomalous cases, these two statements contain the essence of the interpretation of optimum search: We consume existing information at the maximum rate in the expectation of gaining still more information, also at the maximum rate.

The anomalous cases are of three types: (1) the case of two search cells, in which maximum information gain is achieved by search in the cell yielding the lower detection probability, regardless of the amount of search effort applied and regardless of the probabilities in the prior ensemble; (2) the case of three or more search cells, in which maximum information gain is achieved by whereabouts search rather than optimal detection search for very large amounts of search effort; and (3) the case of three or more search cells, in which maximum information gain is achieved by searching the cell that yields the second highest detection probability, but only for small amounts of search effort, and only for special regions in the space of prior probabilities.

The anomaly in the first two cases appears to be caused by the two competing methods of increasing information: by actual detection in the searched cells, or by inference that the target is in the other cell, based on failure to detect in the searched cells. These two mechanisms are at work regardless of the number of cells. When there are only two cells, however, failure to detect in one leads to a large increase in the probability that the target is in the other, and, hence, much more information about the location of the target.

This mechanism of inferred information is dominant in the two-cell case, so that actual detection is not necessary to achieve the maximum information gain, even with small amounts of search effort. The two-cell search is thus an exemplary case of the whereabouts search. For three or more cells, the inferred information mechanism of the whereabouts search is not normally dominant; it becomes important only at large values of the search effort.

In the third case, the anomaly appears to be more dependent on the mathematical structure of the ensemble average information -- on the form $p \ln p$. In those areas of the (p_1, p_2) plane where the anomoly exists (figure 13), p_1 is large, and the amount of self-information to be gained by a detection in cell 1 is small. Detection in cell 2, although less likely, provides more self information. When incorporated in the ensemble average, the greater but less likely expected information contribution from cell 2 -- $p_2 \ln p_2$ -- dominates the smaller but more likely contribution from cell 1 -- $p_1 \ln p_1$.

Future Research. Finally, we need to set out the areas in which further research may prove most fruitful.

The principal case treated here (uniform conditional detection probability) has broad practical applications. Its basic structure has been laid out, and only minor holes remain. Clearly, there are unsolved questions relating to the transition between the small and large search effort limits in the case of I_E. These are important theoretical questions, but less so in a practical sense.

The major extensions needed for use in real world search theories are, first, to false targets, second, to multiple targets, both real and false, and finally, to moving targets. Each of these cases is of immense practical significance, and any theory that does not include them can have little claim to completeness.

Relaxation of the assumption of exponential detection functions, and any further considerations of the case of varying conditional detection probabilities can be deferred on the grounds of lower practical priority, despite their considerable theoretical interest and complex mathematical challenge.

Acknowledgement

The author is indebted to Dr. L. D. Stone of Daniel H. Wagner, Associates, and Dr. R. W. Johnson of the Naval Research Laboratory for careful review of this paper and for helpful comments.

Ms Pamela Thompson patiently retyped the manuscript several times.

This work was done while the author was on the staff of the Center for Naval Analyses, Arlington, VA. Support was provided by the Office of Naval Research under contract No. N00014-76-0001.

References

1. B. O. Koopman, "Search and Information Theory: Part of Final Report on Stochastic Processes in Certain Naval Operations," Columbia University, Dept. of Mathematics Report, 1967. AD #687543.

2. D. F. Mela, "Information Theory and Search Theory as Special Cases of Decision Theory," Operations Research 9, 907-909 (1961).

3. S. M. Pollock, "Search Detection and Subsequent Action: Some Problems on the Interfaces," Operations Research 19, 559-585 (1971).

4. H. R. Richardson, "ASW Information Processing and Optimal Surveillance in a False Target Environment," Daniel H. Wagner, Assoc. Report, October 1973.

5. W. H. Barker, II, "Information Theory and the Optimal Detection Search," Operations Research 25, 304-314 (1977).

6. R. G. Gallagher, "Information Theory and Reliable Communications," John Wiley and Sons, Inc., New York, 1968, p. 16.

7. Gallagher, op. cit., p. 21.

8. B. O. Koopman, "Search and Screening," OEG Report No. 56, Center for Naval Analyses, 1946.

9. L. D. Stone, "Theory of Optimal Search," Academic Press, New York, 1975.

10. J. B. Kadane, "Optimal Whereabouts Search," Operations Research 19, 894-904 (1971).

11. W. J. Browning, "Maximum Information Gain Search Policies," Daniel H. Wagner, Assoc. Report, July 1976.

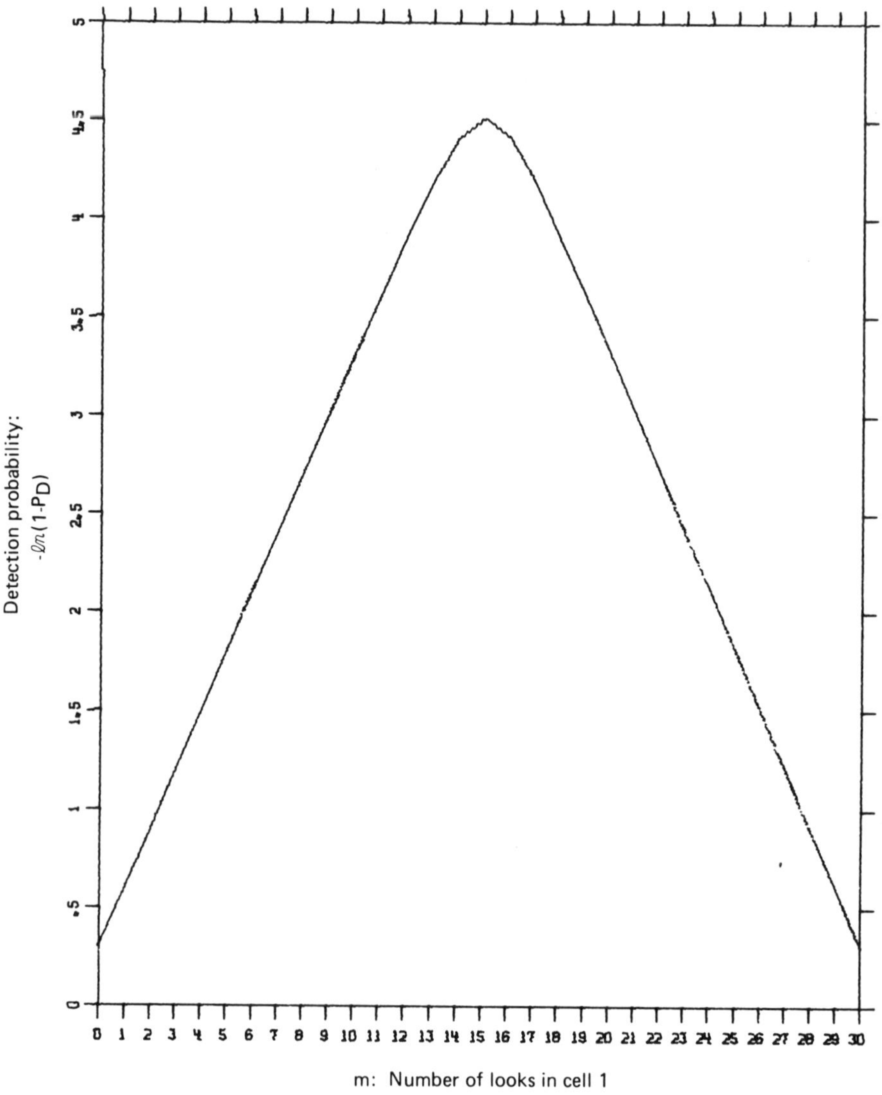

Fig. 1: Mela's First Example
Detection Probability vs Search Allocation

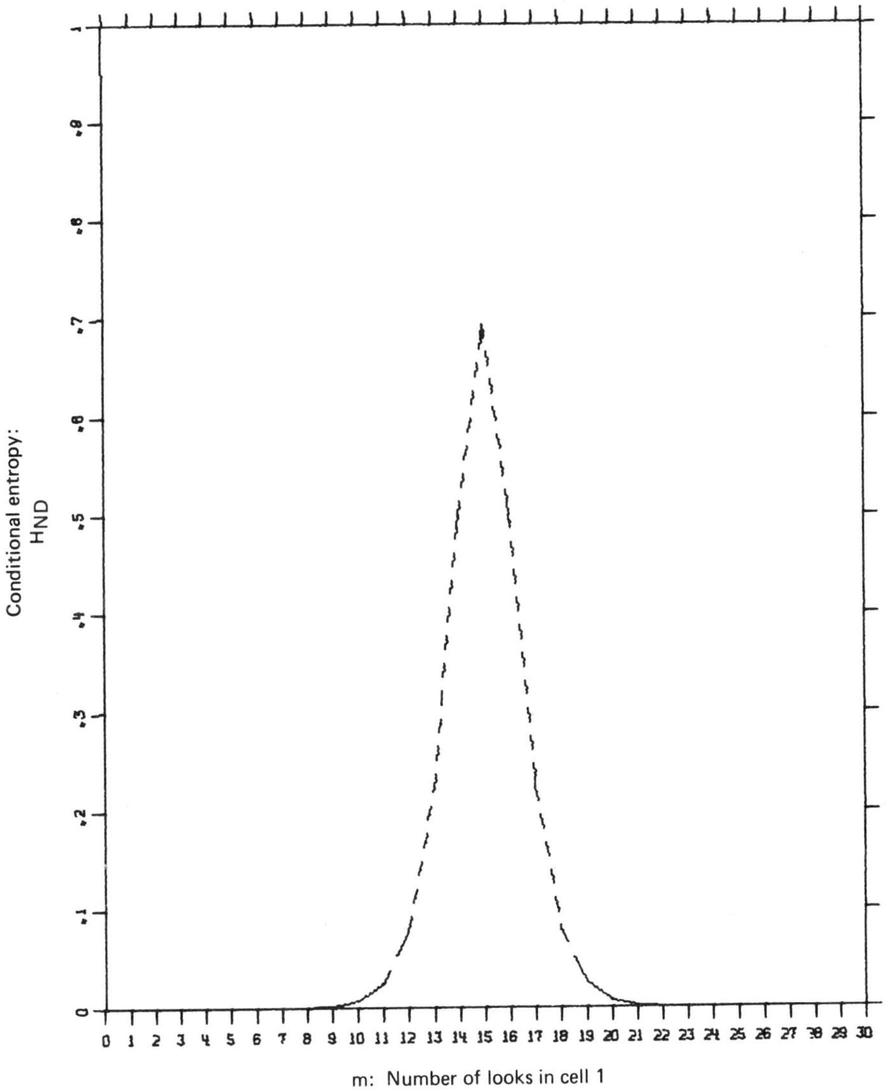

Fig. 2: Mela's First Example
Conditional Entropy vs Search Allocation

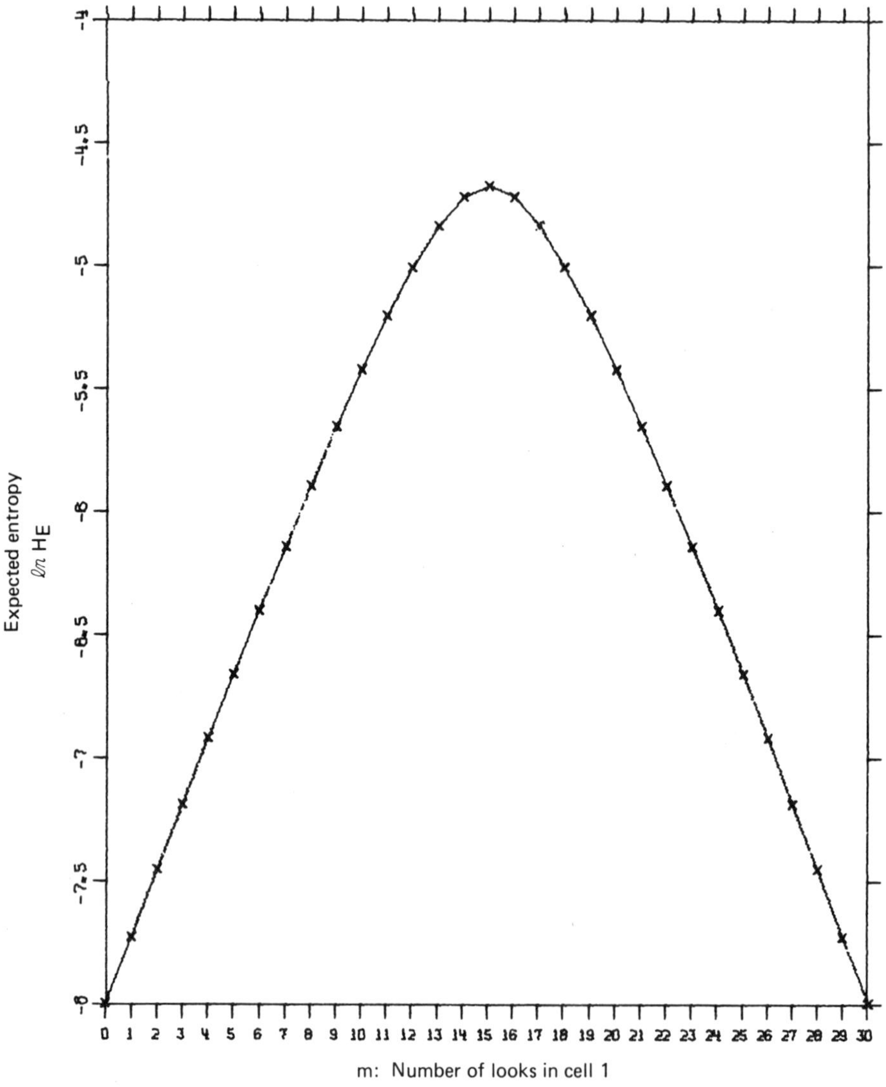

Fig. 3: Mela's First Example
Expected Entropy vs Search Allocation

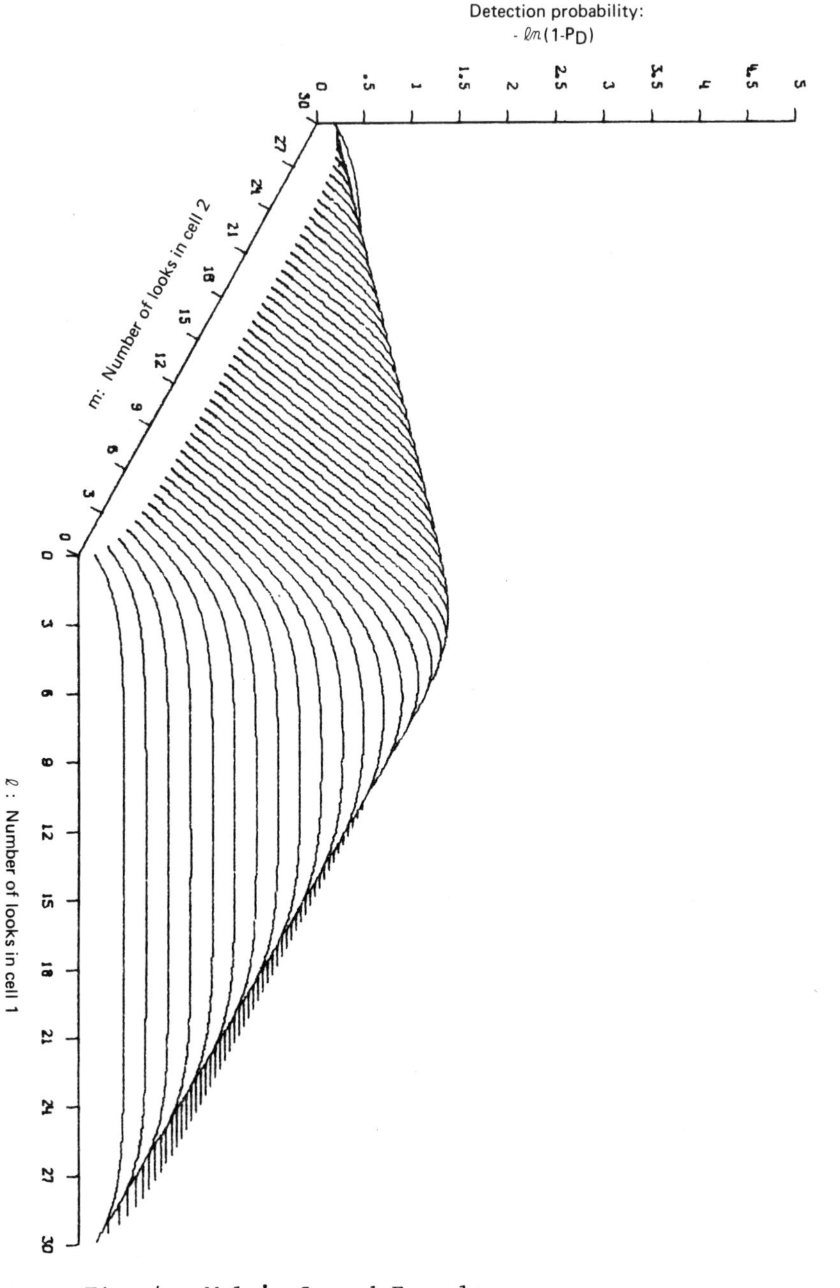

Fig. 4: Mela's Second Example
Detection Probability vs Search Allocation

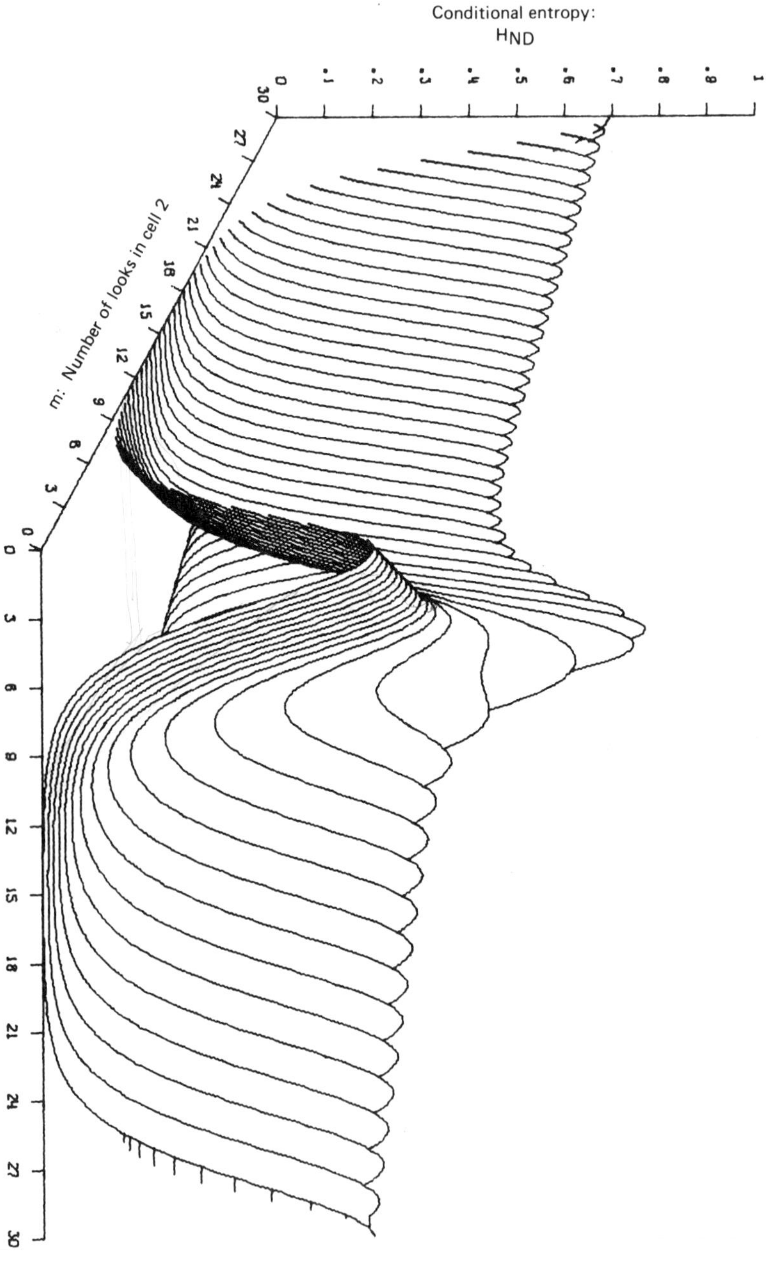

Fig. 5: Mela's Second Example
Conditional Entropy vs Search Allocation

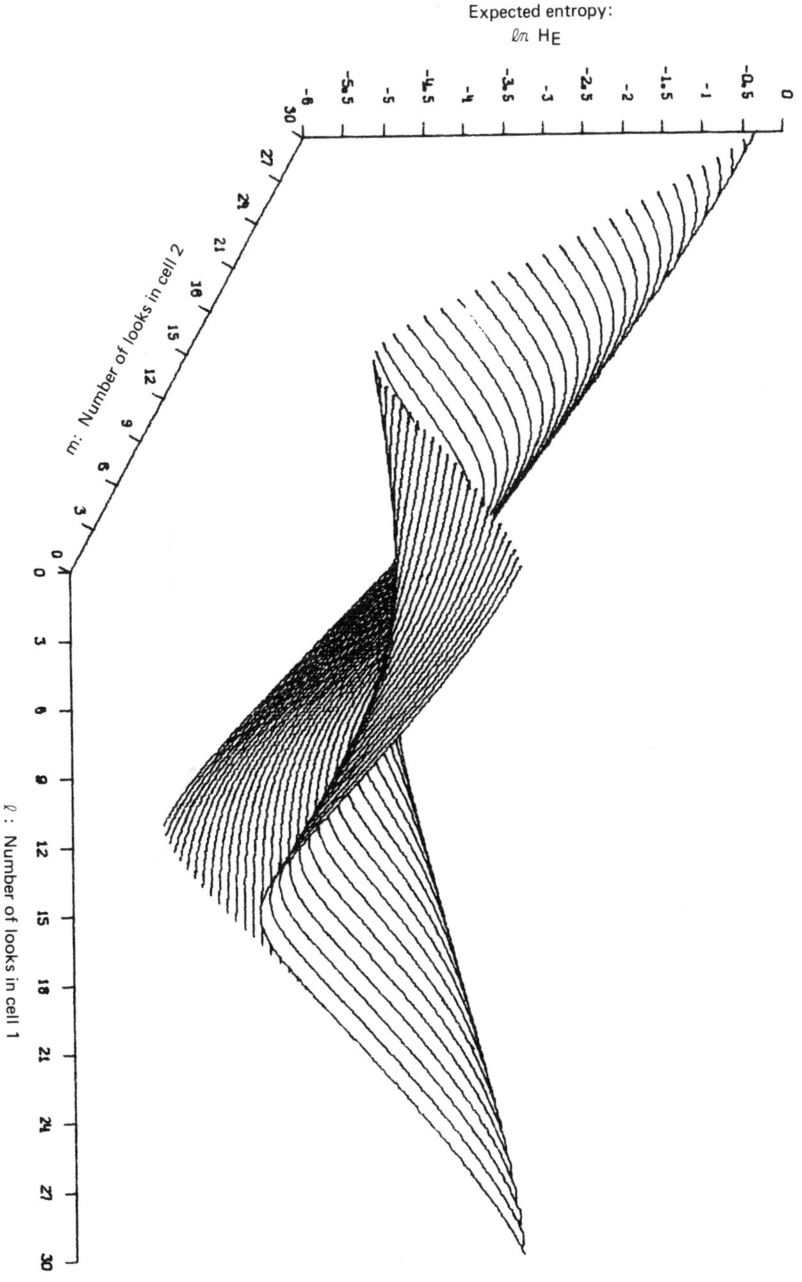

Fig. 6: Mela's Second Example
Expected Entropy vs Search Allocation

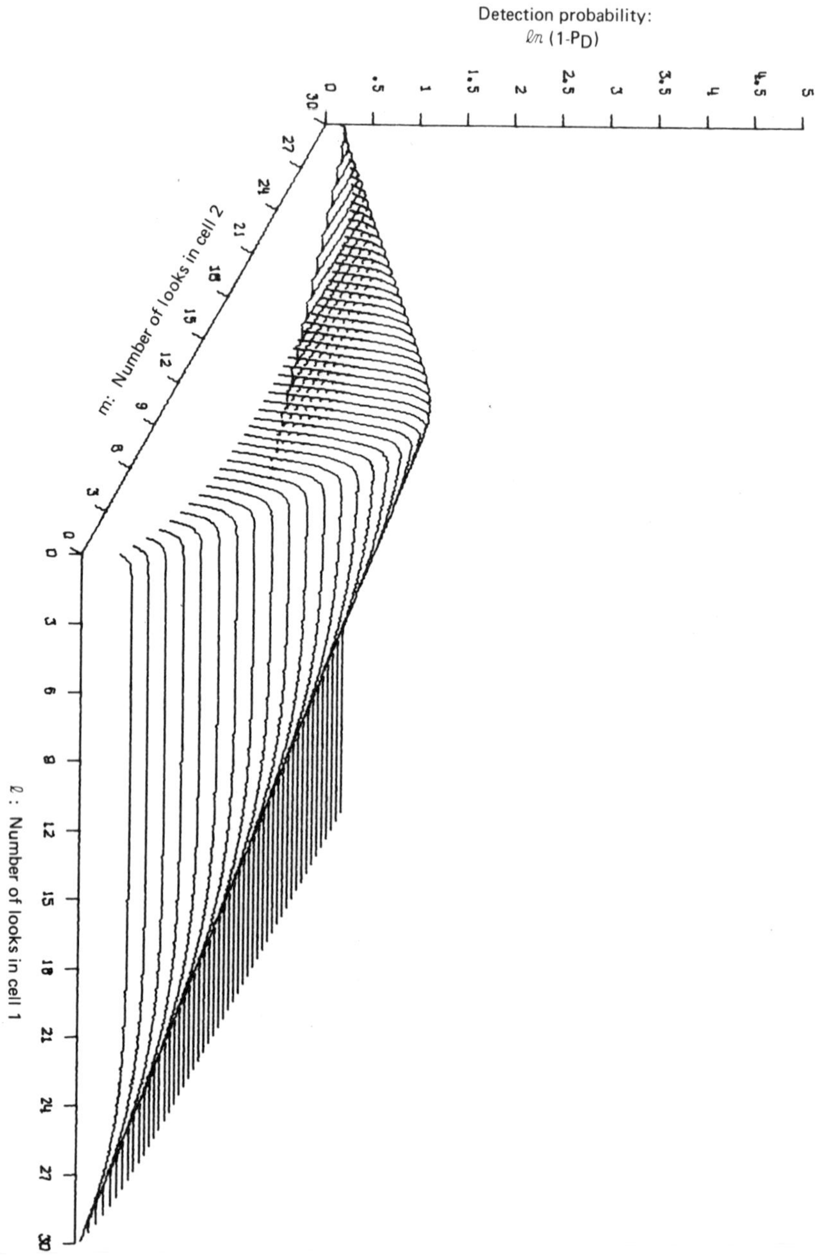

Fig. 7: Pollock's Example - Modified Detection Probability vs Search Allocation

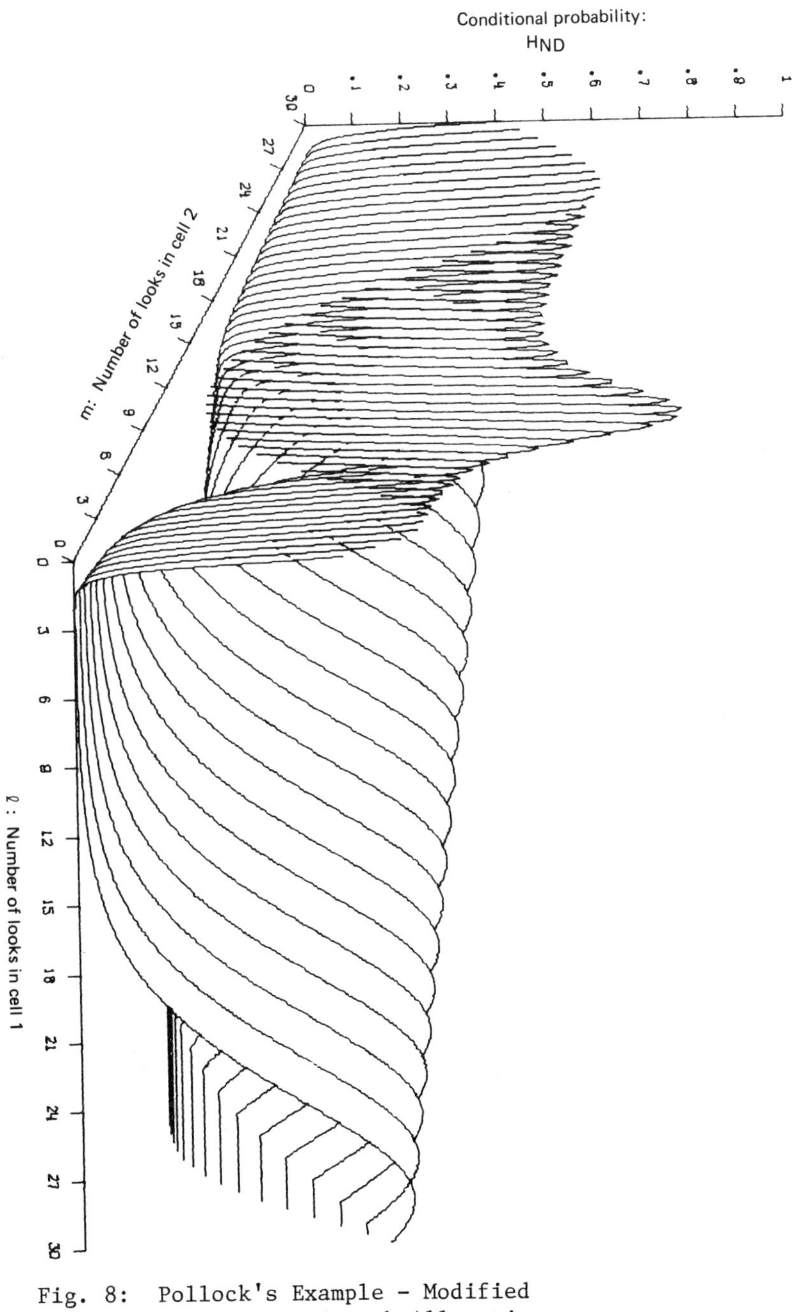

Fig. 8: Pollock's Example - Modified Conditional Entropy vs Search Allocation

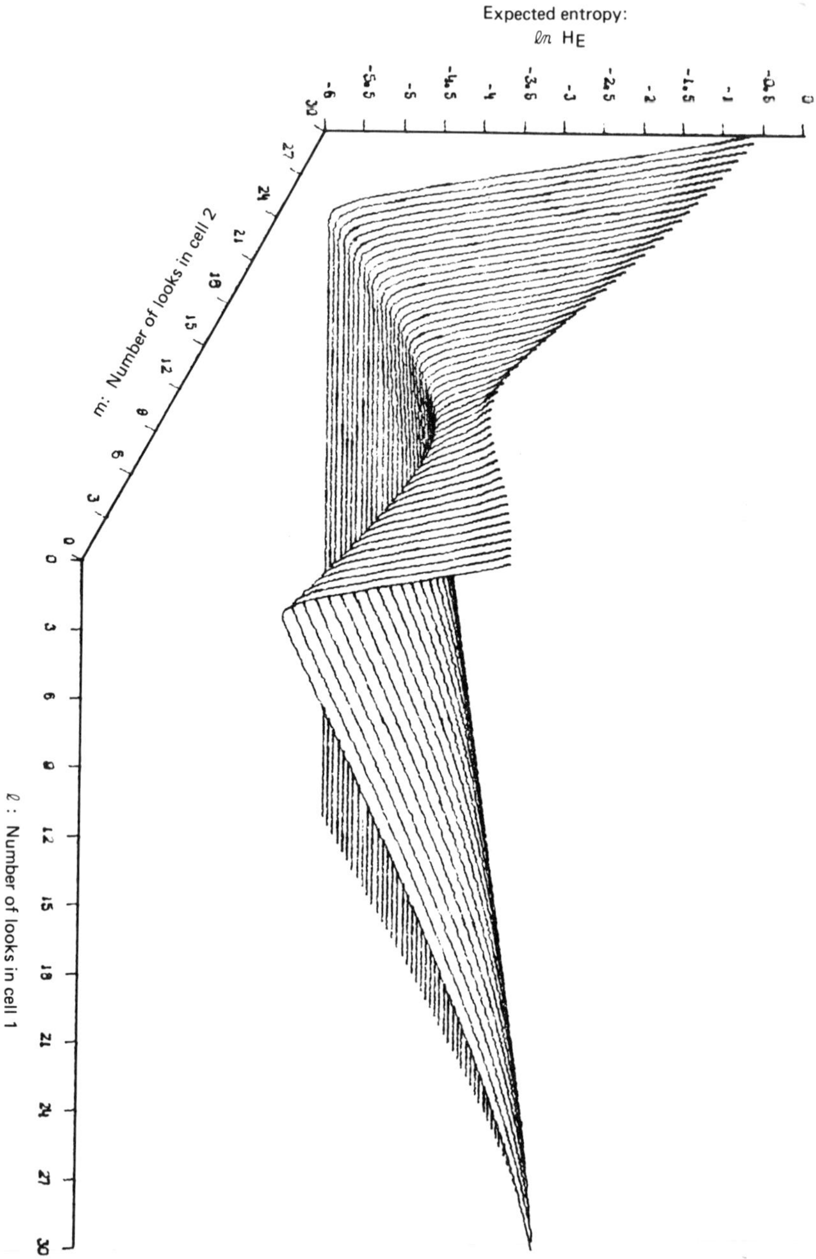

Fig. 9: Pollock's Example - Modified
Expected Entropy vs Search Allocation

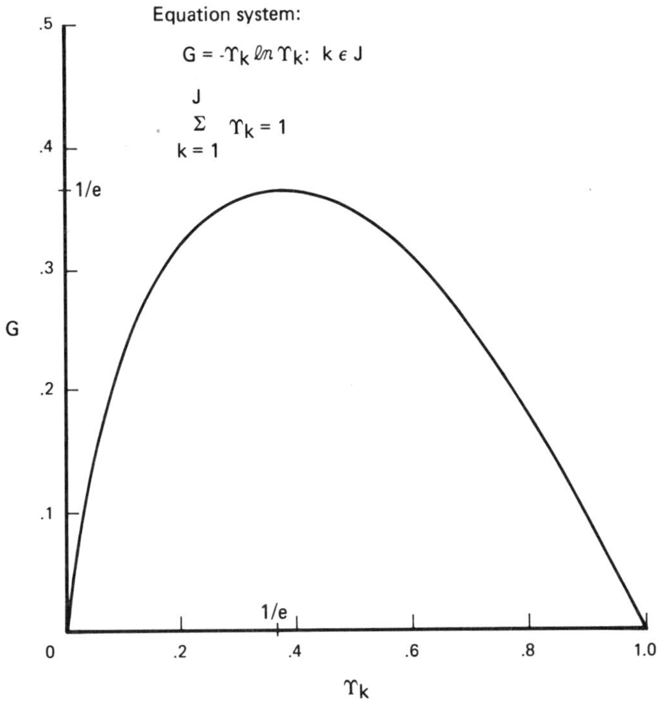

Fig. 10: Graphical Illustration of Equations 4.37 and 4.38

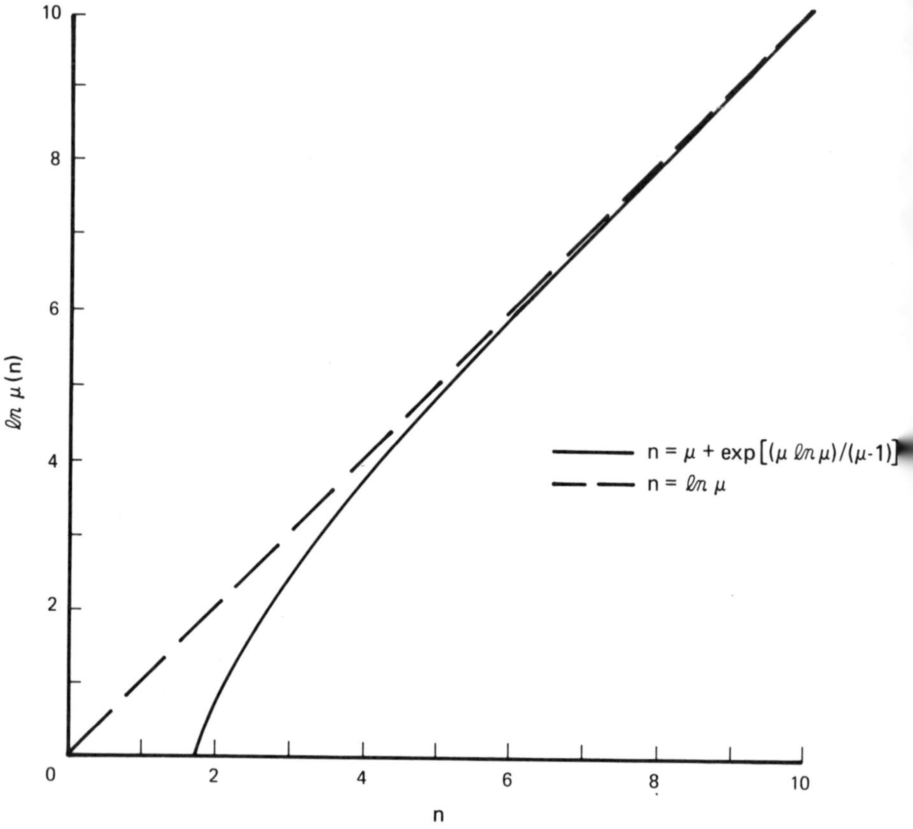

Fig. 11: Graphical Illustration of Equation 4.42

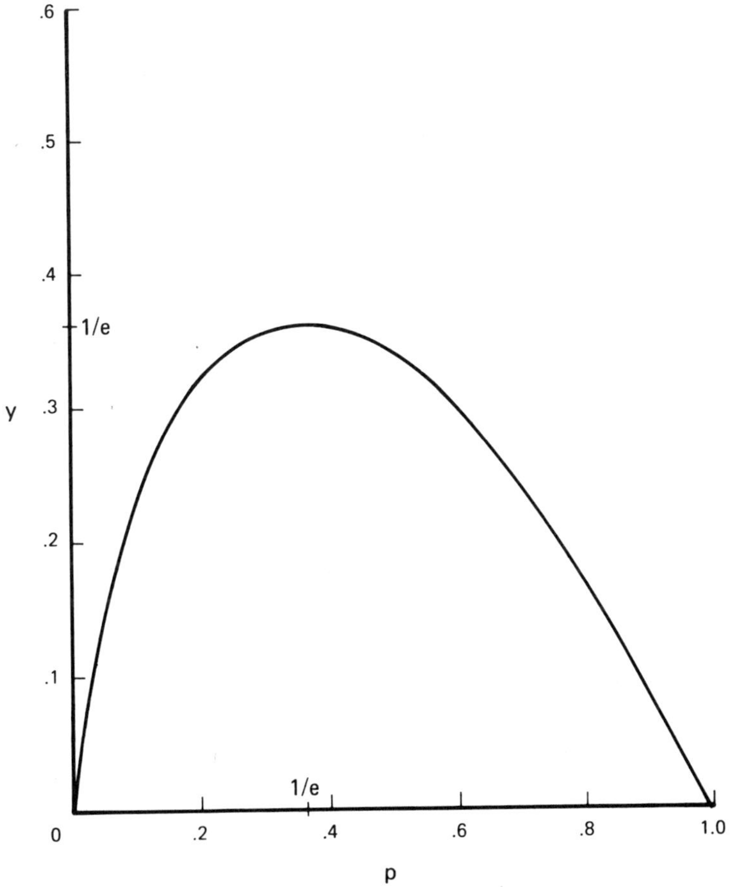

Fig. 12: Graphical Illustration of the Function: $y = -p \ln p$

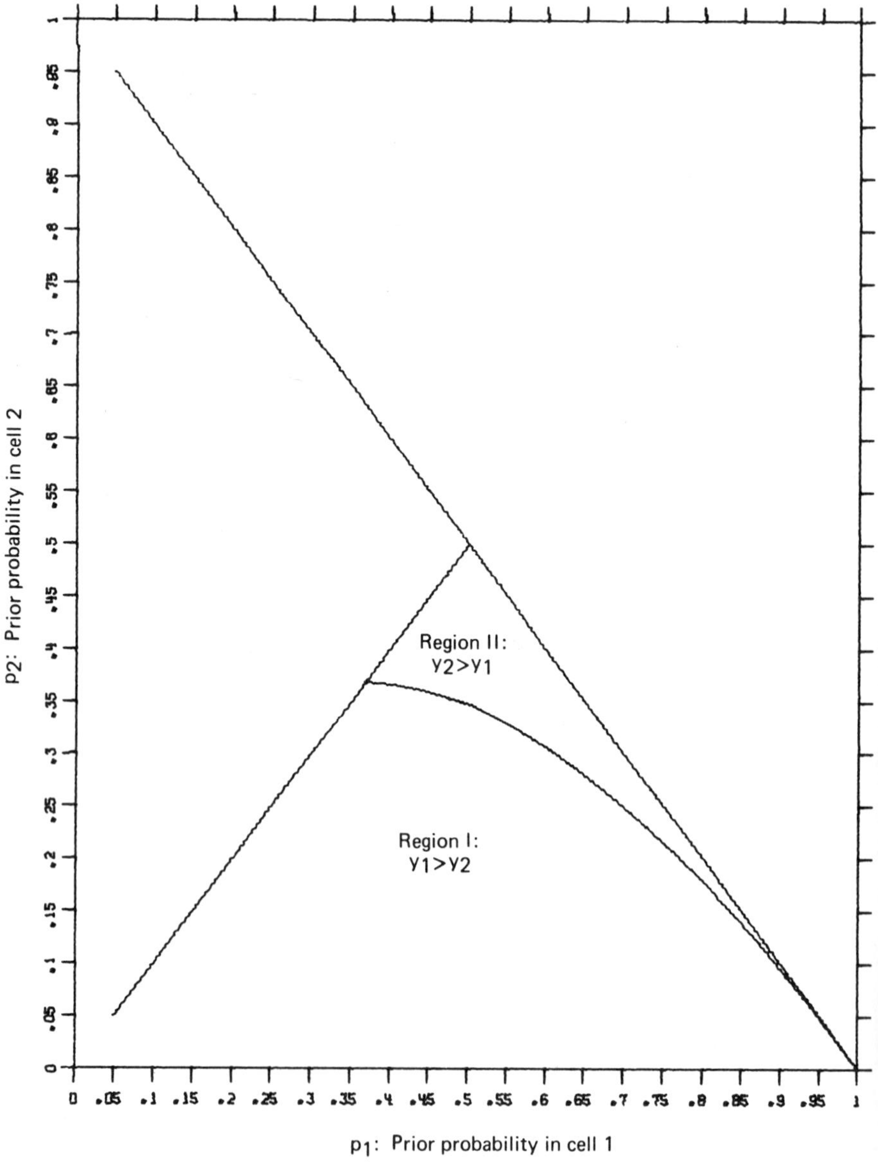

Fig. 13: Delineation of Regions in the Space of Prior Probabilities

ENTROPY INCREASE AND GROUP SYMMETRY

Bernard O. Koopman

The method of entropy maximization -- or information minimization has been applied for a variety of semi-intuitive reasons: entropy represents disorder and in the course of common events tends to increase; and when there is ignorance in a situation, it seems natural to exploit this by minimizing the numerical information. These ideas have had many fruitful applications, but have also led to some paradoxes. Moreover, a mathematically precise formulation of general physical conditions implying the validity of these methods seems to be needed: only very special cases have been treated rigorously; those that are general enough for the purposes of statistical physics are often treated by seemingly harmless approximations that may beg the principle of irreversibility (cf. Mazo's [6] critique of Boltzmann's reasoning).

The present paper approaches this question -- the validation of entropy maximization -- by formulating, in terms of symmetry properties (group invariance) the class of "randomizations" (stochastic processes) which actually increase entropy, as distinguished from the many that do not. And further, it validates, under precisely stated general conditions, this maximization under the constraint of fixing certain expected values. The physical picture behind such cases is given. The conceptualism is broad enough to cover statistical mechanics, both in equilibrium and irreversible (but slow) changes [3],[5]. To convey the ideas in their simplest form, most of the discussion will apply to physical systems having a finite set of possible states.

1. The Case of the Finite Set of States of a Physical System

The states of such a system may be labeled with the integers $(1,2,\ldots,n)$, and on any given occasion -- at any given epoch, t -- the statistics are given by the n probabilities, p_i, that the system will be in the state i. It is assumed that it has reached its state at the epoch considered after passing through various changes, some "deterministic" and others "random." Any such change in states and probabilities amounts to a stochastic process, having a matrix (r_{ij}) of transition probabilities $(i\to j)$ -- i.e., r_{ij} is the conditional probability that the system, after the process, will be in state j, given that at its start it was in state i. This will convert any set of initial probabilities p_i to the final set $p'_i = \sum_j p_j r_{ji}$.

The information (or "negentropy") $I[p]$, as usually defined as

$$I[p] = \sum_i p_i \log p_i, \tag{1.1}$$

will go into $I[p']$. The first question is whether this change increases, decreases, or does not alter the value of I. In spite of the "random" nature of our stochastic process, this question has no answer. Thus if r_{ij} tends to "favor" certain final states, it could take a uniform distribution $p_i = 1/n$ (for which I[p] has its least value, -log n) into a non-uniform one, with unequal p'_i, and therefore increase the information. On the other hand, if we restrict our stochastic process to being symmetrical in the sense that $r_{ij} = r_{ji}$, (the probability of a system in state i going into j is the same as that of its going from j to i), we can easily show that I never increases. More precisely, we have

Theorem 1. *If every $r_{ij} = r_{ji}$, then $I[p'] \leqq I[p]$; and $I[p'] = I[p]$ only when either I[p] is minimal (all p_i equal) or when the process merely permutes the values of the initial probabilities.*

The proof is based on the strict convexity of the function f(x) defined as x log x for x>0, and as its limit, f(0) = lim (x log x) → 0 as x→0, x>0. On the interval x≧0 it is continuous and always concave upward. Hence it satisfies the convex function inequality [4]: if Mu denotes any weighted (arithmetic) mean of a set of values of u, then firstly $f(Mx) \leqq Mf(x)$ -- in this case

$$(Mx) \log (Mx) \leqq M (x \log x). \tag{1.2}$$

Secondly, if this (≦) is replaced by (=), all the weights attached to more than one of the values (of x or f(x)) vanish.

For any stochastic process, we have by elementary probability that $r_{ij} \geqq 0$, and that when r_{ij} is summed over the second index, the value is unity. By our hypothesis of symmetry, this will also be true for summation over the first index: $\Sigma_i r_{ij} = \Sigma_i r_{ji} = 1$. Hence the transformed probabilities $p'_i = \Sigma_j p_j r_{ji}$ are a set of weighted means of the original ones: $p'_i \doteq M_i p$, the index in M_i indicating the particular set of n weights. By the convexity inequality (1.2) we have for each i,

$$(M_i p) \log (M_i p) \leqq M_i (p \log p) \tag{1.3}$$

and hence

$$\begin{aligned}\Sigma_i p'_i \log p'_i &= \sum_i (M_i p) \log (M_i p) \\ &\leqq \sum_i M_i (p \log p) \\ &= \sum_i \sum_j r_{ij} p_j \log p_j = \sum_j p_j \log p_j,\end{aligned} \tag{1.4}$$

the last, by interchanging the order of summation and applying the fact that $\Sigma_i r_{ij} = 1$. Thus $I[p'] \leqq I[p]$: The first part of the conclusion of Theorem 1.

Turning to the second part, (1.3) and (1.4) show that if $I[p'] = I[p]$, then (1.3) applies with ($\leqq$) replaced by (=) for every i. In the case of uniform $p_i = 1/n$, both equations are obvious, since $p_i' = \Sigma_j r_{ji}(1/n = 1/n$. In the case that no two values of p_i are equal, the second part of the convex inequality theorem shows that every $r_{ij} = 0$ or 1. Since their sum in each row is unity, each row has a single 1, the rest 0. Similarly for the columns. Hence the stochastic matrix merely permutes the original probabilities.

In the intermediate case in which the n probabilities p_i fall into k>1 sets of equal values, we proceed as follows: Consider any column in the transition matrix, e.g., the first, of elements r_{j1}. By the second part of the convexity inequality, the only $r_{j1} \neq 0$ correspond to one of the sets of constant p_j. After a possible re-indexing (a permutation), they can be taken as the first set of s: $p_1 = p_2 = \ldots = p_s$ ($s \geqq 1$). Then every $r_{j1} = 0$ when j>s, whereas their sum over $j \leqq s$ is unity. Putting these facts together we see that $p_i' = p_1$ for all $i \leqq s$. A corresponding process applied to each column shows that each of the other sets of equal p_i go into some p_j' unchanged. This completes the proof.

Note that the hypothesis $r_{ij} = r_{ji}$ has been used only to prove that $\Sigma_i r_{ij} = 1$: the latter is a more general property than symmetry.

Theorem 2. A stochastic process (i→j) will leave the value of I[p] unchanged for every choice of p_i if and only if it is essentially deterministic, i.e., every element (r_{ij}) is zero or unity, its determinant is ± 1, and it represents a permutation of states (with unity probability).

That such a permutation does not change I[p] is evident from (1.1). If, conversely, I[p] is never changed, we may take $p_1 = 1$, $p_2 = \ldots = p_n = 0$. Then $I[p] = 0$, so that $I[p'] = 0$. But $p_i' = \Sigma_j p_j r_{ji} = r_{1i}$, and $0 = I[p'] = \Sigma_i r_{1i} \log r_{1i}$. Since $0 = I[p'] = \Sigma_i r_{1i} \log r_{1i}$; and since $0 \leqq r_{1i} \leqq 1$, each term in this sum $\leqq 0$; so that its vanishing implies that its terms all vanish individually. Hence every element in the first row $r_{1i} = 0$ or 1. Since their sum is 1, just one $r_{1i} = 1$, the others, 0. Furthermore, the place of the unit term must be different in each row. Otherwise there would be at least one column, e.g., the first, of all zero elements: $r_{j1} = 0$ for all j. Take as initial probabilities the uniform distribution $p_j = 1/n$, giving $I[p] = -\log n$. Since $p_1' = 0$, I[p'] is a sum of at most n-1 non-vanishing terms, and it cannot be less than $-\log(n-1)$. This contradicts the hypothesis that I is unchanged. Hence the matrix (r_{ij}) has a unit in each row and each column, the other n^2-n elements being zero. Its determinant is ± 1 and it corresponds to a permutation of states, as was to be proved.

We note that no assumption of symmetry has been needed in the hypothesis of this theorem.

2. The Stochastic Process as a Random Transformation in a Group

The chief objective of the present paper is to show how deterministic transitions on the microscopic level can lead to information diminishing ones on the macroscopic one. In this section we continue to confine our attention to the case of a finite set of possible states, as the simplest example of the ideas involved.

The totality of deterministic transitions among n different states is the symmetric group Σ of all n! permutations of states. Suppose that, on a particular occasion, the system is known to make one of these transitions, but all that is known about which particular one is their set of n! probabilities: $P(\xi)$, where ξ is a permutation in Σ. Such an assignment of values amounts to that of an element of the group algebra over Σ (a special one, of non-negative values adding up to unity).

The transition probability r_{ij} is found by elementary probabilistic reasoning in terms of the above data: Let Σ_{ij} be the subset of permutations in Σ that carry i into j; then

$$r_{ij} = \text{sum of } P(\xi) \text{ over all } \xi \text{ in } \Sigma_{ij}. \tag{2.1}$$

The elements of group theory tell us that Σ_{ij} is a left co-set of Σ_i: the subgroup that leaves i fixed. Indeed, if ξ and η are two permutations that each carry i into j, then $\xi^{-1}\eta$ leaves i invariant, since the first operation carries i into j, and the second one, ξ^{-1}, carries j back to i. This is, of course, the criterion of left co-sets: $\xi^{-1}\eta\varepsilon\Sigma_{ij}$ means that $\eta \varepsilon \xi \Sigma_i$, etc.

As explained early in §1, a transition matrix is not necessarily information-decreasing: further assumptions are needed. The simplest and perhaps the most natural from a physical point of view is the assumption of symmetry of probabilities over Σ: that, for every $\xi\varepsilon\Sigma$, $P(\xi^{-1}) = P(\xi)$. Since Σ_{ji} is the set of inverses ξ^{-1} of all the permutations in Σ_{ij}, (2.1) shows that $r_{ji} = r_{ij}$, so that Theorem 1 applies to the decrease of I.

It is necessary to deal with the case of a subgroup Γ of Σ: a group formed by some but not necessarily all of its permutations. Everything said above for Σ applies equally well to Γ; but new possibilities arise. First, Γ may act transitively: given any two i and j, Γ contains at least one permutation carrying i into j, so that Γ_{ij} is non-empty. A formula like (2.1) applies, but with fewer terms, possibly only one (simple transitivity). Again

$P(\xi^{-1}) = P(\xi)$ leads to the symmetry of r_{ij} and thus to non-increasing information. A concrete physical example is that of card shuffling; each operation in shuffling is a random selection of a permutation of the 52 different cards, and obviously the probabilities are symmetric. This shows that this one assumption leads, by minimizing the information, to the reliable view that the probabilities of the various cards being in the various stated places in the pack are essentially equal: a conclusion not requiring knowledge of the actual values of the probabilities of the permutation.

If on the other hand Γ is such that certain pairs i and j cannot go into one another by any permutation contained in Γ, this subgroup is said to act intransitively on the states. Then the set of n states falls into $k>1$ sets, Γ acting transitively on each set, but not carrying any state in one into a state in a different set. Their construction is obvious: apply all the transformations of Γ to the state 1, obtaining a set of states, e.g., $(1,2,\ldots,n_1)$. Then apply them to state n_1+1, obtaining another set such as $(n_1+1,\ldots,n_2)$. If there are any further states not included in $(1,2,\ldots,n_2)$, repeat the process. The result will be a required set of k mutually exclusive subsets; we write

$$\{1,2,\ldots,n\} = S_1 + S_2 + \ldots + S_k = \sum_{h=1}^{k} S_h. \tag{2.2}$$

It is easily seen that each S_h is acted on transitively by Γ. Again r_{ij} is given by (2.1); but since when i and j are in different sets, e.g., S_1 and S_2, there are no elements in Γ_{ij} and hence no terms in the sum, such an $r_{ij} = 0$. Again the assumption $P(\xi^{-1}) = P(\xi)$ leads to non-increase of information; but since our present stochastic process cannot change the total probability $P(S_h)$ (the sum of p_i for $i\epsilon S_h$), the minimal $I[p]$ to which this process leads will in general be given by k separate uniform distributions, such as $p_i = P(S_1)/n_1$ for $i = 1,\ldots,n_1$, etc. This will fail to be true only when initially $P(S_h) = n_h/n$ for all h.

3. Stochastic Processes Preserving Expected Values

Let a quantity ϕ (having any real or complex value) be defined over the set of n states of our system. We write its value at the i state as $\phi_i = \phi(i)$. It is a "microscopic" physical quantity determined by the physical nature of the state itself (just as the Newtonian energy of a particle in a conservative field is determined by its state: position and velocity). Hence the permutations $\xi\epsilon\Gamma$ (transitive or intransitive) will not change the values of ϕ_i: they only change those of p_i. Consider the expected value $E\phi$:

$$E\phi = \sum_{i=1}^{n} p_i \phi_i \tag{3.1}$$

Under the stochastic process of matrix r_{ij} it will go into the new expected value $E'\phi$:

$$E'\phi = \sum_{i=1}^{n} p'_i \phi_i = \sum_{i=1}^{n} \sum_{j=1}^{m} p_j r_{ji} \phi_i \ . \tag{3.2}$$

Of particular interest is the case of invariance of expected values: $E'\phi = E\phi$. Equating the last sums in (3.1) and (3.2), we obtain ($\delta_{ij} = 0$ for $i \neq j, \delta_{ii}=1$):

$$\sum_{j=1}^{n} p_j \sum_{i=1}^{n} (r_{ji} - \delta_{ji}) \phi_i = 0 \ . \tag{3.3}$$

Such an invariance is of little physical interest if it does not include as initial probabilities those corresponding to a known initial state (all p_i zero or 1). Then (3.3) has the force of an identity in these and yields the n homogeneous equations in the ϕ_i:

$$\sum_{i=1}^{n} (r_{ji} - \delta_{ji}) \phi_i = 0 \ , \ j = 1,\ldots,n \ . \tag{3.4}$$

The determinant of $(r_{ji} - \delta_{ji})$ vanishes, as we see by adding the first n-1 columns to the n'th, obtaining $\Sigma_i r_{ji} - 1 = 0$. Hence (3.4) has solutions that are not all zero. The above calculation shows also that the constant function $\phi_i = c \neq 0$ is a solution of (3.4); but this has no physical interest. By an elementary general theorem, the homogeneous system (3.4) has n-r linearly independent solutions, where r is the rank of the matrix $(r_{ji} - \delta_{ji})$. We have seen that $r \leqq n-1$. When $r = n-1$, only the above trivial one exists, and thus in general a stochastic process will change the expected value of every non-constant function ϕ_i.

Hence only special classes of stochastic processes will maintain the expected values of one or more non-constant functions, ϕ_i, ψ_i, etc. Let us consider the information changes that they produce.

The set of n probabilities p_i may be represented as a point $(p_1,\ldots p_n)$ in n-dimensional space, but since $p_i \geqq 0$ and their sum is unity, this point is on the (n-1)-dimensional plane $\Sigma p_i=1$ and, indeed, on the piece of the plane in the first quadrant (when n=2), octant (n=3), and, in general "n-tant."

In the present case, a further set of n-r-1 independent equations must be satisfied; viz., $\Sigma p_i\phi_i = a$, $\Sigma p_i\psi_i = b$, etc., which express the fixed expected value conditions. These, together with $\Sigma p_i = 1$, form a set of n-r independent equations, i.e., represent n-r different (n-1)-dimensional planes. If our problem is to be possible, they must have an intersection on the first "n-tant"; and since they are independent, this will (in general) be of n-(n-r) = r dimensions and "a flat" triangular "simplex" R. The information I[p] is defined over R, over which each function $f(p_i) = p_i \log p_i$ is convex. By reasoning precisely similar to that of §1, we show that each time that our system experiences a symmetrical stochastic process (leaving R invariant) I[p] decreases -- unless it has already reached its minimum, or unless the change of states is deterministic.

We may obtain this constrained minimum in either of two ways. The first is to assume that it can be obtained by formal differentiation with the use of Lagrange multipliers λ, μ, etc.; this gives

$$p_i = C \exp(-\lambda\phi_i - \mu\psi_i - \ldots), \ i=1,\ldots,n, \tag{3.5}$$

and then to show that any different set of probabilities, q_i, which also satisfies the expected value conditions -- so that $(q_1,\ldots,q_n) \in R$ -- makes I[q] > I[p]. The latter fact is shown by noting that the <u>cross-information</u>

$$\Sigma_i q_i \log (q_i/p_i) = I[q] - \Sigma q_i \log p_i. \tag{3.6}$$

Now we have from (3.5) and the common expected values of the two distributions,

$$\Sigma q_i \log p_i = \log C - \lambda a - \mu b - \ldots = \Sigma p_i \log p_i = I[p].$$

Hence the cross-information in this case is I[q] - I[p]. But the cross-information can be written as $\Sigma_i p_i f(z_i)$, where $f(z) = z \log z$ and $z_i = q_i/p_i$: being a weighted mean of a convex function, it has a value greater than $(Mz_i) \log (Mz_i)$; and since $Mz_i = \Sigma q_i = 1$, it is zero. This is the simplest proof of the non-negativeness of the cross-information of two distributions. The second part of the convexity theorem of §1 shows that it is zero if an only if every $q_i = p_i$. This completes the proof that (3.5) gives the minimum of I[p] subject to the expected value constraints. It is understood, of course, that C is a normalizing constant and that λ, μ, etc. have been determined by the expected value requirement. The usual method is to use the partition function Z, equal to $\Sigma \exp(-\lambda\phi_i - \mu\phi_2 - \ldots)$.

The second method of proof is particularly instructive in the explanation it gives of certain problems of generalization to the case of systems with infinitely many states. It focuses attention on the (r+1) - dimensional solid S of points (P,z) in the space of the n+1 variable: $(p_1,\ldots,p_n,z)$, where $P:(p_1,\ldots,p_n)$ is on R, and z is any number $\geqq I[p]$. (S is the "hypergraph" of $\Sigma p_i \log p_i$ in the Cartesian product space R x {z}). This is easily seen to be convex, and bounded below by a strictly convex surface, the graph of I[p]. Hence a plane parallel to R, when moved so that its "ordinate" z increases from the least value - log n indefinitely, it will reach a point on the above convex boundary at which it will be tangent, and in fact be a <u>support plane</u> of the convex S, in the sense of the Hahn-Banach Theorem [1]. The $(p_i,\ldots,p_n)$ corresponding to this point of tangency will, of course, be the minimizing distribution sought. Having proved its existence inside R, the usual formal process described above gives the solution

Things are very often different in the case of a system with infinitely many possible states. Thus in the case of the sequence (1,2,...), the information $I[p] = \Sigma_i p_i \log p_i$ is an infinite series having no lower limit, even under the constraint $\Sigma_i p_i = 1$: each symmetric stochastic process decreases I[p], but it $\to -\infty$. The added expected value constraint $\Sigma_i p_i \phi_i = a$, where $\phi_i = i$, introduces a lower bound and the minimum exists: the familiar exponential distribution. On the other hand, in the case of a doubly infinite set of states, (0, ±1, ±2,...) the above expected value constraint does not produce a lower bound to the information which has no minimum. Only the stronger requirement that $E(i^2) = b$ gives such a bound, and hence a minimum information.

A more subtle case of failure of a constrained minimum is that of a continuum of possible states, well illustrated by the case of a single variable x with the probability density p(x) and the formula

$$I[p] = \int p(x) \log p(x)dx, \quad -\infty < x < +\infty. \tag{3.7}$$

replacing (1.1). If no constraints are imposed beyond those applying to every probability density, I[p] has no lower bound. The same is true if we require that $Ex = a$. There will be a lower bound under the constraint $E(x^2) = b$; and if both of these are imposed (with $b > a^2$), we obtain the Gauss distribution for the minimum. But now suppose that in addition to these we impose the third moment constraint $Ex^3 = c$ (where a, b, c satisfy the moment inequalities). Then if there were a p(x) minimizing I[p], occurring, as required by the convexity, <u>inside</u> the region R of function space, it would be computable by the classical process, and be the exponential of a third degree polynomial. Such a function would have an infinite integral and hence not be a probability density.

The explanation of the failure to give a support plane parallel to R, as in the finite case, is the special form the Hahn-Banach Theorem applies in the non-compactness of our function space S. The support plane exists in the sense that it is closer to S than any other parallel plane not cutting through S; but is has no point in common with S.

The situation just described is important in irreversible processes in statistical mechanics, where the flow of heat is governed by a thermal conductivity that involves a third moment (of momenta): attempts to find a distribution maximizing the entropy under such a constraint are bound to fail. Such facts have led the present author to extend the definition of information of a set of data, as a greatest lower bound of I[p] in (3.7) for all p(x) corresponding to the data.

Returning to the simpler case of finite sets of states, it is important to find a plausible physical basis for the mathematical assumption, in various important cases, that the random changes under which the system is subjected do in fact preserve certain expected values. We must return to the concepts of the theory of groups, and the physical symmetries that they express.

4. On Transforms of State-Transformation Groups

One way of picturing our group Γ of state transformations is by representing each of its elements ξ by a set of n arrows in the plane. If the n states are represented by points labelled 1,2, ...,n, then the pair $(i, \xi i = j)$ is represented by the arrow from i to j; the totality of n arrows, traced from each state i to $j = \xi i$, depicts the permutation ξ. If ξ leaves some state unchanged, the arrow reduces to a point. No two ξ-arrows have the same tail, no two the same head. But there are as many systems of n arrows as there are elements in Γ. The inverse ξ^{-1} is represented by reversing the sense of each earrow of ξ. The product $\eta\xi$, or result of applying η to the point obtained by the application of ξ, represented by adding vectorially: the tails of the η system being placed at the heads of the ξ system.

Suppose next that λ is a new permutation of our n states, belonging, of course, to Σ, but not in general to the subgroup Γ. We wish to apply λ simultaneously to each of the n points in the above representation, and see what effect it has on the n arrows representing ξ. The arrow (i,j) from i to $j = \xi i$ has its extremities replaced by $(\lambda i, \lambda j)$, the result of transforming (i,j): denote these points by (i',j'); clearly

$$j' = \lambda j = \lambda\xi i = \lambda\xi\lambda^{-1}\,\lambda i = \lambda\xi\lambda^{-1}\, i'.$$

Thus j' is obtained from i' by the transformation $\lambda\xi\lambda^{-1}$; and since not only i but $i' = \lambda i$ is an arbitrary one of our n "state" points, our transformed system of arrows is the system representing $\lambda\xi\lambda^{-1}$. If ξ is an arbitrary element of Γ, $\lambda\xi\lambda^{-1}$ is an arbitrary element of what is evidently a subgroup $\lambda\Gamma\lambda^{-1}$ of Σ, in general different from Γ.

Such $\lambda\xi\lambda^{-1}$ and $\lambda\Gamma\lambda^{-1}$ are called the transforms of ξ and of Γ by λ. They are often introduced in their special application to changes of coordinates: if λ represents such a change, $\lambda\xi\lambda^{-1}$ is described as the same geometrical transformation as ξ, but written in a different coordinate system. While this particular, conception is important in many applications, it is not the one that we must use: we shall denote by $\lambda\xi\lambda^{-1}$ an actual geometrical, viz. physical change in the transition ξ of states--as in the picture of its effect on the n arrows representing the latter.

The physical applications that we have in view involve the "motion" of a system, starting at some given state, and passing from state to state along a definite path, which may be a closed orbit. In actuality, this path is followed continuously; but in our simplified model, jumps discontinuously from one state to the next--determined by the one before, as in the deterministic motion of a dynamical system (in phase space or in Hilbert space of wave functions).

Consider the cyclic group $\{\xi\}$ generated by ξ: it is formed by the permutation ξ and all its iterates (powers); and since there are at most n! permutations, a certain lowest positive power, $\xi^m = 1$, the identity (no change). Thus the cyclic group has the m elements: $\{\xi\} = \{1,\xi,\ldots,\xi^{m-1}\}$. The transform by another permutation λ carries this into an iso-morphic cyclic group generated by $\lambda\xi\lambda^{-1}$.

If the system is in any initial state i, the repeated application of ξ will make it move along the path of states $(i,\xi i,\xi^2 i, \ldots,\xi^{m-1})$, and the arrows representing these transitions will form a closed polygon, of m sides if all the states are different, otherwise, of a number m_1 which divides m exactly. This path is the "orbit" of the motion of the system. In the general case, which we shall assume, there are other states -- most others, let us say -- not on this particular orbit. Then, starting from some other state, a different orbit is obtained; and thus the n states lie on a certain number, $k > 1$, of different orbits. Clearly $\{\xi\}$ acts intransitively upon the set of n states.

When λ is a permutation not in $\{\xi\}$, the transformed orbits corresponding with $\lambda\{\xi\}\lambda^{-1} = \{\lambda\xi\lambda^{-1}\}$ into which λ takes the original ones will in general be altogether different--although, of course, with many points of intersection with the former. In this transformation, the lengths of orbits are preserved, since λ is a one-to-one correspondence of the states: a set of m in one orbit cannot go into a set of m' in its transform unless m' = m.

Guided by certain situations in Hamiltonian dynamical systems and their integrals, to be outlined later, we shall be particularly interested in those transformations λ that carry each $\{\xi\}$ orbit into a $\{\xi\}$ orbit -- in general, different. This means that $\lambda\{\xi\}\lambda^{-1} = \{\xi\}$: our cyclic group is self-conjugate under λ; it is a normal subgroup in the group $\{\xi,\lambda\}$ generated by the two elements ξ and λ (all their products of powers, in any order). The class of ξ-orbits is not necessarily changed transitively by λ; certainly a class of orbits interchanged transitively would have to be all of the same length, as we have seen; but this condition may not be sufficient.

5. Physical Reasons for the Preservation of Expected Values

This property, which was examined mathematically in §3, will now be based on a physical model, and this in turn is suggested by the picture of a molecular system enclosed in a vessel, the boundary of which is itself composed of molecules in random motion about their positions of equilibrium. Through such motions, stochastic changes are induced in the paths of the molecules of the system within the container: this is how irreversibility and entropy increase are produced in this system which, if the container were replaced by a "mathematically" perfectly smooth unyielding surface, or fixed field of force, would evolve deterministically.

In actuality, the container acts macroscopically as such a smooth, etc., vessel; but microscopically in addition to the randomization mentioned, it maintains control over certain expected values. In the case of thermodynamic equilibrium, for example, the mean flow of energy across the boundary is zero, in spite of random fluctuations on the microscopic level.

Returning to the simplified picture of the system capable of being in just n states, we shall let ξ represent as in §4 the state-to-state transformation. In this finite case it plays the part of the velocity (or velocity and acceleration) in mechanical motions. Upon each orbit--set of states of the form $(i, \xi i, \xi^2 i, \ldots, \xi^m i)$--we now pick a point, by a method supposedly based on physical considerations, and call it the "point of contact with the boundary." Tus there is just one point on each of the k orbits into which the n states are subdivided. In the picture of the states in the plane, we may think of the boundary contact points as relatively close together, and connect them with a line B, which thus cuts each one of the orbits, in k distinct points in all.

Suppose that the "physical effect at the boundary" is to produce a set of permutation among these k points (k! at most). Let λ be one of them, having the following "length-correspondence" property: if $j = \lambda i$, the orbits through i and through j have the same lengths. If there are orbits of different lengths, evidently λ, or the group $\{\lambda\}$ which it determines, will not act transitively on the k points on B. The set of all such B transformation obviously forms a group.

Evidently any λ having this length correspondence property can be extended to all the n representative points, simply by taking any orbit of given length and superimposing it point-by-point on another orbit into which λ transforms the original boundary point. To put the process algebraically: let s be any point; it is on a unique orbit cut in a unique point i by B; let $s = \xi^r i$, and let $j = \lambda i$. We define λs as $\xi^r \lambda i = \xi^r j$. This means that not only is λ a defined one-to-one correspondence between the set of n states, but that $\lambda \xi^r \lambda^{-1} = \xi^r$. This construction resembles the change of integral curves of a differential system produced by changing their initial values. In both cases the change of orbits will in general be intransitive.

As in §2, we regard the randomizing physical effects of the boundary as produced by a random selection of permutations λ from a group of those described above, and satisfying the symmetric condition $P(\lambda^{-1}) = P(\lambda)$. In any given distribution p_i, the sum of the probabilities of states on a particular orbit will be the probability that the system be moving in that orbit. As long as no random transition λ takes place, this sum remains constant. Under random transitions, these probabilities of being on an orbit are transformed by the symmetric transition probabilities corresponding to λ.

Fundamental to the theory of dynamical systems (Newtonian or quantum-theoretic) are the first integrals of the equations of motion. In classical mechanics, they are physical functions that are constant along each path. In the simplified picture of n states, they have the same value for each state $\xi^r i$ on a given orbit, but except in trivial cases, not the same for all different orbits. One example is, of course, the length of an orbit. A more significant one is the probability of an orbit, as described above. But in cases of more physical importance, there are others. These would correspond in the dynamical case to the energy (Hamiltonian) of a system in steady motion, and under many cases, to the total momentum.

Suppose in our finite case that we have a system having a certain number of such "first integrals," i.e., functions $\phi(i)$, $\psi(i)$, etc. that are constant along each path. The boundary effects of the λ will in general change them; i.e., $\phi(\lambda i)$ is not in general equal to $\phi(i)$, etc. We now postulate on physical grounds that our set of randomly chosen λ do not change the expected values of $\phi(i)$, $\psi(i)$, etc. on the boundary B--and have throughout the whole set of states of the system. We may then apply the convexity arguments of §§1 and 3, obtaining the canonical types of distribution (3.5).

This is the model of the system in a steady thermodynamic state. Both it and its generalization to the case of irreversibility will be outlined in the next section, in terms of classical Hamiltonian theory.

6. Entropy Increase in Hamiltonian Systems

The conventional Newtonian model in the kinetic theory of matter is the system with n degrees of freedom: its position space is a manifold of n dimensions. Therefore, it can be given parametrically (at least locally) in terms of n generalized coordinates $(q) = (q_1,\ldots,q_n)$. If the system consists of N particles of individual coordinates (x,y,z), n = 3N. If there are more degrees of freedom, smooth constraints, etc., n is altered accordingly. The dimensionality of phase space is 2n, its "points" represented (locally) as (q,p), where p_i is the momentum component conjugate to q_i. Globally, the phase space is the position space augmented by the momenta to form a "cotangent bundle" (in the terminology of modern mathematics).

The Hamiltonian is a regular function* H defined on the phase space, and (locally) of the form H(q,p) in the case of steady motion, and H(t,q,p) in general. The corresponding canonical differential equations of motion are

$$dq_i/dt = \partial H/\partial p_i, \quad dp_i/dt = -\partial H/\partial q_i, \quad i=1,\ldots,n \tag{6.1}$$

The solutions of (6.1) express (q,p) as 2n functions of t and the initial values (q^o,p^o) at t^o. These trajectories may be interpreted as a 1-parameter family of transformations of phase space onto itself $(q^o,p^o,t^o) \to (q,p,t)$ (t being the parameter) defined by regular functions and having regular inverses. They form a group in phase space if H does not contain t explicitly.

Any regular function F on phase space varies along each trajectory. If F = F(q,p), its rate of change, calculated by applying (6.1), is dF/dt = [F,H], where the Poisson bracket of any functions f and g of (q,p) is

$$[f,g] = \Sigma_i[(\partial f/\partial q_i)(\partial g/\partial p_i) - (\partial g/\partial q_i)(\partial f/\partial p_i)] \tag{6.2}$$

When F contains the time t explicitly, we have an extra term: $dF/dt = \partial F/\partial t + [F,H]$. These facts show that if H does not contain t explicitly, dH/dt = [H,H] = 0, so that its value is constant along each path: it is a first integral of (6.1); this

---

*Regular in the sense of being single-valued and continuously differentiable to any order required by the theorems applied.

expresses the conservation of mechanical energy.

A function $F = F(t,q,p)$ will be a first integral if it satisfies Liouville's equation

$$\partial F/\partial t + [F,H] = 0. \tag{6.3}$$

It is a consequence of the connection between logic and probability that if F is a probability density of the various possible states of the system, it must satisfy (6.3). For if the system is at epoch t^0 in an elementary region of 2n-dimensional volume dq^0dp^0 containing (q^0,p^0), it will be at t in the volume dqdp, the image of the former under the transformation induced by (6.1); and conversely. The probabilities of these events are $F(q^0,p^0,t^0)\, dq^0dp^0$ and $F(q,p,t)\, dqdp$. Since when two events are logically equivalent (either one implying the other) they have the same probabilities, these two quantities are equal. By Liouvilles's Theorem, all 2n dimensional volumes are preserved under (6.1). Hence $dq^0dp^0 = dqdp$. Therefore $F(q^0,p^0,t^0) = F(q,p,t)$--which, by definition, is the property of first integrals of (6.1)--thus proving (6.3).

The information in the probability distribution of density F is (with an obvious abbreviation of symbols)

$$I[F] = \int\int F \log F\, dqdp, \tag{6.4}$$

the integration being over the whole phase space. The transformation defined by (6.1) leaves its value constant. This is easily seen by making a change of variables of integration, using (q^0,p^0) instead of (q,p) (remembering that the Jacobian $\partial(q,p)/\partial(q^0,p^0) = 1$ in virtue of Liouville's Theorem), and applying Leibniz's rule for t-differentiation under the integral sign. Since F is a first integral, so is F log F; hence $dI[F]/dt = 0$.

As in the earlier sections, we must establish precisely stated conditions under which the information decreases, and, moreover, with certain expected values held constant or made to change in some given manner. Again we shall regard the source of this behavior of I[F] as the set of external disturbances, produced e.g., at physical boundaries, and we shall formulate their action in terms of randomly chosen transformations S belonging to a group. But now the elements of the latter are not only infinite in number, but cannot be identified by a finite set of parameters as in the Lie theory. They will be what Lie termed infinite continuous groups -- identifiable only by the choice of an arbitrary (regular) function.

As in the earlier sections, our transformations S must not only map phase space into itself: $(q,p) \leftrightarrow (q',p')$, but must carry trajectories of (6.1) into trajectories, viz., integrals--in general different--of the same equations. Furthermore, we are concerned with the effect of S on the expected values of certain funcrions, such as G, defined on phase space: the behavior of EG $= \int \int FG\, dqdp$, where F is the probability density, satisfying (6.3). There are two radically different possibilities, according to whether G is an "arbitrarily chosen" function of (q,p,t) (e.g., $G = q_1$ or $= p_3$, etc.) or belongs to a special class. Let us suppose that the point (q',p') through which a particular trajectory passes at the epoch t' is displaced slightly to the position (q",p") on a different trajectory. The latter, while close to the original trajectory at times close to t', will in general separate from it widely with increasing t. Consequently, the value of the arbitrarily chosen G will have values on the trajectories that were close only at t', and become very different thereafter. This is the first and general situation.

The second possibility is illustrated most vividly when G is a first integral of (6.1): then, being constant along each of the two trajectories described above, and having very slightly different values when they were close, will continue to have this property even after the trajectories have widely separated. This case requires that G satisfy Liouvilles equation. Another illustration of the second possibility is the case in which G, without satisfying Liouville's equation, satisfies the equation $[G,H] = 0$. Then G varies along each trajectory according to the law $dG/dt = \partial G/\partial t$. This will always be true of H. Physically, the meaning of this situation is that $H = H(q,p,a,b,\ldots)$, that is, it depends on certain parameters, a,b, etc., such as those specifying the positions of attracting masses, the physical boundaries of the system, e.g., the volume containing a gas, etc., and that we can induce a time dependence by varying these parameters. Since their rate of variation, as produced, e.g., under laboratory conditions, is much slower than the order of velocity of the individual particles of the system, the fact that $dH/dt = \partial H/\partial t = (\partial H/\partial a)\,(da/dt) + \ldots$, means that H varies slowly with time and in a manner which we can either control or understand in terms of macroscopic principles.

The same is true if G is the component in a fixed direction of the total momentum of the system: it would be a first integral if we left the system alone; but it can be changed, e.g., by moving the container as a whole. As in the case of H, the change is relatively slow--but not necessarily "infinitely slow" as in reversible thermodynamics. It may be fast enough to give rise to heat flow, viscosity, and similar entropy increasing processes.

It is from functions described in the last two paragraphs, illustrating the second possibility, that we chose those whose expected values are to constrain the increase of entropy. But obviously another restriction must be imposed: indeed if G is a first integral, so is any function U(G), such as G^2, sin G, exp G, etc.; and the maximum entropy formalism would lead to a mass of absurdities without a further principle of selection of constraints.

The usual (and somewhat loose) term for the restriction is that our functions G must be sum functions, or, since as first integrals they are invariant, additive invariants. This concept presupposes that the system can be regarded as the combination of several sub-systems, and defines a sum function as one whose values for the whole system is the sum of those for its parts. Thus in a molecular gas, the total momentum in a fixed direction, as well as the number of molecules, are additive invariants. For a not too dense gas in which inter-molecular forces are of short range, the energy is very approximately a sum function. Evidently if G is such a function, none of the above functions of G can be: only when U(G) is linear will this property be maintained.

Thus in the classical applications, the sum property is a good physically based slection principle. But we may replace it by a more general one based on group properties.

7. Canonical Transformation Groups

The physical applications involving entropy increase demand that time variation of H, the boundaries B, and other factors be taken into account. For mathematical reasons we wish to retain the conceptualism of groups. But as stated in §6, when $H = H(t,q,p)$, (6.1) do not in general define a group in the 2n dimensional phase space of (q,p). On the other hand, by raising the dimensionality by including t among the "position variables" (q) -- using "space-time" and hence "phase space-time" -- the group property is restored: then (6.1) may be written in the form (subscripts denoting partial derivatives)

$$d\alpha = \frac{dt}{1} = \frac{dq_1}{H_{p_1}} = \dots = \frac{dp_1}{-H_{q_1}} = \dots \tag{7.1}$$

The parameter α thus defined is absent from the 2n+1 coefficients of the system of differential equations, so that, according to the basic theory of such systems, the transformation from initial (t^o,q^o,p^o) to final (t,q,p) during the α change forms a 1-parameter group in the parameter α.

Along with the new spatial variable $q_o = t$ there is its conjugate momentum p_o: this is the negative Hamiltonian, satisfying the equation

$$p_o + H(t,p_1,\dots,p_n,q_1,\dots,q_n) = 0. \tag{7.2}$$

It is necessary to consider such transformations in (q,p) as are defined by (7.1) from a general point of view, as set forth in [2] and [7], according to which the *canonical transformations* are defined in terms of the *Pfaffian differential form*

$$\omega = \sum_{i=0}^{n} p_i \delta q_i \tag{7.3}$$

Here δ is used as a more general differential than the *d* acting along the path. For any 2n+2 variables satisfying a given relation such as (7.2) (hence 2n−1 independent), a *canonical transformation* $(q,p) \rightarrow (Q,P)$ is any 1-to-1 regular transformation preserving the integral $\int\omega$ around any closed curve. Equivalently, it is one for which

$$\sum_{i=0}^{n} (P_i \delta Q_i - p_i \delta q_i) = \delta W \tag{7.4}$$

where W is a function of the variables (2n−1 independent ones among the 2(2n−1) total number).

The usual method (cf. [7], §126) is to regard W as expressed in terms of the 2n+1 spatial variables $(q,Q),t$, and then equate the coefficients of the 2n+1 independent differentials. It is more useful for many physical and mathematical reasons, as pointed out by various authors (e.g., A. Sommerfeld and G. D. Birkhoff) to alter this procedure slightly, taking the old spatial and the new momentum variables as independent. Evidently (7.4) is equivalent to

$$\sum_{i=0}^{n} (Q_i \delta P_i + p_i \delta q_i) = \delta V = \delta\left(\sum_{i=0}^{\infty} P_i Q_i - W\right) \tag{7.5}$$

Equating the coefficients of the differentials (subject to (7.2), etc.) gives the required canonical form. In our case, we are restricting our transformations to *absolute time*: $Q_0 = q_0 = t$; then (7.5) becomes, on using (7.2),

$$\left(\sum_{i=1}^{n} Q_i \delta P_i - \mathfrak{H}\,\delta t\right) - \left(\sum_{i=1}^{n} p_i \delta q_i - H\delta t\right) = \delta V \tag{7.6}$$

where $V = V(t,q,P)$, 2n+1 independent variables. Therefore,

$$Q_i = \partial V/\partial P_i, \quad \mathfrak{H} = H + \partial V/\partial t, \quad p_i = \partial V/\partial q_i. \tag{7.7}$$

These transformations form a group. Evidently the identity is produced when we take $V = \Sigma q_i P_i$, since (7.7) then shows that $\mathfrak{H}$ =

H and that the large and small letters are respectively equal. The inverse is obtained by applying the process that led from (7.4) to (7.7) but with the roles of the large and small letters interchanged, using (t,p,Q) as the independent variables (W appears with its sign reversed).

It is useful to show explicitly the departure of our transformation from the identity: we set $V = \Sigma q_i P_i + U$, so that (7.7) becomes

$$Q_i = q_i + \partial U/\partial P_i, \quad \mathfrak{H} = H + \partial U/\partial t, \quad p_i = P_i + \partial U/\partial q_i. \tag{7.8}$$

First application: let U be replaced by $U\Delta\beta$ and $Q_i - q_i = \Delta q_i$, $P_i - p_i = \Delta p$. Then if in U we replace P_i by $p_i + \Delta p_i$, expand in powers of the Δ - quantities, divide by $\Delta\beta$, and then let $\Delta\beta \to 0$, we obtain (writing $u = U(t,p,q)$)

$$d\beta = \frac{dt}{1} = \frac{dq_1}{\partial u/\partial p_1} = \ldots = \frac{dp_1}{-\partial u/\partial q_1} = \ldots = \frac{dH}{\partial u/\partial t} \tag{7.9}$$

If we take $u = H$ and $\beta = \alpha$, (7.9) reduces to (7.1): this is the classical theorem that the transformation from initial to final values produced by the natural motion (in which d is along the trajectory, as in (6.1), is canonical--in the above sense, defined by (7.4) [with (q,p,Q,P) replaced by (q_0,p_0,q,p) etc.].

Second application: by combining (7.4) (with the above replacement of letters) with two other equations: (7.4) as it stands, and (7.4) with zero superscripts on all its letters, we see that a canonical transformation preserves the canonical property of any canonical transformation. Thus if S is the transformation (7.6) and T is that defined by (6.1) or (7.1), the transform of T by S : $STS^{-1} = T'$, where T' is a canonical transformation, in general different from T. As in §4, two interpretations are possible: S could be used to denote a change of coordinates (reference frame): T' would then be the same transformation (physically and geometrically) as T, merely expressed in different coordinates. But we shall, on the contrary, denote by S an actual physically caused change in the physical transformation T, producing a physically different one, T': our coordinate system is not changed.

Third application: the inverse of the canonical transformation determined by the function U in (7.8), or equivalently by u in (7.9), is determined by -U (after appropriate replacements of letters); or equivalently, by -u.

The fourth application is to the study of the effect of exterior factors such as are produced by random effects at the boundary. Let H be the Hamiltonian of the system when such unpredictable effects are absent. As the representative point (q,p) on a trajectory passes through a "thin" shell-like neighborhood of the boundary, the effect may be expressed by adding another part K = K(t,q,p) to the original H, which then becomes $\mathfrak{H}$ = H + K, the K vanishing outside of the shell. But if we take the partial indefinite integral V = ∫ Kdt, so that K = ∂V/∂t, (7.7) shows that this is the effect of the V-based canonical transformation on the H-based one. Furthermore, the inverse of the former is based on -V: when this is applied to the H-based (7.1), we have $\mathfrak{H}$ = H-K.

We now, as in earlier sections, postulate the equality of the two conditional probabilities: that on a given occasion, disturbance V act; or that its inverse -V act -- conditioned on the assumption that one or the other take place on that occasion. Whichever it is, the value of H becomes unknown, but not its expected value, since $E\mathfrak{H} = (H + V_t)/2 + (H - V_t)/2 = H$. Moreover, if our system is acted on by a random set of disturbances, but if all have the above symmetric property, the expected value of H remains unchanged.

By the same postulate of symmetrical pairs of disturbances it shows that, if dqdp and dq'dp' represent two elementary 2n dimensional regions in phase space of equal 2n-volume, the transition of the state from the former to the latter, and the inverse transition, are events of the same probabilities (all on a given occasion t). As in the case of the finite set of states, the cumulative effect of such disturbances is to lower the numerical information--increase the entropy; but subject to the constraint of given expected H, as explained above.

It is clear that the function H, being the one that generates the group (7.1) of physical motion along the trajectories--the transformations T = T(α)--cannot be replaced in the given physical problem by a function U(H), e.g., H^2, since, while not affecting the trajectories in phase space, it would altogether change them in phase space-time. A similar remark applies to the choice of further constraints in entropy maximization: the functions (such as G, the total momentum component in a fixed direction) generates a group: we could obtain it by writing (7.9) with u replaced by G. This group is that of translations of the system as a whole along the direction in question. Since in important cases of only internal forces, this leaves H unchanged, it is an integral of the system; and furthermore, by the earlier arguments, its expected value is unchanged by the symmetric ensemble of external disturbances. Again, it cannot be replaced by any function of G or of G and H without destroying the group property, except by a linear function with constant coefficients.

Thus the selection of constrained expected values in the maximal entropy formalism is determined by group theory as well as by physics.

In closing, we note that our postulated symmetry of disturbances justifies the assumption of "micromolecular chaos" used in

[3] and [6], and gives a basis for our own treatment [5].

References

[1] For an account of this theorem in its general setting required, see N. Bourbaki, Espaces Vectorielles Topologiques, Hermann, Paris.

[2] Cartan, E., Lecons sur les Invariants Integreaux, Chapters I and XII, Hermann & Fils, Paris, 1922.

[3] Chapman, S. & Cowling, T.G., The Mathematical Theory of Non-Uniform Gases, Cambridge University Press, 1953.

[4] Hardy, Littlewood, & Polya, Inequalities, Chapter III, §3.6-§3.11, Cambridge University Press, 1934-1973.

[5] Koopman, B.O., Relaxed Motion in Irreversible Molecular Statistics, pp. 37-63, in Stochastic Processes in Chemical Physics, K.E. Shuler, Editor; Interscience, 1969, John Wiley & Sons.

[6] Mazo, R.M., Vol. 1: Statistical Mechanical Theories of Transport Phenomena, Pergamon Press, 1967.

[7] Whittaker, E.T., Analytical Dynamics, Chapters X and XI, Second Edition, 1917, Cambridge University Press.

Generalized Entropy, Boundary Conditions, and Biology

Jerome Rothstein
The Ohio State University
Department of Computer and Information Science
Columbus, Ohio

Though most scientists feel no need to assume the existence of special "biotonic" laws for biology, consistent with, but essentially independent of physics, few claim fundamental physics currently gives an adequate account of biology. We propose that generalized entropy (including the information-organization-measurement complex of ideas treated in our earlier papers) provides an adequate framework for <u>constructing</u> an essentially endless variety of biotonic "laws". It corresponds to freedom in <u>design</u> of complex systems, which is essentially like that of Turing machines or computers. Physical law underlies all designs, as with real computers. Their behavioral diversity reflects the different internal constraints, initial conditions, and boundary conditions characterizing the different systems and subsystems. Irreversibility of metabolism or evolution reflects irreversibility of measurement, subsystem preparation procedures or setting up of subsystem boundary conditions, as well as conventional thermodynamic irreversibility; the former class involves the generalized entropy concept. Metastability, constraints, memory, information transfer, feedback and other concepts of physics, biology, cybernetics, and engineering have harmonious roles within this framework, which has essentially no ad hoc characteristics. All essential concepts have long been implicit or explicit in physics itself or in operationally formulatable physical methodologies. The fundamental lack in previous biophysical theorizing can be characterized as inadequate appreciation of the fundamental importance of boundary conditions (in a generalized sense) for biology; dynamical laws, rather than boundary conditions, have historically been given most attention. Goedel's incompleteness theorems and the algorithmic unsolvability of the halting problem for Turing machines lead to incompleteness theorems and undecidability results for the behavior of complex systems. These are discussed in relation to evolution, the second law of thermodynamics, and other topics.

I. Introduction

The construction of an adequate theory of biological phenomena presents a standing challenge to modern physical science which seems to persist despite the significant progress made almost daily. The more we unravel details the more complexity we uncover, and the harder it seems to become to find physical mechanisms able to generate such fantastic complication from an initially disorganized, relatively simply specified, physico-chemical environment. Some have compared the behavior of living matter to that of Maxwell's demon, while vitalists deny the possibility that adequate physical explanations of biology can ever be given. Neo-vitalists, like Elsasser, hypothecate the existence of special "biotonic" laws, consistent with, but essentially independent of physics, which govern some kinds of biological behavior.

The thesis of this paper is that generalized entropy (including the information-organization-measurement complex of ideas treated in our earlier papers) in the framework of a correspondingly generalized thermodynamics, provides an adequate framework for constructing an essentially endless variety of "biotonic laws". These laws correspond to the variety of designs, and the corresponding behaviors, of physically constructible complex systems. Freedom in design of such systems is like that of Turing machines or computers, and has long been familiar in engineering. As engineering systems are designed with specific purposes in mind, the nagging problem of teleological mechanisms or explanations is not how specific purposes or goals can be implemented by tangible physical mechanisms, but how they can arise in a physical context. As we shall show, there is much more than superficial resemblance between this problem and the perennial one of understanding how irreversible processes can occur in a universe governed by completely time-symmetrical dynamical laws.

Physical laws underlie the behavior of all complex systems, including those that contain subsystems capable of measuring (acquiring information), computing, and performing operations on the environment and on various subsystems. We refer to the latter class as "well-informed heat engines", abbreviated {WHE} , and will use them to model any gross behavior admitting macroscopic, operationally definable, specification or description. Individual engines will be denoted as WHE, WHE-1, -2, etc.

The behavioral diversity possible in the class {WHE} reflects the effects of

(1) different internal constraints (including stored information and metastability of its configurations),

(2) different initial conditions (both starting states of active components and memory states, including both stored or acquired data and initially stored, or subsequently modified programs), and

(3) the complex external and internal boundary conditions

characterizing different systems and their subsystems (including flow of information and exertion of control as well as mass or energy flow).
The class {WHE} will be shown to be capable of realizing any operationally defined behavior. It should thus realize biological behavior in principle, insofar as it can be given an operational specification. Accepting {WHE} as a physical model for the class {living things} then implies that explanations of many biological phenomena must be sought in facets of the generalized entropy concept.

In particular, metabolism (including development, differentiation, reproduction, and regeneration) and evolution are accompanied by irreversible processes not readily understandable in thermodynamic terms. The generalized entropy concept, which takes into account the entropy costs of obtaining information (by measurement), or preparing a subsystem in some special set of states, exerting control actions, and of setting up subsystem boundary conditions, is proposed as powerful enough to make those processes understandable. Metastability, constraints, boundary conditions, memory, information transfer, control, feedback, and other concepts of physics, biology, cybernetics, and engineering have harmonious roles within this framework, which has essentially no ad hoc characteristics. All essential concepts have long been implicit or explicit in physics itself or in operationally formulatable physical methodologies.

We belive that a fundamental lack in previous theorizing about biology from the viewpoint of fundamental physics (say quantum and statistical mechanics) can be characterized as follows. Attention has been given almost exclusively to dynamical laws, with comparative neglect of the fundamental importance of boundary conditions, in a generalized sense, for biology. Whenever a constraint is applied or removed, whenever a measurement is made or information transmitted or recorded, whenever a subsystem is prepared, or indeed, whenever anything irreversible occurs, we must deal in essence with a differently specified system, i.e. one subject to altered initial and boundary conditions. Indeed, these almost jump-like changes in generalized boundary conditions can dominate the course of events analogous to events in digital computers, where continuous dynamical variations in currents and fields are of virtually no interest. A computer-like model is often preferable to the trajectory-cum-equations-of motion descriptions of conventional physical theory, as will become abundantly clear in the sequel.

The most general computer, in a logical or mathematical sense, is the Turing machine. It was created originally to clarify, for the foundations of logic and mathematics, the meaning of the intellectual bombshells dropped by Goedel. He showed that in any logical formalism complicated enough to include arithmetic, there were true theorems which could not be proved, and that if a proof could be formulated within the formalism that the formalism was consistent, that the formalism would be inconsistent! Turing made what was in effect a computer capable of processing data in a manner simulating the proof process in any given formal system, and also showed that a "universal" machine could be constructed. That machine, given a properly encoded description of an arbitrary machine, and an encoded version of the data on which the latter was to work, would produce an encoded version of what the arbitrary machine would produce. In short, the universal machine T_u, can simulate an arbitrary machine processing arbitrary data. Goedel's results now imply a number of results for Turing machines, the most famous of which is the algorithmic unsolvability of the halting problem for Turing machines. A machine, given data to process, will either eventually halt, giving a "result" or "decision", or (neglecting things like wear-out or power failure) run on forever. The theorem is that it is impossible to build a Turing machine, say T_u, which given an encoded description of an arbitrary machine and its input, will always correctly decide whether that machine will eventually halt after processing its input data. As real computers differ from Turing machines only in that they have been made faster and more convenient by many clever inventions, and suffer from practical economical limitations, the theoretical limitations of Turing machines are a subset of their limitations. The same is true of {WHE}, for a complete description of a WHE plus its environment can be taken as a Turing machine input; the "next" state of such a description represents the next stage of processing of that input. The active subsystem of the WHE corresponds to the finite state control, the rest of WHE and the relevant part of the environment correspond to the tape, tape plus control (which reads, writes, and moves on the tape) being the essential parts of the Turing machine. The existence of undecidable questions and the corresponding impossibility of constructing some Turing machines implies the existence of undecidable questions about the behaviors of members of {WHE}, and the impossibility of realizing a WHE exhibiting some kinds of behavior. The very phrasing of the last few sentences suggests their kinship with thermodynamic laws appropriate to the generalized entropy concept needed to discuss {WHE}. This topic has many aspects, of which only a few can be treated in this paper.

Undecidability results assume many forms, and for {WHE} the "thermodynamic connection" extends them beyond the logical types of Turing and Goedel. A construction for a WHE which refutes theories of its behavior is given. A kind of "free will" exists for a sufficiently complex WHE. A kind of thermodynamic motivation can be given for something suggestive of evolution of {WHE} under permanently maintained non-equilibrium conditions. Paradoxical as it may sound, it is possible to have a deterministic evolutionary process without being able to decide whether some specified situation will ever evolve from a given situation, or, similarly, whether some specified situation has evolved from a given situation.

Organic evolution is unthinkable, on current theories, without prior chemical evolution, and the complexities of the latter must be very close to realizing {WHE} on a macromolecular scale. It is therefore important to show that generalized statistical thermodynamics allows this. It turns out that precisely the kinds of properties needed to design a molecular WHE do exist, and they do not differ from others long known to exist (primarily exclusion principle).

It therefore appears justified to conclude that biology can be "explained" by physics in a sense not very different from the senses in which chemistry and metallurgy are explained by physics. The architectural principles all seem to be there, but a detailed prediction of all the properties (say the structure-sensitive ones) of even a single ingot of metal is impossible and likely to remain so. The success of our search for simple general principles should not make us forget that changing internal boundary conditions in the history of the ingot have made it impossible for us to predict the precise evolution of its structure-sensitive properties.

II. Generalized Entropy

Schroedinger (1) stressed the importance of "negative entropy" for biology. His terminology emphasizes the non-equilibrium nature of living matter, for entropy increase accompanying approach to equilibrium entails a negative prior value with respect to the equilibrium value. Entropy as a measure of missing information, ignorance, disorder, or disorganization was informally appreciated by the founders of thermodynamics, statistical mechanics, and kinetic theory, particularly Boltzmann and Maxwell. Boltzmann introduced the statistical expression for entropy, later adopted in information theory by Shannon (2), Wiener (3), and many since. Maxwell introduced his "demon", discussed later by Szilard (4), Von Neumann (5), Demers (6), Brillouin (7), Jacobson (8), Gabor (9), and Rothstein (10).

Information is communicated when doubt about which of a set of alternatives, from which a choice is to be made, is resolved. Initial uncertainty is measured by a function identical in mathematical form to entropy, namely

$$I = - \Sigma p_i \log p_i \qquad (1)$$

where

$$\Sigma p_i = 1 \qquad (2)$$

and the p_i are the a priori probabilities of the various alternatives. I is entropy (actually S/k) in statistical mechanics if the set of alternatives are cells in phase space, natural logarithms are used, and the reference level (zero entropy) corresponds to having the representative point of the system localized in one cell (for which p=1, the occupation probabilities of all other cells being zero). In information theory I reduces to zero when the actual choice becomes known, for the new probabilities are now zero, except for one. The entropy thus measures the information conveyed as a consequence of making the choice. In information theory it is the custom to write H for I, to take logarithms to base 2 in the discrete case (in the continuous case natural logs are often used) and to refer to the set of alternatives as messages. In general, messages are strings of symbols; the set of symbols from which messages are formed by concatenation is called an alphabet. Almost always alphabets are finite sets, while the cardinality of the message set is unbounded. It is thus usually convenient to talk of a message as a sequence of selections from an alphabet and to work on a "per symbol" basis.

The problem of putting the relation between physical and informational entropies on a sound basis, and of extending and <u>then</u> applying the generalized entropy which results to a variety of fundamental questions was attacked by Rothstein in a long series of papers, which includes the present one and much work not yet published. The logical and mathematical identity of measurement and communication was shown (11), whence, from an operational viewpoint, the similarity between the entropies of statistical mechanics and information theory ceased to be mere analogy. Some consequences for the Einstein-Podolsky-Rosen paradox and notions of physical reality were also drawn. An operational analysis of macroscopic thermodynamics was then made (12) to see if macroscopic thermodynamic entropy had an inherently informational nature (stemming from the measurements and operations involved). This turned out to be so, the path going through Caratheodory's principle and also giving a "macroscopic ignorance" characterization of heat as a form of energy. This agreement made it clear that nothing incorrect would be deduced from informational generalizations of entropy as long as the information had an operational basis, i.e. it was produced by measurement, or operations like imposing constraints, exerting control actions, or carrying out preparative procedures.

The foundation was thus firm enough to permit research to go in many different directions. These included (a) further generalization of the entropy-measurement-information concept, (b) methodological and philosophical questions, including problems of interpretation of formalisms, (c) resolution of paradoxes of various kinds, (d) applications to complex systems, including living and other non-equilibrium systems. All of them are needed for biology.

Of particular importance is the concept of organization, which can be quantitatively measured in entropy diminution terms just as information can (13,14,15). An organization consists of interacting units whose individual behaviors are not free as they would be in the absence of the constraints, correlations, or couplings in which the organization consists. Let $\{X_i\}$ denote the ensemble of alternatives associated with the ith unit, and $H(X_i)$ the entropy (information-theoretical) of that ensemble. Let $H(X_1, X_2,\ldots)$ be the entropy of the set of compatible complexions, where each complexion is an n-tuple, for n units, consisting of one alternative for each unit, and the meaning of compatible is that all complexions inconsistent with the constraints, etc., are excluded. Were they not, the units would be independent and we would have

$$H_{indep}(X_1, X_2,\ldots) = H(X_1) + H(X_2) + \ldots \quad (3)$$

In general we have

$$H(X_1, X_2,\ldots) \leq H(X_1) + H(X_2) + \ldots \quad (4)$$

with equality only for case (3). We now define the amount of organization (or simply organization) ΔH by

$$\Delta H = \Sigma\, H(X_i) - H(X_1, X_2,\ldots)$$

It is zero for independent units, positive if they are correlated, and attains its maximum value for the pure case (only one complexion admissable). It reduces, in the communication case, to information transmitted (two ensembles, where $\{X_1\}$ say is sent, $\{X_2\}$ is received). In that case, it is often called mutual information, or transinformation. In configurational discussions, amount of organization is a measure of information coded in structure. The "disorganized" state, where all couplings between units are dissolved, is clearly the state of maximum entropy, and this is true for thermodynamics as well. Condensation, crystallization, polymerization and aggregation entail formation of more highly organized states, and we find it attractive to assume that biological materials can be regarded in the same way.

Organization admits hierarchical structure, for the X_i can be tuples individually,

$$\{X_i\} = \{(y_1, y_2, \ldots, y_{m_i})\} \qquad (6)$$

with no explicit change in the preceding definitions. The constraints involved in defining admissible (sub)complexions of m_i-tuples are simply included among those taken into account when computing (5). Some of the complexities which arise include feedback as represented by having some y_k of (6) also being an X_j of (5), where X_j need not be the same as X_i. If we interpret ΔH as a $\Delta S/k$ the above use of the term "complexion" agrees with its conventional usage in statistical mechanics. For time-varying systems we need to discuss the set of complexion-pairs, where the dynamics of the system determines the "next" complexion in terms of the current one. The feedback relation just mentioned then usually has varying times for various occurrences of the same X_i or y_i in different argument positions. For well-known communication theory cases, e.g. the literature on Markov sources beginning with (2), time is discretized, source states are correlated with symbols received, and transitions are probabilistic. The next state of the channel, for example, depends on its current state and the current input symbol to the channel from the source; the next output symbol (supplied by the channel to the receiver) depends either on state alone or on both state and input (in automata theory the two cases are called Moore and Mealy machines respectively). The finite state model of the channel, in the noiseless case, is identical, theoretically, to the finite state automaton, and the general Turing machine is a finite state automaton which both reads its inputs from, and writes its output on, a tape (16). In theory of machines and computation the bulk of the literature deals with the deterministic case (all probabilities 0 or 1), but there is an appreciable literature on nondeterministic machines and stochastic automata (17).

From the terse discussion of organization and entropy so far given it is obvious that traditional entropy surely, and information-entropy also, unless it be generalized, are not able to cope with great organizational complexity in physical systems (including those in engineering or biology) without powerful simplification techniques. A favorite one is analysis into subsystems, with simple flows of energy, heat, matter or information across boundaries between them (compartmental models), and it is indeed useful. It is also the justification for using essentially equilibrium concepts within small subsystems, taken to be at "local" equilibrium, but between which there are appropriate thermodynamic constraints e.g. differences in pressure, temperature, or thermodynamic potential. When "small" becomes infinitesimal we pass over to the usual treatments of non-equilibrium thermodynamics, particularly those related to Onsager's pioneering work (18). But when phenomenological descriptions involve non-local interactions with time delays, as occurs when signal propagation plays a key role or if the system is spatially inhomogeneous and localized sequential reactions can occur (say stages where substances produced at locations A_i must diffuse to B to undergo other reactions with non-linear kinetics) and/or there is a wide spectrum of relaxation times, etc., then one can easily encounter hysteresis. The system displays "hereditary" phenomena, or develops a "memory". If often needs integro-differential-difference equations to treat it. Even very simple cases can be intractable and non-linearity is frequent (19).

The usual phenomenological relations between conjugate forces and fluxes X_i, J_i respectively, are

$$J_i(t) = \Sigma L_{ij} X_j(t) \qquad (7)$$

They become, for media with memory

$$J_i(t) = \sum_j \int_0^t dt' M_{ij} (t-t')X_i (t') \qquad (8)$$

an equation of a form used in theory of viscoelasticity (with local response assumed; Gross) and for which the (local) entropy production rate

$$\sigma(t) = \Sigma X_i(t) J_i(t) \qquad (9)$$

can be momentarily negative. McLennan shows that the time integrated entropy production is always positive, however (20), when the representation of M_{ij} as a correlation function is used:

$$M_{ij}(t-t') = < j_i(t)j_j(t') > \qquad (10)$$

Here the j's are microscopic fluxes whose time dependence is determined by the microscopic equations of motion.

The "momentary" negative values for $\sigma(t)$ are important, for if metastability (13) is also present in the proper way we can "freeze" some "negentropy", to use a term popularized by Brillouin (21), into a metastable subsystem, whereby its organization is increased. This is clearly important for biological evolution, as we have shown elsewhere by very different reasoning (22). Brillouin also makes a distinction between "free" and "bound" information, the former being abstract, the latter being physical in the usual thermodynamic sense. We have not found this distinction necessary in the same sense that we do not feel any need to belabor the difference between the use of "three" in number theory and in the phrase "three cows". Formal concepts admit an infinitude of interpretations in fields to which they can be applied, and once the formal concept is understood one should have no difficulty in knowing whether one is talking about an abstraction or a concrete application! But there is often real difficulty in deciding what ensemble of alternatives is involved, particularly if one jumbles different interpretations together uncritically; in that case it is easy to talk utter nonsense. For example, if one says x bulls plus y cows equals (x+y) "bows" and assumes "bows" always to be well-defined animals, he/she (at last, a case where this hermaphroditic "pronoun" is appropriate!) is due for many a surprise down on the farm. We leave it as an exercise for the reader to straighten out footnote 4, page 157 of (21) on the basis of this remark, and to do the same for the ubiquitous howlers he may encounter in the "applied" information-theoretical literature (particularly psychology, the social sciences, often biology - and even physics!)

The problem of determining just what set of alternatives we mean when we talk about an operationally defined set of physical alternatives is by no means simple. When statistical mechanics deals with precisely defined microstates it is not dealing with the operationally defined distinguishable state provided by macrscopic measurement. The latter is represented by an ensemble in microstate language, the former is a primitive logical notion in the operationally specified situation contemplated. When we specify the operational definition we select a particular set of conditions out of the ensemble of conditions we might have selected, and this is reflected in dependence of the entropy of the system prepared on the preparation conditions. As these are generally set up initially and kept constant, the dependence of entropy on boundary and initial conditions is often overlooked, along with the entropy price paid for the preparation or measurement procedure involved. Note that this price depends on the operations performed, not on the state of mind of the physicist contemplating the results of carrying out those operations (much nonsense has been written because this simple fact was neglected). The language of operational description of a preparation and measurement procedure is a metalanguage for discussing statements in the description language intrinsic to the prepared system. Information or entropy for the former refers to the combination, system of interest plus means for preparation or measurement, considered as a single system. If one were to prepare a system of interest in a pure quantum state its entropy would be zero, but the entropy of the combined system would have increased in the preparation process enough to compensate for the local decrease.

The interaction between a system of interest and the means for preparing or measuring it can be viewed as a relaxation process in an initially metastable compound system. Constraints are then relaxed to let the interaction occur. Constraints are imposed again to end it and decouple the system of interest from the means of measurement or preparation. Though historically associated with quantum mechanics (EPR paradox etc.) this is a purely macroscopic phenomenon. It involves an entropy increase, and elucidates irreversibility both of measurement and of preparation (including manufacture, of which a special case is manufacture of a replica, or reproduction).

Another consequence of the generalized entropy concept is a thermodynamic limit on the precision of any measurement, i.e. a "phenomenological uncertainty principle" (23). For example, if a quantity equiprobably between 0 and u is measured to fall within Δu a choice has been made from $u/\Delta u$ alternatives. To avoid conflict with the second law the minimum entropy expenditure ΔS must satisfy

$$\Delta S \geq k \ln (u/\Delta u) \qquad (11)$$

or

$$\Delta u \geq u \exp (-\Delta S/k) \qquad (12)$$

It would take infinite entropy generation to achieve absolute accuracy in measurement of u, so a "pure case" can not be prepared, operationally.

In quantum mechanics, if u is an observable this means we can never prepare an eigenstate of u, only a mixture. Equivalently, we can never prepare a state of zero entropy, though both can be approached arbitrarily closely. We can view this as one expression of the third law of thermodynamics, the controversial aspects of which have to do with frozen-in metastability, in the main. Tolman (24) takes the content of the third law as the assignment of zero entropy to a pure state. It is also often taken as unattainability of absolute zero in a finite number of operations (25). This harmonizes nicely with eq. (12), as only at absolute zero would thermal fluctuations cease setting a limit on accuracy of measurement. Also harmonious is the adoption of

$$S = 0 \quad \text{at } T = 0 \qquad (13)$$

as a (frequently criticised) statement of the third law, It takes the (maximal information) - (minimal uncertainty) equilibrium state as the only one never to be "spoiled" by fluctuations (cf. (12)). Schroedinger's discussion of the meaning and admissibility of setting the entropy constant equal to zero (26) in

$$S = k \log \Sigma \exp -(\varepsilon_i/kT) + U/T + \text{const.} \qquad (14)$$

should be required reading for all interested in the third law. The core of the objection to equ. (13) is two-fold (a) frozen-in metastable equilibria and (b) degeneracy of the ground state. The entropy equivalent of specified constraints, as in ref. (13) answers objection (a). Even if we did not know them, we could, in principle, use increasingly "gentle" probes, as T approaches zero (e.g. longer and longer wave-length photons) to get increasingly finer information about details of the frozen state. In the limit we would be describing a new ground sate subject to the constraints maintaining the metastable equilibrium. To answer (b) note that either it is possible to resolve the degeneracy by a perturbation (it can be made vanishingly small as T goes to zero), or it is impossible to resolve it by any procedure. If the former, we obtain a new non-degenerate ground state. If the latter, the multiple count is spurious, as in the Gibbs paradox, which can only arise if complexions differing only by permutations of identical particles are not clumped into a single appropriately symmetrized state. This discussion supplements an informational treatment of the third law (27) and a formulation of the third law as the outlawing of Laplace's demon (10).

We end this discussion of absolute entropy (i.e. defined by eq. (13)) and the third law by taking issue with the footnote on page 204 of Fowler (28) and the whole discussion starting with the last paragraph of that page, to the end of section 6.8 on page 205. Planck's position, cited in the footnote, is eminently sound. Indeed we consider his discovery of the quantum through analysis of radiation entropy as implicitly using its informational-operational nature. We were led, among other things, to come close to retracing his steps in working out (27)! Fowler's objection to absolute entropy, that there is no cut-off for analyzing new levels for counting configurations, is answered by a slight paraphrase of the discussion of ground state degeneracy of the previous paragraph. Either the new configurations are "hidden" (forget them) or they are observable in principle and should be included in the complexion count.

Because of the key role played by constraints as elements of operational descriptions or specifications, as boundary conditions, as control elements (e.g. trapdoors, switches, valves) and their effect on entropy, we now describe and analyze a thought experiment clarifying some of these concepts as they relate to each other (29). We were led to devise it, when in 1960, K.R. Popper told the writer he had a thought-experiment refutation of Szilard's thought experiment purporting to establish the entropy equivalent of a bit of physical information (4).

Szilard considers a piston dividing a cylinder into two equal volumes, and puts a molecule into one of them. If one knows into which of the two volumes it was put, one can extract work via the piston even though the whole system is at a uniform temperature. He shows that the second law of thermodynamics is saved if one ascribes an entropy equivalent of k ln 2 to the information in which knowledge of the molecule's whereabouts consists.

Popper argues that the information itself is irrelevant, the essential point being that the molecule is confined to a smaller volume. The situation is then simply the case where an expanding gas can do work. He suggests that according to the "rules of the game" used in designing thought experiments, one can easily arrange a linkage such that the molecule performs work no matter which side of the piston it "sees".

Figure 1 shows a thought experiment to substantiate Popper's contention. The piston has connecting rods on both sides, and these are connected to a rack and pinion arrangement such that a shaft is driven counterclockwise whichever way the piston moves, and can be used to raise a weight. It is thus quite clear, as Popper maintains, that one does not need to know on which side of the piston one has put the molecule. For either choice the same external work is produced. He thus concludes that information on the whereabouts of the molecule is irrelevant to the entropy calculation, that only the volume available to the molecule counts, and that the informational interpretation of entropy is therefore nonsense.

We now show that Popper has imposed a constraint which "organizes away" the uncertainty relevant to Szilard. In Figure 2 we give a pair of two-piston versions of Szilard's and Popper's thought experiments, in which a separate cylinder and piston corresponds to each side of the single piston of their experiments. They are, of course, equally valid thought experiments in themselves, but have the advantages of having the same linkages and configurations in the two cases, and of showing how Szilard's example corresponds to the case of relevant information, Popper's to irrelevant information. In each case a dot indicates the possibility of placing the molecule in that volume.

Consider (a) first. If the molecule is in the lower right volume it can do work against the piston and raise mass M against gravity. If it is in the upper left one this is not so. This is Szilard's case, essentially, differing only by the introduction of the pair of undotted and unused volumes. The distinction between the two dotted volumes is physically relevant in the sense that their occupation by the molecule corresponds to two thermodynamically distinguishable states; if one knows which volume is occupied, one can raise the weight either as shown or by using a rope and pulley. The same situation exists if the two volumes discussed are the two undotted ones, or the two left ones, or the two right ones. The distinction between right and left is irrelevant, only that between upper and lower is significant. The former distinction is a "hidden parameter" bearing no thermodynamic information.

It is now clear from (b) what Popper has done. His linkage has introduced a constraint whereby there is no longer a choice between upper and lower, but only between right and left. The latter is irrelevant in this case just as it is in (a), so there is no longer any paradox. Indeed, one need not have two pistons at all; one clearly has the same situation if the two pistons are considered to be two halves of one side of the same piston.

It is amusing to invert both Szilard's and Popper's experiments, using their set-ups to decide in which of the two volumes a molecule has been placed. Szilard's clearly gives us this bit of information. Popper's does not, for his constraint forces the two situations to give the same indication.

For an analysis of Szilard's experiment and the connection between information and entropy which manages to come to the wrong conclusions despite much correct mathematics, see (30). That both physicists and philosophers can go wrong in these matters points up the need for semantic clarity in handling them, so we will return to further discussion of methodology, foundations, formalisms and language before continuing with our main concerns.

All formalisms used in science (which are essentially specialized languages) evolve inductively from experience. Experience is largely, though not exclusively, measured by the information we gather. Informational-organizational formalism thus has a particularly fundamental role to play. As logic also has a peculiarly basic role their relationship was explored (31) in the sense of seeing what operational justification there might be for believing in the applicability of logic to the world. Logic survived and the treatment later led to characterizing information and organization as intrinsic to the language of the operational viewpoint in a philosophically neutral sense (32). The role of theory as providing more complete information about, or organization of experience than one has without it, and resultant criteria for deciding between rival theories is treated in (33), along with related topics. The fundamental roles of time and space as informational and relational manifolds was examined in (34), leading uniquely to (3+1) dimensionality for space-time with surprisingly few additional assumptions. Application to an informational resolution of the irreversibility paradoxes of statistical mechanics is made in (35) and (36), where the first shows how spin echo experiments realize a Loschmidt reflection and so suspend entropy increase by storing phase information, while the second, by analyzing the operational requirements of realizing the paradoxes of Loschmidt and Zermelo, shows that they do not exist in the sense that they merely represent the illegitimate use of words in multiple meanings. Most of the foregoing is reviewed in (37). It also relates these ideas to notions of reality and objectivity, time and irreversibility, the legitimacy of hidden variables and physical continua from an operational viewpoint, and how generalized entropy is relevant to biology. The discussion is supplementary to this paper, with moderate overlap. Some novel forms of the generalized second law, including their statement as undecidability theorems, are given in (38). As measurement and preparation are very close, operationally, one is not surprised to find that statistical quality control theory can be formulated in the same general terms as statistical communication theory (39). A critique of information theory and generalized entropy as applied to a broad interdisciplinary spectrum of applications is given in (40). It also presents a common informational-organizational characterization of theory, concept, pattern and meaning.

We close this chapter on generalized entropy with a few remarks on amplification. It is now commonplace to view amplifiers as devices containing subsystems at a negative temperature (41); these are "hotter" than infinite temperature in a sense explained by standard equations like

$$(\partial S/\partial E)_x = 1/T \qquad (15)$$

Here x is usually volume, but can be any other thermodynamic coordinates contributing terms of the form $X_i\ dx_i$ to dE. The amplifier is metastable, i.e. its "input" energy "unlocks" the metastability holding some stored energy in the subsystem. It then pours out as the amplified "output". No changes other than formal ones are needed in (15) to take this into account.

III. Well Informed Heat Engines

This section deals with physical systems capable of exhibiting behavior of essentially unbounded complexity. Their design is now almost commonplace in engineering (many fields) and relevant theory is particularly well developed in computer science. Essentially any behavior whose description is not logically contradictory or physically impossible can be realized in principle and modelled on a computer. In theory this could include modelling biological or even intelligent behavior, if we could but describe them with sufficient precision and rigor. Hope for devising such descriptions is rejected by vitalists but viewed as a distant goal, or at least one which cannot be ruled out as impossible in principle, by others. We outline some of the implications for generalized thermodynamics of the existence of such systems. We also seek to understand how they could come into existence, in the course of events natural for a large non-equilibrium system, as subsystems of the large system. While this really requires generalized statistical mechanics rather than macroscopic generalized thermodynamics, a subject to be touched on in IV, the macro theory can at least show why such evolution is more to be expected than to be a source of astonishment.

The well informed heat engine (22, 37, 38) differs from the ordinary heat engine in that it possesses measuring equipment (sensory, input, or monitoring subsystems for gathering information about the environment or its subsystems), a computer system (data processor, decision maker, "brain"), and effector equipment (output, "muscles", means for performing physical operations on its environment or on its subsystems). Its data storage system (memory) can be both internal (part of the computer) and external (to be read by the measuring equipment and "written on" by the effector equipment; in computer jargon we permit interactions with external memory through the I/O interfaces). We set no fixed bounds on memory size, though at any one time WHE interacts with but a finite memory (often called "tape"; one often talks of input and output tapes, storage tapes, etc.). We also permit WHE to read computer programms, add to its "library" of stored programs and data, and control its input and output (its effector apparatus can couple or uncouple part of its sensory apparatus or a part of its effector apparatus from a part of its environment or itself).

It is hardly necessary to dwell on the fact that WHE, as characterized above, includes essentially all engineering systems and models of complex systems, living or not. Theoretical models are represented by the programs which carry out the procedures by which the calculations used in the underlying theory are made. Physical models can either be similarly represented by their computer models, or a system itself can be viewed as its own simulator or analog computer. We therefore pass on to generalized thermodynamical considerations.

In (10) and (38) we showed that a number of philosophical questions could be reformulated to apply to well informed heat engines, with the impossibility of answering them showing up as aspects of a generalized second law. For example, a WHE with all the dynamical laws of its universe coded in memory could not predict the ultimate fate of the universe by extrapolating from data it could measure. The second law "spoils" the sharpness of the predictions as time goes on, the time for the initial information to "relax" to utter uselessness being an informational analog of approach to equilibrium. The same is true for retrodiction; the remote origin in time of the universe would be similarly inaccessible (note the formal similarity to a Loschmidt reflection). This time-symmetric informational generalization of Boltzmann's H-theorem is aesthetically satisfying.

Though "self-reproducing" automata (and thus WHEs) can be constructed (42, 43, 44), they can never fully follow the philosophical admonition "know thyself". Proof: Define self-knowledge for a WHE as a complete specification of its state coded in its memory. But total specification of the state of the memory alone exhausts the information storage capacity of that (finite) memory, no matter how large, leaving no room to store the state of the rest of WHE. Remark: Note that limited cases of self-knowledge are possible, in a sense, e.g. those in which the memory is largely "blank" in an informational sense, e.g. a stored program exists permitting computation of memory contents of a far larger block than that required to store the program. But then the state changes if the program is ever executed! Also note that nothing prevents having a complete description in an external memory. But then the very act of reading it, by WHE, would change the state of WHE in a manner serving either to invalidate that external description, or, if it remains valid, the WHE state is unchanged. But this means that WHE has "learned" nothing about itself by reading its complete description! Either it is unable to assimilate the "truth" about itself or the act of assimilating it makes it no longer the truth.

This result has a flavor like Goedel's theorems, for if WHE is given a complete description of its state it could never verify (decide) by operational-computational means, that it was, in fact, a complete description. It also has a thermodynamic flavor (third law), for a complete description of a physical system corresponds to S=0 for that system (pure state), i.e. T=0, the attainment of which is prohibited by one of the conventional statements of the third law. That it also has a quantum mechanical flavor emerges both from the fact that WHE's state is changed by interaction with another system ("absorption of quanta" of information) and by the third law remark above. The undecidability aspect just mentioned is a logico-mathematical one in the sense that the discussion could have been framed entirely in terms of abstract machines. Indeed, a continuing attempt at "proving" the correctness of a description looks very much like the hunting of a Turing machine for a proof of the unprovable. This discussion reinforces the conjecture in note 10 of (40), that in a future formulation of the statistical thermodynamics of well informed heat engines, the laws of thermodynamics will show up as Goedelian theorems.

The dual logico-thermodynamical nature of the second law also shows up in a kind of "free will" for a WHE. By this we mean non-self-predictability (45, 22, 38). The WHE could only predict what it is predetermined to do by the state and dynamical laws of its universe, including itself, if it can get the data by measurement. But getting the data increases entropy, i.e. invalidates state data in a region containing any subregion about which information is obtained. It follows that the second law implies no self-predictability for a WHE. There are other cases mentioned in (22) and (38), but we pass on to one which is even more intriguing, as it smacks of an "orneriness" in behavior not ordinarily associated with physical systems, even those with "bugs".

We now show that it is possible in principle to design a WHE, which, in a non-trivial sense refutes all theories of its behavior (46). We set up the following game between a theorist T, who devises well-formed theories T_1, T_2... in his attempts to describe and predict the behavior of a WHE for all time. With no essential loss in generality let the behavior B(t) consist of a sequence of choices

$$B(t_i) = B_i \qquad (16)$$

at discrete times t_i, where

$$B_i \in B = \{B^1, B^2, \ldots B^k\} \qquad (17)$$

with B the "alphabet" (finite set) of possible behaviors WHE is capable of exhibiting at any one time, the specific choice B_i at time t_i being a member of that set. Let

$$B_{mn} \in B \qquad (18)$$

be the behavior predicted by T, using theory T_m at time t_n. The entire set of possibilities can be displayed as an infinite matrix

(19)

	B_1	B_2	B_3	...	B_n...
T_1	B_{11}	B_{12}	B_{13}	...	B_{1n}...
T_2	B_{21}	B_{22}	B_{23}	...	B_{2n}...
T_3	B_{31}	B_{32}	B_{33}	...	B_{3n}...
⋮	⋮	⋮	⋮		⋮
T_m	B_{m1}	B_{m2}	B_{m3}	...	B_{mn}...
⋮	⋮	⋮	⋮		⋮

where the columns are labelled by the B_n at t_n, the rows by theories T_m. The theorist knows everything about the structure and operation of WHE, but does not know how it is programmed, and seeks to discover the program, i.e. the specific "laws" peculiar to the individual (WHE + program) which make it exhibit the observed behavior ... B_i, B_{i+1},.... His theories T_m are programs (this is no loss in generality), which he must "publish". The designer seeks to give WHE a program such that no theory devised by T will ever be correct. Who wins? The designer! He gives WHE a program which refutes all theories T_m in the following way. He makes WHE consult the "library" which T generates and choose for its behavior at t_k some B_i other than B_{kk}. This can be done with any programmed permutation P of the set B into itself

$$P: \left\{ \begin{matrix} B^1 & B^2 & \dots & B^k \\ B^{P1} & B^{P2} & \dots & B^{Pk} \end{matrix} \right\} \qquad (20)$$

which satisfies

$$P_i \neq i \qquad (21)$$

Such a permutation is called a derangement; their number, for k objects, is given by

$$d_k = k! \left(1 - \frac{1}{1!} + \frac{1}{2!} - \dots + (-1)^k \frac{1}{k!}\right) \qquad (22)$$

For large k this is approximately k!/e. For k=6 the approximate value of the expression in parentheses (1/e) is 0.36788, compared to the exact value to six figures, 0.36806 (47). The strategy is fiendishly effective (demonic?) and is essentially Cantor's classic diagonalization argument.

Of course the game is "unfair", but it shows very dramatically how different a WHE is from systems whose behavior does not depend on incoming information, which do not contain amplifiers, metastability, feedback, cannot behave selectively, etc. Whenever information is received we must deal with what can be called, with equal justification, a new boundary condition, a new initial condition, or an altered specification of the system. Computers and WHEs can also be viewed as fantastically multiply metastable, for each change in internal boundary condition, say when a flip-flop flips, is driven by a controlled energy source, and such changes increase entropy when the driving energy involved is degraded. Frequently the energy minima corresponding to initial and final metastable states differ negligibly in energy. The switching action begins with raising the initial energy level relative to the final one and (often) lowering the energy barrier between them (opening a "trapdoor"). It terminates with essential restoration of the initial relative levels and of the barrier, and with dissipation of at least some of the work done in level raising and trapdoor operation.

The richness of possible WHE behaviors is as vast as inventive talents can make it, and can only be hinted at here. Automation of process control, manufacturing, and laboratory analyses, computer animation, the automated library, global networks of on-line computers for the space program, image processing and pattern recognition techniques, even progress in "artificial intelligence", are some of the burgeoning areas involved. It does not seem far-fetched to say that the limits of modeling by WHEs are those of well-formulated theory. It seems likely that undecidability results within well-formulated physical theories will ultimately have as their analogs "principles of impotence" or impossibility, of which thermodynamics, quantum mechanics, and relativity have provided many examples.

Before we give a kind of thermodynamic motivation for spontaneous evolution of WHEs under long term maintenance of non-equilibrium conditions, we reinterpret some undecidability results on Turing machines for WHEs. The result is that no well-defined general theory can be devised which will always decide whether an arbitrary WHE, started up in a well-defined state in a "world" completely described by specification A, say, would ever, by its actions produce a world situation completed described by a given specification B. Similarly, if WHE is presently operating in situation B, it is generally impossible to tell if A is one of the situations which preceded it. We discuss only the first case, as the second follows from the first by a very simple argument - simply "reverse" the machine and take B as the initial "tape". As the worst that can happen for Turing machines under such a reversal transformation is to replace a (possibly) deterministic finite state control by a non-deterministic one (making it a non-deterministic Turing machine) and because the class of non-deterministic Turing machines is known to be equivalent in computational capability to the class of deterministic ones, the asserted result follows (48).

To prove the main result, simulate WHE and its world on a universal Turing machine T_u. T_u is specified by the triple (K, Γ, M). Here

$$K = \{q_o, q_1, \ldots q_n\} \tag{23}$$

is the set of states of the finite control. The set of tape symbols is

$$\Gamma = \{\sigma_1, \sigma_2, \ldots \sigma_m, B\} \tag{24}$$

where the σ_i are the symbols used to form coded groups of symbols representing the states, program, and world description, and B is often used for convenience instead of a blank square. In general it is conventional to view the tape as having a finite connected region as containing all the information on the tape, connection being maintained by representing erasure of a symbol by printing B in place of it. The infinite stretch(es) of blanks after (and before) the non-blank symbols can be represented by another symbol, e.g. $\square$. Moves to another square are symbolized by

$$M = \{R, L, S\} \qquad (25)$$

where R,L, and S respectively mean that the control moves to the square immediately to the left, to the right, or remains stationary. A move of T is a selection from a set of quintuples representing desired mapping

$$K \times \Gamma \rightarrow K \times \Gamma \times M \qquad (26)$$

The meaning of (26) is that when the finite control is in the state q_i scanning symbol σ_i at the square where it is currently located, it prints σ_f at its current square in place of σ_i, goes into a new state q_f and moves right or left or stays put. The move S is dispensable and is often omitted. The mapping (26) is often given as a table, sometimes called a functional matrix, rather than an ordered list of quintuples, its rows and columns being labeled by the elements of K and Γ and representing q_i and σ_i, the cells containing the triple representing q_f, σ_f, and the move. Data is processed or computations are made by starting the machine in some state, say q_o, looking at an initial symbol of an input word α on the tape, operating successively on symbol after symbol until some specific condition (halt state) is encountered. The result then can be read on the tape. It is proved in many places, e.g. Trakhtenbrot, ch. 13 (48), no procedure exists, i.e. no program can be found for T_u, guaranteed to halt <u>ultimately</u>, which can decide whether an arbitrary <u>particular</u> Turing machine (representing WHE), will, when "started" with $\alpha(A)$ as its input, process it and ever produce $\alpha(B)$ or not. We suggest that this and earlier results for HWEs remove the sting of the reproaches often leveled at theories of evolution, psychoanalysis, social science and "historical" subjects generally, that they have no predictive value. They can still have <u>explanatory</u> value. Similarly the functional matrix of T_u "explains" the string of symbols $\alpha(t_i)$ at any t_i, but it can not predict, in general, whether it will ever encounter a string α at <u>any</u> t_i. When a physical system is complicated enough to contain something like a WHE, we can expect to encounter these undecidable questions.

We now give the promised thermodynamic plausibility argument for the evolution of {WHE}. Assume permanently maintained non-equilibrium, taken below as steady state, even though periodically varying constraints are actually used. This is not thought to be serious for the following reasons. The deviation from equilibrium contemplated is not violent, so Fourier expansions of observables, as a function of time, and enough linearization of the problem to legitimize application of steady state analysis to the coefficients of the periodic terms, seem reasonable. The discussion follows (22) and (46).

Assuming a tendency to evolve towards a state of minimum rate of generation of entropy, one can show that there will be a thermodynamic motivation towards complicated configurations constituting subsystems of a large thermodynamic system, in such a way that the subsystems will tend towards perfect Carnot engines. The argument below applies to the kind of open system which might have been constituted by the Earth before and during the period during which life evolved.

Consider a system, N, alternately in thermal contact with heat reservoirs R_1 and R_2, at absolute temperatures T_1 and T_2, respectively, with T_1 less than T_2. After contact with R_1, the temperature of N is less than T_2, so that, after contact with R_2, a quantity of heat, Q_2, say, will flow into N. On the next contact with R_1, N will reject a quantity of heat, Q_1, to R_1, as its temperature was raised by contact with R_2 to a value greater than T_1.

Consider a steady state in the sense that the same thing happens on each successive cycle of contacts with the two reservoirs. The entropy change per cycle is that of the reservoirs, for the system comes back to its initial state. R_2 decreases in entropy and R_1 increases. The overall entropy change is given by

$$\Delta S = Q_1/T_1 - Q_2/T_2 \qquad (27)$$

The second law is satisfied if ΔS is positive. Take the unit of time as the interval required per cycle and then eq. (27) is the rate of entropy generation.

It is a result of nonequilibrium thermodynamics that a steady state of entropy production is characterized as the state of minimum rate of entropy production. Equilibrium is the special case in which the rate of entropy generation is identically zero.

The maximum value ΔS can have occurs when $Q_1 = Q_2$ and is given by

$$(\Delta S)\ \text{max.} = Q_2\ (1/T_1 - 1/T_2) \tag{28}$$

The minimum value is

$$(\Delta S)\ \text{min.} = 0 \tag{29}$$

which occurs when

$$Q_1/T_1 = Q_2/T_2 \tag{30}$$

ΔS being a minimum implies that an amount of energy

$$W = Q_2 - Q_1 = Q_2\ (1 - T_1/T_2) \tag{31}$$

remains in the system. This is high-grade energy available for doing work; in fact, equations (30) and (31) describe a reversible engine working between T_2 and T_1 and doing work at a rate W per cycle.

If we now decompose the system, N, into a heat engine M, and a third reservoir of matter, R_3, which is isolated except that system M can interact whith it mechanically (not thermally), doing work, W, on it per cycle, then this system would be operating in the preferred steady state of minimum rate of entropy generation. (28) describes the least preferred case -- the heat flowing in merely flows out again. One begins to suspect that the system described by equ. (29) - (31) is more highly organized than that described by equ. (28) and that in a permanently maintained nonequilibrium condition, a system, N, initially unorganized, tends to evolve into a highly organized state --and this is the appropriate generalization of the classical trend to equilibrium.

Let us spell this out in more detail. In the following discussion, one can think of the sun as R_2, the depths of space as R_1, and the earth as N (M plus R_3). As a preliminary, we show that reversibly storing W in a system organizes it, compared to its state if the same energy is added to the system and it is allowed to come to equilibrium. The entropy of a system at equilibrium is determined by the energy and volume

$$S = S\ (E, V) \tag{32}$$

and the temperature by

$$\left(\frac{\partial S}{\partial E}\right)_V = \frac{1}{T} \tag{33}$$

As temperature is positive, so is the derivative, and so S must increase with E. If reversible work, W, is done on the system, the work is stored and thus available. The internal energy has increased from E_0, say to $E_0 + W$, but the entropy is unchanged. Were the constraints (organization) keeping the energy stored relaxed so that the system could come to equilibrium, the entropy change (if the volume were constant) would be

$$\Delta S = S(E_0 + W, V) - S(E_0, V) \quad (34)$$

This is positive, and gives a quantitative measure of the amount of organization required to store energy W at constant V. To take care of variable V, we obtain, by differentiating equ. (32), multiplying by T, and using equ. (33),

$$T\,dS = dE + T\left(\frac{\partial S}{\partial V}\right)_E dV \quad (35)$$

This can be compared with the first law of thermodynamics for reversible processes

$$T\,dS = dE + P\,dV \quad (36)$$

whence we conclude that

$$\left(\frac{\partial S}{\partial V}\right)_E = \frac{P}{T} \quad (37)$$

P/T is positive. It then follows that increasing the volume available to the system at constant internal energy increases entropy, and decreasing the volume decreases the entropy. This and the earlier conclusions are intuitively obvious, because the system has more states to "choose from" if one gives it more energy or more volume. Consider equ. (36) again. If work is done on the system by compressing it, and all of the energy is stored and available, the entropy change must vanish. The increase in E balances the decrease in V to maintain constant S.

Starting from an equilibrium configuration and storing energy (reversibly) in the system, in the general case means starting from a point in a suitable (E, V) plane and going along a curve of constant S, which is a curve in this plane, to another point. A differential path element has E and V components, which have been discussed above, whence it follows that the organization is monotonically increased with each path element, and hence, likewise by integration, for the finite displacement. A simple discussion might have been given in terms of the Helmholtz free energy F or Gibbs free energy G, defined by

$$F = E - TS$$
$$G = E - TS + PV \quad (38)$$

but we preferred the above because of the more direct connection with the organization concept. One can see immediately that F and G increase when energy is stored in a system, or when its entropy is decreased (organization increased); it is well known that F and G represent the maximum work obtainable from the system at constant volume and pressure, respectively.

Return now to our system N, cyclically interacting with R_1 and R_2. If N stores no energy, equ. (28) applies. After contact with R_2, states are accessible to N which were not accessible to it after the previous contact with R_1. The heat Q_1, transferred to R_1, depends on the state of N when it is in contact with R_1. By the theorem of minimum rate of entropy increase, when the cycle repeats, N will tend, for fixed Q_2, toward the state corresponding the lowest Q_1, where the state is to be taken from those accessible to N after contact with R_2. If there is even one state that corresponds to Q_1 less than Q_2 by an amount W, which can be arbitrarily small, then the trend is away from the situation of equ. (28). Each cycle sees an amount of energy, W, stored in N, and, as the energy of N increases, more states become accessible to it. If one of these corresponds to still smaller Q_1, N tends toward it, otherwise energy continues to accmmulate, still more states become accessible, and so on. The free energy of N continually increases, with energy storage each cycle corresponding to a change in the V component of the discussion above and a change of state to a change of the E component. The organization of N thus increases monotonically; it "evolves". The culmination of evolution would be when equ. (29), (30), and (31) hold. For these equations to hold in the steady state, one would either have to have Q_2 vanish ($T_2 = 0$, excluded by the third law) or else permit N to perform work W per cycle indefinitely. The last seems to require that N be decomposable into two parts, M and R_3, with R_3 able to store the energy, i.e., become organized when M performs work on it. This alternative seems more reasonable - the earth, for example, after cooling, had an outer shell corresponding to M; the rest of the planet, to R_3.

We believe the foregoing not only shows organic evolution to be compatible with thermodynamics, but that something akin to it is actually required under the set of conditions obtaining in the solar system when life evolved. One suspects that "living" or other highly organized matter may be far more ubiquitous in the cosmos than heretofore thought possible, for the above considerations would apply to cases where life as we know it is impossible (49).

IV. Toward Well Informed Heat Engines at a Macromolecular Level

Life had to evolve chemically, for there were no shelves stocked with motors, wire, electrical components, etc., from which WHEs could have been assembled initially. It is therefore necessary to verify that physical and chemical properties of chemical species which could exist or be formed eventually in the primitive earth environment were capable of carrying out a repertoire of possibly very simple functions from which complex functions and behaviors can be synthesized. We will concentrate primarily on the communication and control side and components other than the "tape memory", for the latter is and has been so penetratingly studied that even high school students know that hereditary information is stored in DNA. The base-nucleotide pairs constitute a four letter alphabet which codes both species and individual programs and initial data for controlling the assembly of an individual from a zygote. There is a "translator", m RNA, which encodes parts of the gene message into "protein language" using a twenty letter alphabet (polymer chains of different amino-acids). We seek principles by which the rest of the "computer" works.

We were led to consider that the principles of quantum mechanics, which predict many characteristic properties of metals, insulators, and semiconductors from first principles, would also predict the existence of macromolecules which, with justification, could be called metal-like, insulator-like, or semiconductor-like. The original motivation (circa 1950) came from the search for new semiconductor materials and ways to miniaturize devices and circuits; the idea of a molecular transistor manufacturable in tank-car quantities seemed intriguing. It turned out virtually all the theory worked out at the time had been anticipated by many workers, usually associated with Mulliken (Platt, Kuhn, Bayliss and others) who had approached the subject from "the other end", i.e. the structure of molecules, particularly as revealed by spectroscopic methods. This disappointment, together with problems of coupling in to molecules electronically (the wiring problem) and other demands on the time available led to dropping it until the announcement of the first biophysics conference. As it had been realized that living systems might well be using "molecular circuitry", this aspect of the work was retrieved and organized for the conference and later published (50). The concepts involved, incidentally, enabled one physicist, deficient even in organic chemistry, to assimilate much biochemistry very quickly. Briefly, the conclusions are that the resources available for charge transfer, resonance, configurational changes, many kinds of tautomerism, reversible polymer reactions, and so on, make it possible to see how chemical systems could exist, which could respond to, and react on, their environment. In a real sense relays, and so logical circuitry, were indeed possible. Furthermore, only common properties and reactions of rather common organic compounds were involved. There was no real reason to believe that these things were exotic or unlikely to be available in the early history of life on earth.

Photochemical and electrochemical input and output were clearly possible. Mechano-chemical interactions have also been studied (51); they are clearly important for building chemical versions of WHEs. As a result we were satisfied that much of the componentry for communication and control was available in principle, but we needed reassurance that whole systems with sufficient stability could exist. As the simplest systems, virus, sometimes consist of only a few long-chain molecules, we wondered whether its ancestor might have been a single chain, and whether a rationale might be found for nature's apparent preference for chains.

We therefore asked the following question. Can statistical thermodynamics say anything of a general nature about the nature of a macromolecule which has the property that it can induce the formation of a replica of itself in an appropriate environment? The environment is assumed to be "simple", e.g. a solution in which component "pieces" of the macromolecule are available, and only "non-exotic" kinds of chemical interactions occur. It turns out that a very simple statistical thermodynamical argument suggests that a self-replicating macromolecule is, with very high probability, a chain molecule (52). The argument makes essential use of the excluded volume property. First we show that the build-up phase requires the assembly process to be exothermic and that only sheets or chains need be considered. Then, for the separation phase (necessary to distinguish replication from crystallization) configurational entropy dominates; sheets are likely to remain "glued" where chains readily separate. It is argued from this that chains could evolve with a self-catalytic property tantamount to reproduction much more easily than other macromolecules.

The simplicity of the argument and the importance of both excluded volume and configuration entropy for it then clearly made it necessary to examine both more closely. This was done in a fairly large investigation, some results of which have been published in a number of abstracts (53) and papers (54, 55), the rest being in reports (56).

The decisive point in the argument of (52) was as follows.

When two polymer sheets separate, the totality of configurations accessible to a thin slab is replaced by those accessible to two flexible sheets. When two chains separate, the set of configurations of a thin ribbon is replaced by that of a pair of flexible chains. We assume, for comparison, the same number of units in sheet and chain, no two units permitted to occupy the same space at the same time (excluded volume requirement), rigid bonds between adjacent units and bends occurring at the units (henceforth called beads) in random places at random times. It then turns out, by a simple calculation, that the increase in configurational entropy consequent to chain separation dwarfs that attending sheet separation. Thermodynamics is thus "more willing" to help chains to reproduce than it is to help sheets, for there is a greater increase in solution entropy (decrease in free energy) available to drive the separation phase. This supports the view that self-reproducing chains would evolve before sheets in solution.

The configurations of chains consisting of beads joined by rigid links was chosen as the simplest model to study with exclusion properties. The "world" in which the configurations occur and change was taken as a square lattice. It was later shown that other lattices and dimensionalities changed nothing essential and only made already intractable mathematics more so.

More specifically, we consider chain molecules consisting of beads held together with rigid links, the beads constituting perfect hinges between links and allowed to occupy the lattice points of a quadratic grid. Chains move by a sequence of elementary flips, occurring at one bead at a time, with no change in link length, and with motion of the bead possible only to an unoccupied lattice site. The chains then consist of zig-zags on the lattice, with an interior bead capable of flipping across a cell diagonal into an empty site if it is at the corner of a bend (it will be immobile as long as it is an interior bead of two or more links in the same straight line), and it may have one or two flip possibilitites if it is a terminal bead (one if it is the end of a bend, two if it is the end bead of a two-link straight portion; occupancy of neighboring sites can reduce the number of flip possibilities to zero).

With this simple set of flip rules to govern the chain configurations accessible from an initial one, one can set up transition diagrams between configurations whose complexity increases exponentially with chain length. It turns out that chains containing less than eleven links have all their configurations mutually accessible from each other by a sequence of elementary flips. At length eleven, one can draw the following configuration on the lattice in which steric hindrance, i.e. excluded volume, prohibits any change in configuration. Draw a 2 x 3 rectangle containing ten lattice points on its perimeter and two in its interior (all occupied). Except for the middle link of one of the longer sides regard the perimeter as constituting links of the chain. From the two ends of the perimeter minus that one link draw two more links to the two interior lattice points. The configuration is now a chain with its ends "tucked in" such that the sites to which the ends might flip are all occupied, and the same is true of all corners. This configuration, though admissible from the point of view of the excluded volume problem, is inaccessible from any other configuration under the simple flip rules described.

It is easy to generate examples ad libitum of configurations which are mobile in the plane (start with the straight shape, say) in the sense that they can wiggle off to infinity by a succession of flips, and it is possible to construct families with properties like the one just discussed, namely of being inaccessible under the flip rules from other families, though configurations within a family are mutually accessible. As the chains grow longer and longer, the number of such families increases, as does their maximum "membership". If one introduces constraints in the form of boundaries, pre-empted sites, or mutual exclusion effects between chains, many configurations mutually accessible in the absence of those constraints become members of mutually inaccessible families. These considerations have been carried much further than the glimpse just given, and disclose a fantastically rich and complicated configuration structure (56).

The foregoing shows the existence of the kind of structure needed for control, but unless there were means for controlling whether or not passage between families were possible, this configurational richness could not form the basis of control actions (and thus for selective behavior). Means for such control are readily available, however. One need only point to the existence of catalysts, which serve to "liberalize" flip rules in the sense of permitting transitions to occur easily that would occur with negligible probability in the absense of the catalyst. A similar situation occurs if the links can be formed or destroyed (snipping even one bond "unlocks" a surprisingly large number of sessile configurations). Analysis of such situations is very complicated, and has not even been completely carried out for the "simplest" case of condensation of a gas. It is sufficient for the purpose at hand to note that in a condensed system in which polymerizaton-depolymerization reactions can occur, in which adsorption of chain molecules on interfaces is a possibility and in which the temperature is not so low as to freeze all flips, that the achievement of configurational complexity and control of accessibility between flip-connected families is indeed at hand.

It is now perhaps not so bold an extrapolation to consider, as an essential step in the evolution of complex chemical configurations, the prior evolution of chains and catalysts (possibly with parts of a chain serving as a catalyst when its configuration is right) which permit access to families of configurations not possible earlier. Similarly, when these are at hand, they can become "launching pads" for the preparation of new, still more complex configurations, and so on. There are essentially two aspects of thermodynamic drive relevant here, either energy decreases or entropy increases (or an appropriate combination, like free energy, decreases). The first, though important (both in its own right and in various thermodynamic potentials, like free energy), yields first place to generalized entropy whenever metastability, amplification and control, or selective behavior are involved.

Control of accessibility of regions of phase space is exactly what the catalyst is needed for. Excluded volume (synonymous here with steric hindrance) or more generally, any kind of generalized selection rule, potential barrier, or exclusion principle, are overcome by a catalyst or other controller, e.g. a "symmetry breaker". Increase in the measure of the relevant phase space implies that previous equilibrium, i.e. previous maximum entropy, no longer obtains. Proof: Let Ω_i be the initial phase volume, and Ω_f, defined by

$$\Omega_f = \Omega_i + \Delta\Omega \tag{39}$$

the phase volume after the additional phase volume $\Delta\Omega$ has become accessible. As the corresponding equilibrium entropies are given by

$$S_i = k \log \Omega_i \tag{40}$$

$$S_f = k \log (\Omega_i + \Delta\Omega) \tag{41}$$

we have

$$S_f - S_i = k \log (1 + \Delta\Omega/\Omega_i) \tag{42}$$

which is always positive. The original equilibrium system is thus no longer in equilibrium until it has evolved to a state of higher entropy S_f. The catalyst has changed the specification of the system, or equivalently, has altered a boundary or initial condition.

This is adequate as one step of a complex behavior, not more (when the new equilibrium is reached nothing further happens). Now energy comes into its own: if in the newly opened region of phase space new reactions (say exothermic ones) can occur, barriers may be built against having the representative point wander into some portion of Ω_f. Let ΔE be available, at temperature T for this "compression" from Ω_f to Ω_f', defined by

$$\Omega_f' = \Omega_f - \Delta\Omega_f \quad (43)$$

We now have

$$S_f - S_f' = - k \log (1 - \Delta\Omega_f/\Omega_f) \quad (44)$$

i.e. entropy decreases. Remembering that ΔE "pays" for it, we conclude that

$$\Delta E \geq -kT \log (1-\Delta\Omega_f/\Omega_f) \quad (45)$$

Complex programmed sequences of physico-chemical events at a macromolecular level are clearly possible by alternating steps corresponding to eqs. (42) and (45) if a suitable succession of catalytic configurations is available. This is clearly suggestive for biology (56) (57). By summing up all such steps we obtain something like a constrained free energy cost of producing the end product configuration. For a cyclic process this is analogous to a metabolic rate.

The viewpoint of equ. (39) - (45) now illuminates equ. (27) - (38) and suggests the following. The earlier discussion is all right as far as it goes, but as in the thermodynamics of perfect gases versus that of real gases, it is an ideal limiting case. It appears to be directly applicable as long as one can assume that there are always catalysts and a smooth succession of Ω's bridging the gap between disorganized initial conditions and the generation of an "ecological" Carnot engine. In reality, though, the path will be thorny, all kinds of missing catalysts, evolutionary dead-ends, long delays at intermediate stages of approximation to a Carnot engine and the like can be expected. Just as no one would expect a quenched ingot to become a perfect crystal near absolute zero, so would one expect to approach minimum rate of constrained entropy generation in the steady state, rather than an absolute minimum. As so many constraints would be microscopic and of the same general nature as the frozen-in metastability of third law fame, it may never be possible to write them down explicitly. If we are lucky there may be sufficiently few classes of constraint, where we can neglect variation within classes, to permit progress. The utility of compartmental models, the existence of functionally and anatomically describable organ systems, reasonably well-formulated physiological and biochemical principles, and the ubiquity of hierarchical structure everywhere in biology all suggest that the case is <u>not</u> hopeless.

Before we consider the tactics of dealing with complexity, a subject more in discrete mathematics, logic and computer science than in customary physics or biology, the following remarkable circumstance should be discussed. It is, that unlike Schroedinger, molecular biologists, biochemists, most biophysicists, and most others in the biological and physical sciences, a number of outstanding physicists have taken an almost neo-vitalist stance, feeling that there are inherent limits to our probings of life and mind set by physics itself.

The dean, if not founder, of this "school" is Bohr (58). He introduced a doctrine of generalized complementarity, based on the uncertainty principle, and applied it to biology. The key idea is that the very act of determining the value of some variable or a property of a system may involve the establishment of conditions incompatible with the very definition or existence of that or some other property or variable. As is well-known this is fundamental to the Copenhagen interpretation of quantum mechanics. To extrapolate it to biology consider the problem of determining by measurement, <u>now</u>, whether a microscopic spore is dead or alive. Probing it with photons or particles to determine whether its current state (in the language of quantum or statistical mechanics) is a member of the class {dead states} or of the class {live states} would literally blast it apart. Determining it to <u>be</u> alive would then be incompatible with its <u>being</u> alive. If all physical observables and properties are defined by the operations used to measure them (a position Einstein found overly positivistic and too tolerant of solipsism (59)), then one can argue against the possibility of ever having a physical definition of life. Elasser (60) sharpened the complementarity argument by considering systems with so many particles that, as in the usual discussions in stastical mechanics, the energy levels of the whole system become very closely spaced. The usual approximation is to consider them as virtually a continuum. Let the energy difference between live and dead states be some very small ΔE. Applying the uncertainty principle to the system as a whole, we have

$$\Delta E \ \Delta t \geq h/2\pi \qquad (46)$$

where Δt is the time to distinguish between the levels in questio without blasting the system apart. Reasonable estimates of ΔE give unreasonably long values for Δt. In later work (61) he advocates the existence of special "biotonic" or "organismic" laws for biology, consistent with, but essentially independent of physics. Polanyi (62), in a paper which uses the phrase "boundary conditions" in a more general and less technical sense than we do, and which says many things in a spirit close to that of this paper, can be counted in this school in view of his concluding three sentence summary. "Mechanisms, whether man-made or morphological, are boundary conditions harnessing the laws of inanimate nature, being themselves irreducible to those laws. The pattern of organic bases in DNA which functions as a genetic code is a boundary condition irreducible to physics and chemistry. Further controlling principles of life may be represented as a hierarchy of boundary conditions extending, in the case of man, to consciousness and responsibility." Wigner (63) also feels the need for special biotonic laws and even ascribes a role to consciousness which seems to put it into physics, rather than the other way round.

How can one answer such a group of "superstars", whose views are presented in their writings far more forcefully than might be inferred from the above? How explain, if they are partly wrong, where and how it happened? Only an overwhelming conviction of the soundness of the generalized entropy concept, even in its applicability at the level of macromolecular automata, permits the writer to suggest that they have omitted a decisive feature of the whole situation, viz. the (essentially macroscopic) informational-organizational and thermodynamic aspect of the problem.

To answer Bohr-Elsasser first most of the detailed information about a system of which they speak (which would certainly explode the system it if could be obtained) is irrelevant. Details about water molecules in cells, for example, need not be known, for if they are disordered, as in fluid water, details are irrelevant. If they are ordered by hydrogen bonding to protein, say, their quasicrystalline structure could be clarified in subsidiary studies and the results would illuminate any particular individual case of interest. It is not clear that more detailed information than the arrangement in the first few shells around hydrophilic groups, say, need be of concern. Similar arguments can be (and have been) made about many other physical and chemical details. Just as crystals can be well understood with the aid of measures far short of destroying them, and one can augment that understanding by gathering "lethal" information from other crystals, so can one clarify the behavior and structure of an individual live organism via mild probing of it and lethal, if necessary, probing of others of its species. To return to the spore earlier mentioned, one can incubate it under conditions where it would grow if alive and lyse if it were dead. The proper conditions to do so would have been learned by experiments with other spores. Whether one can make a particular determination (decision) with respect to a particular attribute is a factual question which varies from case to case. There clearly must be questions whose answers cannot be obtained, but it is not clear that they are, in fact, of such a fundamental nature that leaving them unanswered makes the field forever incapable of being understood. The proper attitude, it seems to me, is that adopted by researchers in the field - forge ahead and worry about barriers to understanding when they are actually blocking progress. Also, one must make sure that questions hard to answer are properly posed. It is very hard to get a sensible answer to a nonsensical question! Lastly, the difficulties all seem to be those of understanding complex systems. Perhaps biology needs help from computer science, where complex systems abound - and where system behavior is far from understood.

But, some readers will say, these answers are far too facile. Given a single virus particle and no knowledge of what kind of cell it needs to replicate, how can one determine whether it has the ability to replicate in an appropriate host ("live" condition) or not (dead)? Given a zygote, can one even tell its species before it actually develops without lethal chromosome staining or the like? Probably not, and if we can get around these difficulties experimentally, there will doubtless be new cases where we can't, and when we later can, new borderline cases will crop up and so on in a game reminiscent of Goedelian leap-frog (22, 40). Perhaps the following is a proper answer in all such cases. As we close the gap in scale between macroscopic and microscopic we are less able, by the same "mild" measurement procedure, to determine properties of a system of interest. More and more we must rely on the statistics of either many repetitions of the same experiment on the same system or simultaneous observation of an assembly of similar systems. The latter is used, for example, in spectroscopic determination of atomic energy levels, the former in both usual studies of macroscopic systems and in studies of interference patterns from a very weak coherent source. We are approaching the limits of resolution of our measurements, so it is not surprising that we can easily end up getting inconclusive information. We cannot determine a diffraction pattern photographically with two or three photons, or even determine the amino acid sequence of one isolated macromolecule of a protein-like polymer (even if it is fairly short). Measurement of species, say, is not an elementary act, but rather a complex one akin to pattern recognition.

Another answer of a very different kind, but related to the last sentence, is that characterization of organization of a system is wanted, not details about the standardized components used in its construction. This is more important for the case at hand than energies, where, very often, all that is needed is that threshold, but not destructive ΔE's be involved.

To answer Elsasser-Polanyi we repeat what was said earlier about machines and their design. If by biotonic or organismic law is meant the consequences of system design for system behavior, then the only objection this writer would have to the phrase is that it is perhaps too exalted for such a hum-drum concept. That "boundary conditions" in the sense of Polanyi are irreducible in the sense of not being derivable from physics is no more surprising than the fact that boundary conditions in the technical sense are not derivable from the equations of motion. Those boundary condition are chosen. They define what specific solutions the equations of motion generate. They are part of the specification of the problem, not results of solving the problem. They are prior to solving the problem, not reducible in any way to the process of solution. No formal system derives its own axioms.

Another answer, which also relates to Wigner, is to stress that undecidable or non-determinable questions or parameters within a language can be decided or determined within a metalanguage. Atomistic physics arose by abstracting away from specific properties of individual macroscopic objects of which atoms are constituents. To recapture the macroscopic objects again one must put back what was thrown away. This added ingredient is essential. The absolute minimum is the "coarse-graining" needed to achieve irreversibility in statistical mechanics (35, 36, 37), and there is no maximum.

In a sense automata theory (in a generalized sense which includes all computers) deals precisely with the constraints in which organization consists. One can say, with some justification, that in the past physical scientists tended to neglect such automata theory aspects of biology, concentrating on quantal and statistical thermodynamic aspects. The newer generation of automata modelers of biology on the other hand, have tended to concentrate on computer science approaches with insufficient attention to the thermodynamic side (in addition to cellular automata models (42, 43, 44) see the contributions of Bremermann et al, Stahl, and Pattee in (64), pp 1-105; this volume contains other useful papers and bibliographics). The thesis of this paper is that {WHE}, which synthesizes both into its very foundations, is the appropriate starting point for physicotheoretical models of biology.

But what kind of computer or automaton can hope to cope with the complexities of biological behavior on the time scale of biological reaction times and be able to evolve? How can we understand patterns of hierarchical structure and their evolution, and relate them to patterning of behavior? How can we model relationships whose complexity is open-ended in an evolutionary sense?

We took our first clue from the analogy between the asembly phase of self-replication of chains and crystallization (52) **and the fact that the "stacking algorithm" for building ice out of** water molecules could make snow crystals with the right boundary conditions. This launched several investigations into how far the simplest possible algorithms could go in generating complex pattern. We chose groupoid strings for study. A finite groupoid is an algebraic system

$$G = \{a_1, a_2 \ldots a_n, *\} \qquad (47)$$

where the a_i are members of a finite set (also frequently designated as G), and * is a binary operation defined for any pair of elements of G, producing a unique element of G, viz.

$$a_i * a_j = a_k \qquad (48)$$

say. Form strings by concatenation of elements of G; these are groupoid strings. Daughter strings are generated algorithmically by successive application of eq. (48) to nearest neighbor element pairs, thereby producing a succession of daughter string elements (65). Even with very simple groupoids (e.g. cyclic group of order two) the most complex and beautiful patterns emerged (take different colors for distinct elements, covering a half-plane with the descendants of an infinite string, say). They often became totally different with even tiny changes in the initial string. It was also immediately obvious that successive generations of daughters could be computed in parallel everywhere along a parent string by a linear cellular automaton, and the whole pattern displayed on a planar cellular acutomaton. The idea of genes as "aperiodic crystals" was very congenial to groupoid strings, so their application to biological structure was natural (66).

Along with this study of the generation of patterns a parallel program studied pattern recognition by planar cellular automata (67) motivated by the problem of modeling visual (and nervous) systems (retina like devices). This provided the insight that the recognition problem resembled the inverse problem of determining parent strings, given their daughters (second clue). For several years this had to be done sequentially, but then the idea of combining the cellular automaton with an iterated switching array under local control by the automata occurred to me (bus automata) whereby separated cells could become nearest neighbors, in effect. This permitted the "parent problem" to be solved completely, in parallel, "immediately". This spurred discovery of a proof that the groupoid formalism was computation universal, that tremendous speed-up was possible in general, going all the way to "immediacy" in a large number of important cases, many surprisingly complex (68). Further studies have shown the utility of modeling skill acquisition with bus automata (69). As computers "speak" formal languages only, it was necessary to view pattern from this point of view. Straight line "language" (67) had been brought under control. The problem of fitting the "best" straight line required new concepts to be developed both in pattern research and statistics (70); that problem is now solved "immediately" (70).

During the initial groupoid studies it was realized that the binary operation could often be generalized to binary relation very easily, and that groupoid structures (e.g. the composition-series and subgroup structures in groups) contained an unlimited repertoire of hierarchical structures on various subsets. This suggested the possibility of devising a new kind of formal language, which we called relational languages, which would admit reductionism, holism, synthesis, open-ended hierarchical complexity - and metalinguistic extensibility - and perhaps be useful in biology. It turned out to be less difficult than it sounds (72); that line of research was shelved only because (a) bus automata seemed even more exciting (b) time was needed to assimilate more literature in many fields.

The present situation now seems to be that all the elements for building bus automata are present at a macromolecular level. The excluded volume property, which is essentially an aspect of the exclusion principle, splits configuration space into an enormous number of mutually inaccessible regions. Bond migration, formation and scission can change accessibility and inaccessibility drastically, and even alter its whole structure. Resonance and tautomerism permit such effects to propagate over macromolecules, and propagation of effects between molecules is possible by these and other mechanisms. Information storage in polymers made of sequences of different monomer residues is now commonplace, and increasing numbers of specific catalytic effects of regions of such polymers are found almost daily. The generalized thermodynamics, based on the information-organization extension of classical entropy, not only permits macromolecular fast bus automata to operate but "pushes" for their evolution.

V. Conclusion

Biology can be "explained" by physics aided by generalized entropy, generalized boundary conditions, and formal concepts like those used in computer science, even including evolution. Much structure-sensitive detail can be explained but not predicted, and some questions are inherently undecidable. These novel features (for physics) are a result of complexity in the systems, rather than specifically biological, and many are logically inherent in language, logic, and mathematics themselves.

References

(1) E. Schroedinger, What Is Life? Cambridge University Press, Cambridge, England (1944).
(2) C. Shannon, Bell System Tech. Journal 27 279, 623 (1948).
(3) N. Wiener, Cybernetics, John Wiley, New York (1948).
(4) L. Szilard, Zeit. f. Physik 53, 840 (1929).
(5) J. v. Neumann, Math. Grundlagen der Quantenmechanik, Ch V, Julius Springer, Berlin (1932); reprinted by Dover Pub., New York (1943).
(6) P. Demers, Can. Jour. Res. 22, 27 (1944), A23, 47(1945).
(7) L. Brillouin, Science and Information Theory, Ch XIII, Academic Press, New York (1956 and 1962).
(8) H. Jacobson, Trans. N.Y. Acad. Sci. 14, 6 (1951).
(9) D. Gabor, in E. Wolf, ed, Progress in Optics 1, North-holland, Amsterdam (1961).
(10) J. Rothstein, Methodos, 11 94 (1959).
(11) J. Rothstein, Science 114, 171 (1951)
(12) J. Rothstein, Phys. Rev. 85, 135 (1952).
(13) J. Rothstein, Jour. Appl. Phys. 23, 1281 (1952).
(14) J. Rothstein, Trans. Prof. Group on Engr. Management, IRE (now IEEE) 1, 25 (1954).
(15) J. Rothstein, Trans. Prof. Group on Information Theory, PGIT-4, 64 (1954).
(16) There are many texts on automata theory, computation and formal languages to which the reader can refer, if necessary e.g. J.E. Hopcroft and J.D. Ullman, Formal Languages and their Relation to Automata, Addison-Wesley, Reading, Mass. (1969).
Z. Manna, Mathematical Theory of Computation, McGraw-Hill, New

York (1974).
A. Ginzberg, Algebraic Theory of Automata, Academic Press, New York (1968).
Z. Kohavi, Switching and Finite Automata Theory, McGraw Hill, New York (1978).
(17) P.H. Starke, Abstract Automata, North Holland, Amsterdam, (1972).
A. Paz, Introduction to Probabilistic Automata, Academic, New York, (1971).
A.A. Lorents, Stochastic Automata, John Wiley, New York (1974).
(18) L. Onsager, Phys. Rev. 37 405 (1931); 38 2265 (1931). The later literature has become too vast and diversified to be cited adequately here. Three samples with numerous references are Non-Equilibrium Thermodynamics, Variational Techniques and Stability, R.J. Donnelly, R. Herman, and I. Prigogine, editors Univ. of Chicago Press, Chicago (1966).
A. Katchalsky and R.F. Curran, Non-Equilibrium Thermodynamics in Biophysics Harvard Univ. Press (1965).
M. Lax, Fluctuations from the Nonequilibrium Steady State, Rev. Mod. Phys. 32, 25 (1960).
A modern general treatment of thermodynamics and statistical mechanics, with one of its five major sections devoted to the thermodynamics of irreversible processes (by J. Meixner and H.G. Reik) is
Encyclopedia of Physics, edited by S. Fluegge, vol. III/2, Principles of Thermodynamics and Statistics, Springer, Berlin (1959).
For a survey of the controversial field of relativistic thermodynamics, with 369 references up to Oct. 1969, see
A. Guessous, Thermodynamique Relativiste, Gauthier-Villars, Paris (1970).
(19) The study of such hereditary phenomena began with V. Volterra, Lecons sur les equations integrales et les equations integro-differentielles, Paris (1913).
See also H.T. Davis, Introduction to Nonlinear Differential and Integral Equations, U.S. Atomic Energy Commission Washington (196
(20) J.A. McLennan, Jour. Chem. Phys. 41 1159 (1964).
Relations like (8) occur in B. Gross, Actualites Scientifiques et Industrielles no. 1190, Hermann et Cie., Paris (1953).
(21) L. Brillouin, Science and Information Theory, Academic Pres New York (1956, 1662).
(22) J. Rothstein, Communication, Organization, and Science Falcon's Wing Press, Indian Hills, Colorado (1958). The publishe included a forword for which the author disclaims all responsibility. Chapter 11 gives a simple thermodynamic argument for the evolution of well informed heat engines.
(23) J. Rothstein, Phys. Rev. 86, 640 (1952) (abstract).

(24) R.C. Tolman, Principles of Statistical Mechanics, Oxford University Press London (1936), pp. 612-13.
(25) R.H. Fowler and E.A. Guggenheim, Statistical Thermodynamics, Second edn. Cambridge University Press, Cambridge (1949) p. 224. Chapter V discusses much of the history and controversy surrounding Nernst's heat theorem and the third law, and Simon's decisive contributions to their clarification.
D. ter Haar, Elements of Statistical Mechanics, Rinehart, New York (1954), Appendix III is a succinct and clear introduction to the third law as is chapter 7 of A.H. Wilson, Thermodynamics and Statistical Mechanics, Cambridge University Press, Cambridge (1957). A more recent, and comparatively elementary treatment is found in J. Wilks, The Third Law of Thermodynamics, Oxford University Press, London (1961).
(26) E. Schroedinger, Statistical Thermodynamics, Cambridge University Press, 2nd ed. Cambridge (1952).
(27) J. Rothstein, Bull. A.P.S. II 1 74 (1956) and unpublished work.
(28) R.H. Fowler, Statistical Mechanics Cambridge University Press, Cambridge (1936).
(29) J. Rothstein, Bull. A.P.S. II 9, 98 (1964).
(30) J.M. Jauch and J.G. Baron, Helvetica Physica Acta 45 220 (1972).
(31) J. Rothstein, Phys. Rev 85, 722 (1952) (abstract) Philosophy of Science 23, 31 (1956).
(32) J. Rothstein, Philosophy of Science 29, 406 (1962).
(33) J. Rothstein, Phys. Rev. 86 634 (1952) (abstract); Revue Internationale de Philosphie 40, 211, Fasc. 2 (1957).
(34) J. Rothstein, Phys. Rev. 86 640 (1952), Bull. A.P.S. II 6, 42 (1964); ibid 7, 238 (1961) (abstracts); Physics Today 15, 28 (Sept. 1962), and unpublished work.
(35) J. Rothstein, Bull. A.P.S. 30 7 (1955), American Journal of Physics 25, 510 (1957).
(36) J. Rothstein, Foundations of Physics 4, 83 (1974).
(37) J. Rothstein, Informational Generalization of Physical Entropy, in Quantum Theory and Beyond, T. Bastin, ed., pp. 291-305 Cambridge University Press, London (1971).
(38) J. Rothstein, Philosophy of Science 31 40 (1964).
(39) J. Rothstein, I.R.E. National Convention Record, Part 4, pp. 3-11 (1956).
(40) J. Rothstein, ch XIX in Communication: Concepts and Perspectives, L. Thayer, ed. Spartan Books, Washington (1967).
(41) N.F. Ramsey, Phys. Rev. 103 20 (1956), M.J. Klein, Phys. Rev. 104 589 (1956).
(42) J. von Neumann, The Theory of Self-Reproducing Automata, A.W. Burks, ed., University of Illinois Press, Urbana (1966).

(43) Essays on Cellular Automata, A.W. Burks, ed., University of Illinois Press, Urbana (1970).
(44) E.F. Codd, Cellular Automata, Academic Press, New York (1968).
(45) J. Rothstein, Phys. Rev. 95, 643 (1954) (abstract).
(46) J. Rothstein, Bull. A.P.S. II 8, 394 (1963) and unpublished work, including pp. E-1 to E-36, final report on Contract No. AF 49(638)-1450, Oct. 30, 1966, Air Force Office of Scientific Research.
(47) C.L. Liu, Introduction to Combinatorial Mathematics, pp. 106-8.
(48) For additional material on Turing machines, in addition to the first two books cited in (16), the following introductory treatments can be consulted: M. Minsky, Computation, Finite and Infinite Machines, Prentice-Hall, Englewood Cliffs, N.J. (1967); A. Yasuhara, Recursive Function Theory and Logic, Academic Press, New York (1971); B.A. Trakhtenbrot, Algorithms and Automatic Computing Machines, D.C. Heath, Lexington, Mass (1963)
(49) J. Rothstein, Second International Biophysics Congress (IOPAB), Vienna, 1966, abstract 664.
(50) J. Rothstein, Heuristic Application of Solid State Concepts to Molecular Phenomena of Possbile Biological Interest, pp. 77-85, Proceedings of the First National Biophysics Conference (1957), H. Quastler and H.J. Morowitz, eds. Yale University Press, New Haven (1959).
(51) For an introductory summary see A. Katchalsky, S. Lipson, I. Michaeli, and M. Zwick, Elementary Mechanochemical Processes, Contractile Polymers, pp. 1-40, Pergamon Press, London (1960).
(52) J. Rothstein, IEEE Transactions on Military Electronics, Vol. MIL-7, Nos. 2 and 3 (special bionics issue), 205 (1963).
(53) J. Rothstein, Bull. A.P.S. II, 9, 662 (1964); 10, 87 (1965); 11, 165 (1966); 13, 179 (1968) (exclusion, chain diffusion). 3, 201 (1958); 9, 255, 414, 681 (1964) (metastability, steady state, evolution, 14, 504 (1969) (reproduction). J. Rothstein, Biophysical Society Abstracts, annual meetings, 1960, 1964, 1966, 1969 (organization, aging, membranes, gen. stat. mech., J. Rothstein, Int. Biophys. Cong. (IOPAB), Vienna (1966); (gen. stat. mech.)
Int. Biophys. Cong. (IOPAB), Cambridge (1969); (non-equil. phases, coop. phenomena).
(54) J. Rothstein, Excluded Volume Effects as the Basis for a Molecular Cybernetics, pp. 229-245, Cybernetic Problems in Bionics, H.L. Oestreicher and D.R. Moore, eds., Gordon and Breach New York (1968).
(55) J. Rothstein and P. James, Jour. Appl. Phys. 38, 170 (1967).

(56) J. Rothstein, with assistance of P. James, R. Donaghey, D.R. Childs, S. Zohn, E. Feuerstein. Basic Principles of Self-Replicative Information Storage in Chemical Systems. Vol. I, Tech. Rpt. AFAL-TR-65-60 (April 1965) 270 pages, Vol. II, Tech. Rpt. AFAL-TR-66-228 (July 1966), 443 pages, Vol. III, Tech. Rpt. AFAL-TR-66-255 (Oct. 1967), 423 pages, Vols. I and II were prepared under Contract No. AF33(615)-1464, Vol. III under Contract No. AF33(615)-3769 with the Air Force Avionics Laboratory, Research and Technology Division, Air Force Systems Command, Wright-Patterson Air Force Base, Ohio.
(57) J. Rothstein, Biophys. Jour. 8 Society Abstracts, PA-34 (1968).
(58) N. Bohr, Atomic Theory and the Description of Nature, Cambridge University Press, London (1934) Atomic Physics and Human Knowledge, John Wiley, New York (1958).
(59) Albert Einstein, Philosopher - Scientist, vol. VII, Library of Living Philosophers, P.A. Schilpp ed; University of Illinois Press, Glencoe (1949). Reprinted in paperback, Open Court Publishing Co., LaSalle Illinois (1973).
(60) W.M. Elsasser, Phys. Rev 56, 987 (1937).
(61) W.M. Elsasser, Jour. Theor. Biol. 1 27 (1961); Proc. N.A.S. 54, 1431 (1965); The Physical Foundation of Biology, Pergamon Press, London (1958).
(62) M. Polanyi, Science 160, 1308 (1968).
(63) E. Wigner, Foundations of Physics 1, 35 (1970); pp. 231-238 in The Logic of Personal Knowledge, Routledge and Kegan Paul Ltd., London (1961); Proc. Amer. Philosophical Soc. 113, 95 (1969).
(64) Natural Automata and Useful Simulations, edited by H.H. Pattee, E.A. Edelsack, L. Fein, and A.B. Callahan, Spartan Books, Washington (1966).
(65) J. Rothstein, Patterns and Algorithms, Proc. IEEE Sympos. on Adaptive Processes (9th), Paper II. 4 (1970). IEEE Pub. 70 C 58-AC.
(66) J. Rothstein, Biophys. Soc. Abstracts 10, 233a (1970).
(67) J. Rothstein and C. Weiman, Computer Graphics and Image Processing 5 106 (1976).
(68) J. Rothstein, Proc. 1976 International Conference on Parallel Processing, pp. 206-212, IEEE Catalog No. 76CH1127-OC; M. Moshell and J. Rothstein, ibid., pp. 222-229.
(69) J. Rothstein, Transitive Closure, Parallelism, and the Modelling of Skill Acquisition, 1977 Proceedings of the International Conference on Cybernetics and Society, pp. 232-236, IEEE Catalog No. 77CH1259-1SMC.
(70) J. Rothstein, loc. cit. ref (69), pp. 572-576.
(71) J. Rothstein and J. Mellby, work in progress (JM's disser-

tation).
(72) J. Rothstein, Formal Languages for Possible Biological Applications, First European Biophysics Congress, Baden (Vienna), 1971; Biophys. Soc. Abstracts 13 40 a (1973).

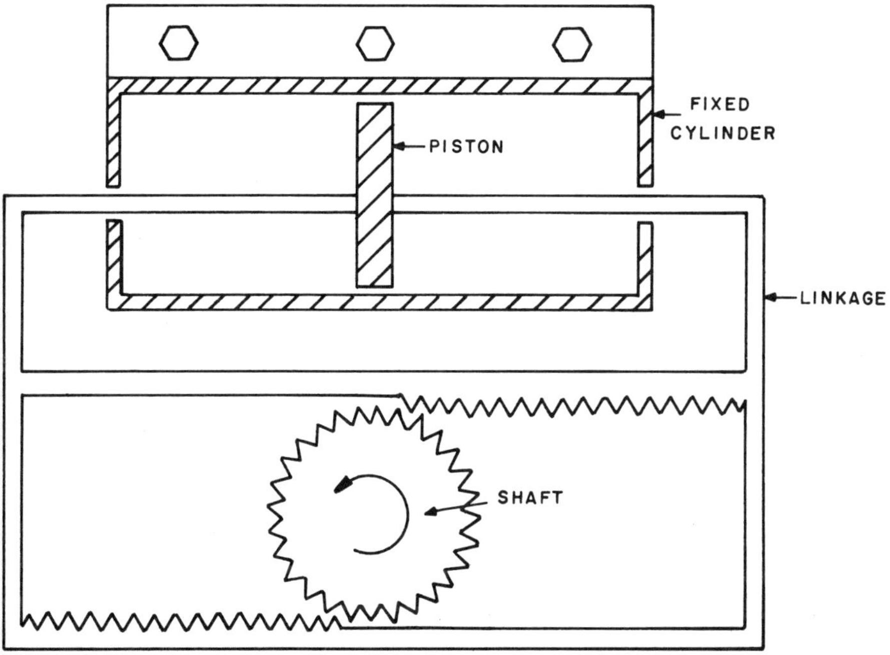

Figure 1 Thought experiment illustrating Popper's refutation of Szilard's assignment of an entropy equivalent to physical information.

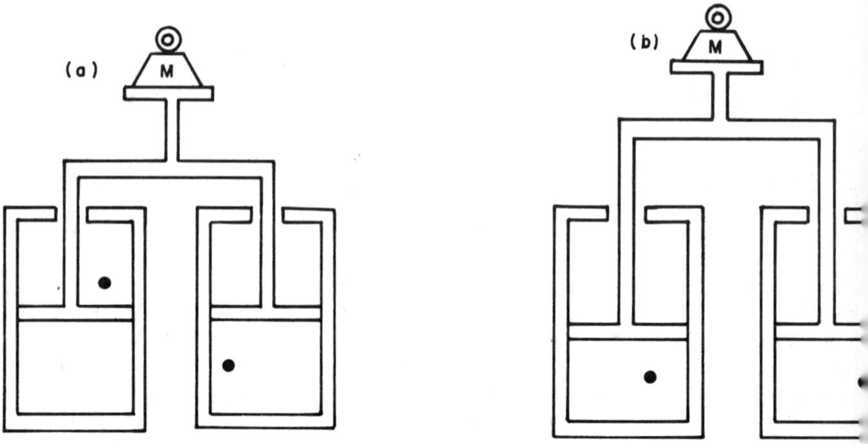

Figure 2 Two-piston representations of (a) Szilard's, (b) Popper's, thought experiments.

THE GIBBS GRAND ENSEMBLE AND THE ECO-GENETIC GAP

Edward H. Kerner

The ground between ecology and genetics has long been recognized as the site of major biological world views, where primary evolutionary mainsprings are actuated. Yet, while the two disciplines have developed their own bodies of facts and principles, the inter-zone has remained noticeably amorphous, shot through with qualitative insights but few quantitative ones, and leaving the eco-genetic synthesis of the evolutionary picture short of consummation. This eco-genetic gap is exposed by questions like the following: How is determined the extent of speciation of a well-defined eco-group? How does such speciation fluctuate over the long term? How, if at all, does fluctuation of speciation connect to fluctuation of population size? The present note remarks a novel and quantitative approach to these matters, basing upon Gibbs-ensemble or maximum-entropy methods, which provides definite if only preliminary or suggestive answers.

The approach is first of all through ecology, where at least at theoretical level the necessary global view for entertaining such questions is at hand, - one can scarcely, in Hutchinson's (1) phrase, visualize the evolutionary play without the ecological theater. In particular, Volterra's (2) ecological theory allows consideration of the unlimitedly large, highly speciated eco-complex. While it has sometimes been criticized as simplistic in its micro-ecological aspects, it stands unrivalled in its scope, and it grasps at its base those essential features of species self-growth and -decay and predator-prey interaction which must remain central to the large view of ecology. At the same time it allows an overview of statistical-mechanical sort (3) accounting sensibly and in detail for a great range of macro-ecological observation: for the swings in amplitude and periodicity of populations: for the competitive exclusion phenomenon: for the relation between 'stability' and 'diversity' in eco-systems: for the species-abundance relationship, etc.

If $N_1, N_2, \cdots N_n$ denote the population sizes of species $1, 2, \cdots n$ locked together in an ecologic network, the Volterra dynamics is

$$\dot{N}_i = \epsilon_i N_i + \frac{1}{\beta_i} \alpha_{i\sigma} N_i N_\sigma \quad ,$$

(summation on repeated Greek indices understood), where ϵ_i is an auto-increase (or -decrease) rate, and $\alpha_{i\sigma}/\beta_i$ a set of coupling coefficients for the interaction of N_i with N_σ, taken to be of generalized predator-prey type, in which the reciprocal nature of predator-gain/prey-loss is expressed in the skew-symmetry $\alpha_{ij} = -\alpha_{ji}$. A simpler and more useful representation of the dynamics (3), preparing the way for Gibbs ensembles, is

$$\dot{v}_i = \gamma_{i\sigma} \frac{\partial G}{\partial v_\sigma} \quad , \qquad G \equiv \tau_\sigma \left(e^{v_\sigma} - v_\sigma \right) .$$

Here $v_i \equiv \log (N_i/q_i)$ is a logarithmic measure of population N_i relative to a reference level q_i equalling the mean population size $\overline{N_i}$, while $\gamma_{ij} \equiv \alpha_{ij}/\beta_i \beta_j$ and $\tau_i \equiv q_i \beta_i$. In this form it is readily seen, owing to the skew-symmetry of γ_{ij}, that the quantity G is conserved, playing the role of a conservation principle similar to energy conservation in physical systems. Secondly, volume elements $dv_1\, dv_2 \cdots dv_n$ in the phase space $v_1, v_2, \cdots v_n$ are conserved, due to $\partial \dot{v}_\sigma / \partial v_\sigma = 0$ (Liouville's theorem).

These two elements of simplicity allow the step to ensemble considerations, wherein detailed knowledge of all the many $v_i(t)$ is given up in favor of probabilistic statements based on the (petit) canonical probability density $\rho \sim \exp(-G/\theta)$ in phase space. Then the *eco-temperature* θ connects to an assortment of mean-values of quantities associated with the noisy population-versus-time curves $N_i(t)$ that are accessible to observation. For example, besides $\overline{N_i} = q_i$, the mean values $\overline{(N_i/q_i - 1)^2} = \theta/\tau_i$ measure the amplitudes of the eco-noise, and low θ (relative to τ_i) behavior describes a slight rippling of $N_i(t)$ about q_i, while high θ ($>> \tau_i$) behavior is telling of strongly fluctuating $N_i(t)$ in a pattern of population surge-and-crash. The single $N_i(t)$, jostled upwards and downwards in the large play of the eco-network in which it is immersed, is akin to a molecular coordinate in Brownian motion in a gas as it is knocked about through impact with the surrounding swarm. Within Volterra dyanamics this eco-noise is intrinsic, not to be confounded with any fluctuations due to effects of varying environmental conditions. The reason for the noise being sustained, without damping, is the omission of self-interaction or self-limiting terms of Pearl-Verhulst type (i.e. $\alpha_{ii}=0$). Under the usual ergodic hypothesis of statistical mechanics, the barred quantities above stand for mean values computable from the ensemble on the one hand, and long time averages available observationally on the other hand.

The central point now is, that G may be partitioned into clusters $G_a + G_b + \cdots + G_s$ with each G_k gathering together eco-types $k, k', \cdots$ of like value of τ, that is

$$G_k = \tau_k\left(e^{v_k} - v_k + e^{v_{k'}} - v_{k'} + \dots \quad \text{to } \nu_k \text{ terms}\right).$$

The *eco-clusters* $a, b, \cdots k \cdots s$ in short are something like genera-like groupings of closely similar eco-types, taken to be $\nu_a, \nu_b, \cdots \nu_k \cdots \nu_s$ in number in the different clusters. It should be noticed that the common value of τ_k for the eco-types in the k'th cluster (meaning a common value within observational error of $\overline{(N_k/q_k - 1)^2}, \overline{(N_{k'}/q_{k'} - 1)^2} \cdots$) does not at all erase the identities of these types, as the dynamics entails not only the τ_k parameters but also the interaction coefficients $\gamma_{k\sigma}, \gamma_{k'\sigma}, \cdots$. More precisely, the similar eco-types comprising a cluster are not merely those having closely similar τ values, but having as well closely similar interaction coefficients γ with other species, so that the types correspond to close neighbors by ordinary ecological criteria, and exhibit as well closely similar eco-noise patterns $N_k(t)$, $N_{k'}(t), \cdots$ (they are then distinguished (3) as to noise, not so much by the structure of the noise patterns, but by these patterns being out of phase with one another). The explicit γ dependence of the noise is well seen for instance in a computation of $\overline{\dot{v}_k^2}$, which works out to be $\theta \gamma_{k\sigma}^2 \tau_\sigma$. Thus G alone and τ alone do not describe all the salient features of eco-typal noise; the full complement of γ's also is required. It could happen, as a sort of 'accidental degeneracy', that two definitely different types k and ℓ, clearly not to be bracketted in a single cluster, might yet have a closely common $\tau_k \approx \tau_\ell$. Owing however to $\overline{\dot{v}_k^2}$ being distinctly different from $\overline{\dot{v}_\ell^2}$, there is no question as to k and ℓ

belonging to different clusters. It suffices then to define types falling into one cluster, as species with closely common γ_k *and as well with comparable* $\gamma_{k\sigma}$ *values.*

It should be noticed here that in principle the cases of odd and even total speciation must be distinguished, as the odd and even Volterra systems have different (and ecologically cogent) characteristics. The odd/even distinction indeed poses quite subtle difficulties, of a kind not elsewhere encountered, which blocked the problem of variable speciation for some time. The full discussion (4) requires a rather long detour. But in the end it turns out that one can virtually 'forget' odd/even distinctions.

The situation then becomes wholly comparable with that in chemical thermodynamics of many chemical species, for which Gibbs' (5) landmark invention of the grand ensemble allows for variable numbers of degrees of freedom. This allowance translates in present context into the grand phase space density

$$\rho \sim e^{\frac{\mu_a \nu_a}{\theta} + \frac{\mu_b \nu_b}{\theta} + \cdots} \; e^{-\left(\frac{G_a}{\theta} + \frac{G_b}{\theta} + \cdots\right)} \Big/ \nu_a!\, \nu_b! \cdots ,$$

describing not merely probability assignments for the population levels v_λ or N_λ, but also probabilities for the degrees of speciation $\nu_a, \nu_b, \ldots$ within clusters. The factors $\nu_a!\ \nu_b! \cdots$ enter as shown, in order that equivalent configurations in the phase space (equivalent under permutations of $v_k, v_{k'}, \cdots$ within the separate clusters) be not counted as distinguishable in ranging over the whole phase space (i.e. over the so-called specific phases). Analogously to the usual chemical potentials of chemical theory, the μ_k here are *cluster-potentials* whose preliminary role is to grant a free specification of the mean cluster sizes $\overline{\nu_k}$, just as at the outset θ enters so as to allow arbitrary specification of $\overline{G}$. It may be observed that, thermodynamically, the chemical (or cluster) potentials rank in importance comparably to the concept of temperature itself. The grand density presumes thermostatic equilibrium in the sense that $\nu_k(t)$, like $N_\lambda(t)$, makes up a *stationary* noise pattern, in which strips of the noise taken over different long time intervals yet are statistically very much alike, so that earlier and later strips cannot be distinguished.

Upon normalizing ρ and integrating over phase space, there comes forth directly the probability law for the ν_k :

$$p(\nu_a, \nu_b, \cdots) = \frac{\omega_a^{\nu_a} e^{-\omega_a}}{\nu_a!} \cdot \frac{\omega_b e^{-\omega_b}}{\nu_b!} \cdot \quad \cdots \quad ,$$

$$\omega_k \equiv e^{\frac{\mu_k}{\theta}} \left(\frac{\tau_k}{\theta}\right)^{-\frac{\tau_k}{\theta}} \Gamma\left(\frac{\tau_k}{\theta}\right) \quad ,$$

with Γ designating the gamma function.

In words, the grand ensemble view crystallizes out very simply the fundamental statement that *the cluster sizes* $\nu_a, \nu_b, \cdots$ *have independent Poisson statistics.*

It is typical of the obtuse strength of the thermalistic approach that up to this point the *mechanism* of speciation need not have been detailed; merely the fact of speciation into clusters is sufficient to set the structure of the probability law for their sizes, as well for eco-species in the present context as for chemical species in ordinary chemical thermodynamics. Moreover, the outcome of Poisson statistics does not depend very strongly on the specific structure of the Volterra dynamics (for example, many modifications of the dynamics will still imply Poisson statistics).

The ω_k are of course the same as the mean speciation levels $\bar{\nu}_k$ within clusters, and to prescribe $\bar{\nu}_k$ is merely to fix the cluster-potential μ_k . This fix, however, is in itself quite empty (as is the detailed structure of ω_k in terms of τ_k/θ), without indeed probing into the mechanism by which the ν_k may change.

Here the chemical analogy is instructive. For a chemical reaction $A \leftrightarrows B$ the stoichiometry dictates that $dN_A + dN_B = 0$. At the same time, maintenance of equilibrium requires $\mu_A^* dN_A + \mu_B^* dN_B = 0$, according to well-known thermodynamic principles. Hence $\mu_A^* = \mu_B^*$ is the cogent characterization of equilibrium, with μ^* denoting the ordinary chemical potential.

On this same path, one can recognize that if a and b are two genetically interconnected eco-clusters, one may expect to be able to describe the connection ultimately as having the nature of monomolecular-type transition within the chemico-genetic apparatus of a and b member elements; the important feature of some degree of geographical separation (6) between clusters may be considered to be subsumed in the labels a or b identifying the clusters. One can then look upon $N_a + N_{a'} + \dots$ and $N_b + N_{b'} + \cdots$ as so many amplified representations (or 'eco-molecules') of substantially the same types of inner molecular complexions within each of the a and b clusters. A simple mean measure of the counts of the eco-molecules will be $q_a \nu_a$ and $q_b \nu_b$ (taking $q_a, q_{a'}, \cdots$ to have sensibly a practically common value, and similarly for $q_b, q_{b'}, \cdots$). The stoichiometry of genetic transition may now be stated as $d(q_a \nu_a) + d(q_b \nu_b) = 0$, signifying that incremental gain or loss of a eco-molecules is accompanied by corresponding loss or gain of b eco-molecules in the course of the eco-genetic flow or counterflow. The flows must of course be understood to be rather slow, and they may be heavily uni-directional but in general not totally so, as the rules of chemical kinetics make clear. Once again, it is a straightforward step to show, as in the chemical case, that thermostatic equilibrium requires $\mu_a d\nu_a + \mu_b d\nu_b = 0$.

Therefrom it follows that $\mu_a / q_a = \mu_b / q_b$ or, from the relation above connecting $\omega_k = \bar{\nu}_k$ with μ_k,

$$\frac{\log \bar{\nu}_a}{\bar{N}_a} + \frac{\bar{N}_a}{\overline{(N_a - \bar{N}_a)^2}} = \frac{\log \bar{\nu}_b}{\bar{N}_b} + \frac{\bar{N}_b}{\overline{(N_b - \bar{N}_b)^2}}$$

after using the Stirling approximation for the Γ function ($\theta \ll \gamma_a, \gamma_b$) for simplicity, and harking back to the meaning of θ/γ as $\overline{(N/\bar{N} - 1)^2}$. All quantities entering here are in principle accessible to observation, with the time-scales for N variation being perhaps months or years, and for ν variation decades-to-centuries.

A principal conclusion is, then, that *the geno-statistics* $\bar{\nu}$ *are now no longer freely assignable, but are necessarily connected to one another, and conjointly to the eco-statistics* $\bar{N}$, $\overline{(N-\bar{N})^2}$ *for clusters genetically linked directly and network-linked ecologically.*

In a word, something like an eco-genetic synthesis flows quite naturally from the grand ensemble view of ecologic-collision on the one hand (as embraced in Volterra dynamics), and monomolecular-type genetic transition on the other hand. The present elemental statement of eco-genetic connection clearly must eventually be amplified to recognize the entire array of genetic and ecologic linkages under which, network-wide, all of the speciation levels and population levels co-determine one another. It is, additionally, quite possible to go beyond the thermostatics of ensembles to a consideration of non-equilibrium eco-genetic states, basing upon the analogy of chemical kinetics, as elsewhere discussed (4), and as well taking into account 'micro-evolutionary' steps of speciation within clusters on the basis of competitive exclusion processes, besides the 'macro-evolutionary' transitions between clusters described above.

It may be remarked in conclusion that the grand ensemble as employed here in classical pattern is not yet grand enough for the entire range of biological purposes. For it has been tacitly presumed that the number of clusters has itself been set, whereas biologically an openness as to number of clusters, virtually infinite in extent, must finally be reckoned with, in some greater-than-grand ensemble without precedent.

I thank the National Science Foundation for its partial support of this work.

References

1. Hutchinson, G. E. *The Ecological Theatre and the Evolutionary Play* (Yale University Press, New Haven, 1965).

2. Volterra, V. *Lecons sur la Théorie Mathématique de la Lutte pour la Vie* (Gauthier-Villars, Paris, 1931).

3. Kerner, E. H. *Gibbs Ensemble: Biological Ensemble* (Gordon & Breach, New York, 1972); *Bull. Math. Biol.* 36, 477, (1974).

4. Kerner, E. H. *Bull. Math. Biol.* (in press, 1978).

5. Gibbs, J. W. *Elementary Principles in Statistical*

Mechanics (Longmans, Green & Co., New York, 1902).

6. Mayr, E. *Animal Species and Evolution* (Belknap Press, Cambridge, 1963).

TOWARD A MATHEMATICAL DEFINITION OF "LIFE"

Gregory J. Chaitin

Abstract: In discussions of the nature of life, the terms "complexity," "organism," and "information content," are sometimes used in ways remarkably analogous to the approach of algorithmic information theory, a mathematical discipline which studies the amount of information necessary for computations. We submit that this is not a coincidence and that it is useful in discussions of the nature of life to be able to refer to analogous precisely defined concepts whose properties can be rigorously studied. We propose and discuss a measure of degree of organization and structure of geometrical patterns which is based on the algorithmic version of Shannon's concept of mutual information. This paper is intended as a contribution to von Neumann's program of formulating mathematically the fundamental concepts of biology in a very general setting, i.e. in highly simplified model universes.

1. Introduction

Here are two quotations from works dealing with the origins of life and exobiology:

"These vague ideas can be made more precise by introducing the idea of information. Roughly speaking, the information content of a structure is the minimum number of instructions needed to specify the structure. One can see intuitively that many instructions are needed to specify a complex structure. On the other hand, a simple repeating structure can be specified in rather few instructions." [1]

"The traditional concept of life, therefore, may be too narrow for our purpose... We should try to break away from the four properties of growth, feeding, reaction, and reproduction... Perhaps there is a clue in the way we speak of living *organisms.* They are *highly organized,* and perhaps this is indeed their essence... What, then, is organization? What sets it apart from other similarly vague concepts? Organization is perhaps viewed best as 'complex interrelatedness'... A book is complex; it only resembles an organism in that passages in one paragraph or chapter refer to others elsewhere. A dictionary or thesaurus shows more organization, for every entry refers to others. A telephone directory shows less, for although it is equally elaborate, there is little cross-reference between its entries... " [2]

If one compares the first quotation with any introductory article on algorithmic information theory (e.g. [3-4]), and compares the second quotation with a preliminary version of this paper [5], one is struck by the similarities. As these quotations show, there has been a great deal of thought about how to define "life," "complexity," "organism," and "information content of organism." The attempted contribution of this paper is that we propose a rigorous quantitative definition of these concepts and are able to prove theorems about them. We do not claim that our proposals are in any sense definitive, but, following von Neumann [6-7], we submit that a precise mathematical definition must be given.

Some preliminary considerations: We shall find it useful to distinguish between the notion of degree of interrelatedness, interdependence, structure, or organization, and that of information content. Two extreme examples are an ideal gas and a perfect crystal. The complete microstate at a given time of the first one is very difficult to describe fully, and for the second one this is trivial to do, but neither is organized. In other words, white noise is the most informative message possible, and a constant pitch tone is least informative, but neither is organized. Neither a gas nor a crystal should count as organized (see Theorems 1 and 2 in Section 5), nor should a whale or elephant be considered more organized than a person simply because it requires more information to specify the precise details of the current position of each molecule in its much larger bulk. Also note that following von Neumann [7] we deal with a discrete model universe, a cellular automata space, each of whose cells has only a finite number of states. Thus we impose a certain level of granularity in our idealized description of the real world.

We shall now propose a rigorous theoretical measure of degree of organization or structure. We use ideas from the new algorithmic formulation of information theory, in which one considers individual objects and the amount of information in bits needed to compute, construct, describe, generate or produce them, as opposed to the classical formulation of information theory in which one considers an ensemble of possibilities and the uncertainty as to which of them is actually the case. In that theory the uncertainty or "entropy" of a distribution is defined to be

$$-\sum_{i<k} p_i \log p_i,$$

and is a measure of one's ignorance of which of the k possibilities actually holds given that the *a priori* probability of the ith alternative is p_i. (Throughout this paper "log" denotes the base-two logarithm.) In contrast, in the newer formulation of information theory one can speak of the information content of an individual book, organism, or picture, without having to imbed it in an ensemble of all possible such objects and postulate a probability distribution on them.

We believe that the concepts of algorithmic information theory are extremely basic and fundamental. Witness the light they have shed on the scientific method [8], the meaning of randomness and the Monte Carlo method [9], the limitations of the deductive method [3-4], and now, hopefully, on theoretical biology. An information-theoretic proof of Euclid's theorem that there are infinitely many prime numbers should also be mentioned (see Appendix 2).

The fundamental notion of algorithmic information theory is $H(X)$, the algorithmic information content (or, more briefly, "complexity") of the object X. $H(X)$ is defined to be the smallest possible number of bits in a program for a general-purpose computer to print out X. In other words, $H(X)$ is the amount of information necessary to describe X sufficiently precisely for it to be constructed. Two objects X and Y are said to be (algorithmically) independent if the best way to describe them both is simply to describe each of them separately. That is to say, X and Y are independent if $H(X,Y)$ is approximately equal to $H(X)+H(Y)$, i.e. if the joint information content of X and Y is just the sum of the individual information contents of X and Y. If, however, X and Y are related and have something in common, one can take advantage of this to describe X and Y together using much fewer bits than the total number that would be needed to describe them separately, and so $H(X,Y)$ is much less than $H(X)+H(Y)$. The quantity $H(X{:}Y)$ which is defined as follows

$$H(X{:}Y) = H(X) + H(Y) - H(X,Y)$$

is called the mutual information of X and Y and measures the degree of interdependence between X and Y. This concept was defined, in an ensemble rather than an algorithmic setting, in Shannon's original paper [10] on information theory, noisy channels, and coding.

We now explain our definition of the degree of organization or structure in a geometrical pattern. The d-diameter complexity $H_d(X)$ of an object X is defined to be the minimum number of bits needed to describe X as the "sum" of separate parts each of diameter not greater than d. Let us be more precise. Given d and X, consider all possible ways of partitioning X into nonoverlapping pieces each of diameter $\leq d$. Then $H_d(X)$ is the sum of the number of bits needed to describe each of the pieces separately, plus the number of bits needed to specify how to reassemble them into X. Each piece must have a separate description which makes no cross-references to any of the others. And one is interested in those partitions of X and reassembly techniques α which minimize this sum. That is to say,

$$H_d(X) = \min\,[H(\alpha) + \sum_{i<k} H(X_i)],$$

the minimization being taken over all partitions of X into nonoverlapping pieces

$$X_0, X_1, X_2, \ldots, X_{k-1}$$

all of diameter $\leq d$.

Thus $H_d(X)$ is the minimum number of bits needed to describe X as if it were the sum of independent pieces of size $\leq d$. For d larger than the diameter of X, $H_d(X)$ will be the same as $H(X)$. If X is unstructured and unorganized, then as d decreases $H_d(X)$ will stay close to $H(X)$. However if X has structure, then $H_d(X)$ will rapidly increase as d decreases and one can no longer take advantage of patterns of size $> d$ in describing X. Hence $H_d(X)$ as a function of d is a kind of "spectrum" or "Fourier transform" of X. $H_d(X)$ will increase as d decreases past the diameter of significant patterns in X, and if X is organized hierarchically this will happen at each level in the hierarchy.

Thus the faster the difference increases between $H_d(X)$ and $H(X)$ as d decreases, the more interrelated, structured, and organized X is. Note however that X may be a "scene" containing many independent structures or organisms. In that case their degrees of organization are summed together in the measure

$$H_d(X) - H(X).$$

Thus the organisms can be defined as the minimal parts of the scene for which the amount of organization of the whole can be expressed as the sum of the organization of the parts, i.e. pieces for which the measure of organization decomposes additively. Alternatively, one can use the notion of the mutual information of two pieces to obtain a theoretical prescription of how to separate a scene into independent patterns and distinguish a pattern from an unstructured background in which it is imbedded (see Section 6).

Let us enumerate what we view as the main points in favor of this definition of organization: It is general, i.e. following von Neumann the details of the physics and chemistry of this universe are not involved; it measures organized structure rather than unstructured details; and it passes the spontaneous generation or "Pasteur" test, i.e. there is a very low probability of creating organization by chance without a long evolutionary process (this may be viewed as a way of restating Theorem 1 in Section 5). The second point is worth elaborating: The information content of an organism includes much irrelevant detail, and a bigger animal is necessarily more complex in this sense. *But if it were possible to calculate the mutual information of two arbitrary cells in a body at a given moment, we surmise that this would give a measure of the genetic information in a cell. This is because the irrelevant details in each of them, such as the exact position and velocity of each molecule, are uncorrelated and would cancel each other out.*

In addition to providing a definition of information content and of degree of organization, this approach also provides a definition of "organism" in the sense that a theoretical prescription is given for dissecting a scene into organisms and determining their boundaries, so that the measure of degree of organization can then be applied separately to each organism. However a strong note of caution is in order: We agree with [1] that a definition of "life" is valid as long as anything that satisfies the definition and is likely to appear in the universe under consideration, either is alive or is a by-product of living beings or their activities. There certainly are structures satisfying our definition that are not alive (see Theorems 3 to 6 in Section 5); however, we believe that they would only be likely to arise as by-products of the activities of living beings.

In the succeeding sections we shall do the following: give a more formal presentation of the basic concepts of algorithmic information theory; discuss the notions of the independence and mutual information of groups of more than two objects; formally define H_d ; evaluate $H_d(R)$ for some typical one-dimensional geometrical patterns R which we dub "gas," "crystal," "twins," "bilateral symmetry," and "hierarchy;" consider briefly the problem of decomposing scenes containing several independent patterns, and of determining the boundary of a pattern which is imbedded in a unstructured background; discuss briefly the two and higher dimension cases; and mention some alternative definitions of mutual information which have been proposed.

The next step in this program of research would be to proceed from static snapshots to time-varying situations, in other words, to set up a discrete universe with probabilistic state transitions and to show that there is a certain probability that a certain level of organization will be reached by a certain time. More generally, one would like to determine the probability distribution of the maximum degree of organization of any organism at time $t + \Delta$ as a function of it at time t. Let us propose an initial proof strategy for setting up a nontrivial example of the evolution of organisms: construct a series of intermediate evolutionary forms [11], argue that increased complexity gives organisms a selective advantage, and show that no primitive organism is so successful or lethal that it diverts or blocks this gradual evolutionary pathway. What would be the intellectual flavor of the theory we desire? It would be a quantitative formulation of Darwin's theory of evolution in a very general model universe setting. It would be the opposite of ergodic theory: Instead of showing that things mix and become uniform, it would show that variety and organization will probably increase.

Some final comments: Software is fast approaching biological levels of complexity, and hardware, thanks to very large scale integration, is not far behind. Because of this, we believe that the computer is now becoming a valid metaphor for the entire organism, not just for the brain [12]. Perhaps the most interesting example of this is the evolutionary phenomenon suffered by extremely large programs such as operating systems. It becomes very difficult to make changes in such programs, and the only alternative is to add new features rather than modify existing ones. The genetic program has been "patched up" much more and over a much longer period of time than even the largest operating systems, and Nature has accomplished this in much the same manner as systems programmers have, by carrying along all the previous code as new code is added [11]. The experimental proof of this is that ontogeny recapitulates phylogeny, i.e. each embryo to a certain extent recapitulates in the course of its development the evolutionary sequence that led to it. In this connection we should also mention the thesis developed in [13] that the information contained in the human brain is now comparable with the amount of information in the genes, and that intelligence plus education may be characterized as a way of getting around the limited modifiability and channel capacity of heredity. In other words, Nature, like computer designers, has decided that it is much more flexible to build general-purpose computers than to use heredity to "hardwire" each behavior pattern instinctively into a special-purpose computer.

2. *Algorithmic Information Theory*

We first summarize some of the basic concepts of algorithmic information theory in its most recent formulation [14-16].

This new approach leads to a formalism that is very close to that of classical probability theory and information theory, and is based on the notion that the tape containing the Turing machine's program is infinite and entirely filled with 0's and 1's. This forces programs to be self-delimiting; i.e. they must contain within themselves information about their size, since the computer cannot rely on a blank at the end of the program to indicate where it ends.

Consider a universal Turing machine U whose programs are in binary and are self-delimiting. By "self-delimiting" we mean, as was just explained, that they do not have blanks appended as endmarkers. By "universal" we mean that for any other Turing machine M whose programs p are in binary and are self-delimiting, there is a prefix μ such that $U(\mu p)$ always carries out the same computation as $M(p)$.

$H(X)$, the algorithmic information content of the finite object X, is defined to be the size in bits of the smallest self-delimiting programs for U to compute X. This includes the proviso that U halt after printing X. There is absolutely no restriction on the running time or storage space used by this program. For example, X can be a natural number or a bit string or a tuple of natural numbers or bit strings. Note that variations in the definition of U give rise to at most $O(1)$ differences in the resulting H, by the definition of universality.

The self-delimiting requirement is adopted so that one gets the following basic subadditivity property of H:

$$H(<X,Y>) \leq H(X) + H(Y) + O(1),$$

This inequality holds because one can concatenate programs. It expresses the notion of "adding information," or, in computer jargon, "using subroutines."

Another important consequence of this requirement is that a natural probability measure P, which we shall refer to as the algorithmic probability, can be associated with the result of any computation. $P(X)$ is the probability that X is obtained as output if the standard universal computer U is started running on a program tape filled with 0's and 1's by separate tosses of a fair coin. The algorithmic probability P and the algorithmic information content H are related as follows [14]:

$$H(X) = -\log P(X) + O(1). \tag{1}$$

Consider a binary string s. Define the function L as follows:

$$L(n) = \max\{H(s) : \text{length}(s) = n\}.$$

It can be shown [14] that $L(n) = n+H(n)+O(1)$, and that an overwhelming majority of the s of length n have $H(s)$ very close to $L(n)$. Such s have maximum information content and are highly random, patternless, incompressible, and typical. They are said to be "algorithmically random." The greater the difference between $H(s)$ and $L(\text{length}(s))$, the less random s is. It is convenient to say that "s is k-random" if $H(s) \geq L(n)-k$, where $n = \text{length}(s)$. There are at most

$$2^{n-k+O(1)}$$

n-bit strings which aren't k-random. As for natural numbers, most n have $H(n)$ very close to $L(\text{floor}(\log n))$. Here floor(x) is the greatest integer $\leq x$. Strangely enough, though most strings are random it is impossible to prove that specific strings have this property. For an explanation of this paradox and further references, see the section on metamathematics in [15], and also see [9].

We now make a few observations that will be needed later. First of all, $H(n)$ is a smooth function of n:

$$|H(n)-H(m)| = O(\log|n-m|). \tag{2}$$

(Note that this is not strictly true if $|n-m|$ is equal to 0 or 1, unless one considers the log of 0 and 1 to be 1; this convention is therefore adopted throughout this paper.) For a proof, see [16]. The following upper bound on $H(n)$ is an immediate corollary of this smoothness property: $H(n) = O(\log n)$. Hence if s is an n-bit string, then $H(s) \leq n+O(\log n)$. Finally, note that changes in the value of the argument of the function L produce nearly equal changes in the value of L. Thus, for any ε there is a δ such that $L(n) > L(m) + \varepsilon$ if $n > m + \delta$. This is because of the fact that $L(n) = n+H(n)+O(1)$ and the smoothness property (2) of H.

An important concept of algorithmic information theory that hasn't been mentioned yet is the conditional probability $P(Y|X)$, which by definition is

$P(<X,Y>)/P(X)$. To the conditional probability there corresponds the relative information content $H(Y|X^*)$, which is defined to be the size in bits of the smallest programs for the standard universal computer U to output Y if it is given X^*, a canonical minimum-size program for calculating X. X^* is defined to be the first $H(X)$-bit program to compute X that one encounters in a fixed recursive enumeration of the graph of U (i.e. the set of all pairs of the form $<p,U(p)>$). Note that there are partial recursive functions which map X^* to $<X,H(X)>$ and back again, and so X^* may be regarded as an abbreviation for the ordered pair whose first element is the string X and whose second element is the natural number that is the complexity of X. We should also note the immediate corollary of (1) that minimum-size or nearly minimum-size programs are essentially unique: For any ε there is a δ such that for all X the cardinality of {the set of all programs for U to calculate X that are within ε bits of the minimum size $H(X)$} is less than δ. It is possible to prove the following theorem relating the conditional probability and the relative information content [14]:

$$H(Y|X^*) = -\log P(Y|X) + O(1). \tag{3}$$

From (1) and (3) and the definition $P(<X,Y>) = P(X)P(Y|X)$, one obtains this very basic decomposition:

$$H(<X,Y>) = H(X) + H(Y|X^*) + O(1). \tag{4}$$

3. Independence and Mutual Information

It is an immediate corollary of (4) that the following four quantities are all within $O(1)$ of each other:

$$\left\{\begin{array}{l} H(X)-H(X|Y^*), \\ H(Y)-H(Y|X^*), \\ H(X)+H(Y)-H(<X,Y>), \\ H(Y)+H(X)-H(<Y,X>). \end{array}\right.$$

These four quantities are known as the mutual information $H(X{:}Y)$ of X and Y; they measure the extent to which X and Y are interdependent. For if $P(<X,Y>) \approx P(X)P(Y)$, then $H(X{:}Y) = O(1)$; and if Y if a recursive function of X, then $H(Y|X^*) = O(1)$ and $H(X{:}Y) = H(Y) + O(1)$. In fact,

$$H(X{:}Y) = -\log\left[\frac{P(X)P(Y)}{P(<X,Y>)}\right] + O(1).$$

which shows quite clearly that $H(X{:}Y)$ is a symmetric measure of the independence of X and Y. Note that in algorithmic information theory, what is of importance is an approximate notion of independence and a measure of its degree (mutual information), rather than the exact notion. This is because the algorithmic probability may vary within a certain percentage depending on the choice of universal computer U. Conversely, information measures in algorithmic information theory should not vary by more than $O(1)$ depending on the choice of U.

To motivate the definition of the d-diameter complexity, we now discuss how to generalize the notion of independence and mutual information from a pair to an

n-tuple of objects. In what follows classical and algorithmic probabilities P are distinguished by using curly brackets for the first one and parentheses for the second. In probability theory the mutual independence of a set of n events $\{A_k : k<n\}$ is defined by the following 2^n equations:

$$\prod_{k \in S} P\{A_k\} = P\{\bigcap_{k \in S} A_k\}$$

for all $S \subset n$. Here the set-theoretic convention due to von Neumann is used that identifies the natural number n with the set $\{k : k<n\}$. In algorithmic probability theory the analogous condition would be to require that

$$\prod_{k \in S} P(A_k) \approx P(\Theta_{k \in S} A_k) \tag{5}$$

for all $S \subset n$. Here $\Theta\ A_k$ denotes the tuple forming operation for a variable length tuple, i.e.

$$\Theta_{k<n} A_k = \langle A_0, A_1, A_2, \ldots, A_{n-1} \rangle.$$

It is a remarkable fact that these 2^n conditions (5) are equivalent to the single requirement that

$$\prod_{k<n} P(A_k) \approx P(\Theta_{k<n} A_k). \tag{6}$$

To demonstrate this it is necessary to make use of special properties of algorithmic probability that are not shared by general probability measures. In the case of a general probability space,

$$P\{A \cap B\} \geq P\{A\} + P\{B\} - 1$$

is the best lower bound on $P\{A \cap B\}$ that can in general be formulated in terms of $P\{A\}$ and $P\{B\}$. For example, it is possible for $P\{A\}$ and $P\{B\}$ to both be $1/2$, while $P\{A \cap B\} = 0$. In algorithmic information theory the situation is quite different. In fact one has:

$$P(\langle A,B \rangle) \geq c_2\, P(A)P(B),$$

and this generalizes to any fixed number of objects:

$$P(\Theta_{k<n} A_k) \geq c_n \prod_{k<n} P(A_k).$$

Thus if the joint algorithmic probability of a subset of the n-tuple of objects were significantly greater than the product of their individual algorithmic probabilities, then this would also hold for the entire n-tuple of objects. More precisely, for any $S \subset n$ one has

$$P(\Theta_{k<n} A_k) \geq \hat{c}_n\, P(\Theta_{k \in S} A_k) P(\Theta_{k \in n-S} A_k) \geq \tilde{c}_n\, P(\Theta_{k \in S} A_k) \prod_{k \in n-S} A_k.$$

Then if one assumes that

$$P(\Theta_{k \varepsilon S} A_k) >> \prod_{k \varepsilon S} P(A_k)$$

(here >> denotes "much greater than"), it follows that

$$P(\Theta_{k<n} A_k) >> \prod_{k<n} P(A_k).$$

We conclude that in algorithmic probability theory (5) and (6) are equivalent and thus (6) is a necessary and sufficient condition for an n-tuple to be mutually independent. Therefore the following measure of mutual information for n-tuples accurately characterizes the degree of interdependence of n objects:

$$[\sum_{k<n} H(A_k)] - H(\Theta_{k<n} A_k).$$

This measure of mutual information subsumes all others in the following precise sense:

$$[\sum_{k<n} H(A_k)] - H(\Theta_{k<n} A_k) \quad = \quad \max \{[\sum_{k \varepsilon S} H(A_k)] - H(\Theta_{k \varepsilon S} A_k)\} + O(1),$$

where the maximum is taken over all $S \subset n$.

4. *Formal Definition of H_d*

We can now present the definition of the d-diameter complexity $H_d(R)$. We assume a geometry: graph paper of some finite number of dimensions that is divided into unit cubes. Each unit cube is black or white, opaque or transparent, in other words, contains a 1 or a 0. Instead of requiring an output tape which is multidimensional, our universal Turing machine U outputs tuples giving the coordinates and the contents (0 or 1) of each unit cube in a geometrical object that it wishes to print. Of course geometrical objects are considered to be the same if they are translation equivalent. We choose for this geometry the city-block metric

$$D(X,Y) = \max |x_i - y_i|,$$

which is more convenient for our purposes than the usual metric. By a region we mean a set of unit cubes with the property that from any cube in it to any other one there is a path that only goes through other cubes in the region. To this we add the constraint which in the 3-dimensional case is that the connecting path must only pass through the interior and faces of cubes in the region, not through their edges or vertices. The diameter of an arbitrary region R is denoted by $|R|$, and is defined to be the minimum diameter $2r$ of a "sphere"

$$\{X : D(X,X_0) \leq r\}$$

which contains R. $H_d(R)$, the size in bits of the smallest programs which calculate R as the "sum" of independent regions of diameter $\leq d$, is defined as follows:

$$H_d(R) = \min [\alpha + \sum_{i<k} H(R_i)],$$

where

$$\alpha = H(R \mid \Theta_{i<k} R_i) + H(k),$$

the minimization being taken over all k and partitions of R into k-tuples ΘR_i of nonoverlapping regions with the property that $|R_i| < d$ for all $i < k$.

The discussion in Section 3 of independence and mutual information shows that $H_d(R)$ is a natural measure to consider. Excepting the α term, $H_d(R) - H(R)$ is simply the minimum attainable mutual information over any partition of R into nonoverlapping pieces all of size not greater than d. We shall see in Section 5 that in practice the min is attained with a small number of pieces and the α term is not very significant.

A few words about α, the number of bits of information needed to know how to assemble the pieces: The $H(k)$ term is included in α, as illustrated in Lemma 1 below, because it is the number of bits needed to tell U how many descriptions of pieces are to be read. The $H(R \mid \Theta R_i)$ term is included in α because it is the number of bits needed to tell U how to compute R given the k-tuple of its pieces. This is perhaps the most straight-forward formulation, and the one that is closest in spirit to Section 5 [5]. However, less information may suffice, e.g.

$$H(R \mid \langle k^*, \Theta_{i<k} (R_i^*) \rangle) + H(k)$$

bits. In fact, one could define α to be the minimum number of bits in a string which yields a program to compute the entire region when it is concatenated with minimum-size programs for all the pieces of the region; i.e. one could take

$$\alpha = \min \{ |p| : U(pR_0^* R_1^* R_2^* \ldots R_{k-1}^*) = R \}.$$

Here are two basic properties of H_d: If $d \geq |R|$, then $H_d(R) = H(R) + O(1)$; $H_d(R)$ increases monotonically as d decreases. $H_d(R) = H(R) + O(1)$ if $d \geq |R|$ because we have included the α term in the definition of $H_d(R)$. $H_d(R)$ increases as d decreases because one can no longer take advantage of patterns of diameter greater than d to describe R. The curve showing $H_d(R)$ as a function of d may be considered a kind of "Fourier spectrum" of R. Interesting things will happen to the curve at d which are the sizes of significant patterns in R.

Lemma 1. ("Subadditivity for n-tuples")

$$H(\Theta_{k<n} A_k) \leq c_n + \sum_{k<n} H(A_k).$$

Proof:

$$H(\Theta_{k<n} A_k) = H(\langle n, \Theta_{k<n} A_k \rangle) + O(1) =$$

$$H(n) + H(\Theta_{k<n} A_k \mid n^*) + O(1) \leq \hat{c} + H(n) + \sum_{k<n} H(A_k).$$

Hence one can take

$c_n = \hat{c} + H(n)$.

5. *Evaluation of H_d for Typical One-Dimensional Geometrical Patterns*

Before turning to the examples, we present a lemma needed for estimating $H_d(R)$. The idea is simply that sufficiently large pieces of a random string are also random. It is required that the pieces be sufficiently large for the following reason: It is not difficult to see that for any j, there is an n so large that random strings of size greater than n must contain all 2^j possible subsequences of length j. In fact, for n sufficiently large the relative frequency of occurrence of all 2^j possible subsequences must approach the limit 2^{-j}.

Lemma 2. ("Random parts of random strings")
Consider an n-bit string s to be a loop. For any natural numbers i and j between 1 and n, consider the sequence u of contiguous bits from s starting at the ith and continuing around the loop to the jth. Then if s is k-random, its subsequence u is $(k+O(\log n))$-random.

Proof:
The number of bits in u is $j-i+1$ if j is $\geq i$, and is $n+j-i+1$ if j is $< i$. Let v be the remainder of the loop s after u has been excised. Then we have $H(u)+H(v)+H(i)+O(1) \geq H(s)$. Thus $H(u)+n-|u|+O(\log n) \geq H(s)$, or $H(u) \geq H(s)-n+|u|+O(\log n)$. Thus if s is k-random, i.e. $H(s) \geq L(n)-k = n+H(n)-k+O(1)$, then u is x-random, where x is determined as follows: $H(u) \geq n+H(n)-k-n+|u|+O(\log n) = |u|+H(|u|)-k+O(\log n)$. That is to say, if s is k-random, then its subsequence u is $(k+O(\log n))$-random.

Lemma 3. ("Random prefixes of random strings")
Consider an n-bit string s. For any natural number j between 1 and n, consider the sequence u consisting of the first j bits of s. Then if s is k-random, its j-bit prefix u is $(O(\log j)+k)$-random.

Proof:
Let the $(n-j)$-bit string v be the remainder of s after u is excised. Then we have $H(u)+H(v)+O(1) \geq H(s)$, and therefore $H(u) \geq H(s)-L(n-j)+O(1) = L(n)-k-L(n-j)+O(1)$ since s is k-random. Note that $L(n)-L(n-j) = j+H(n)-H(n-j)+O(1) = j+O(\log j)$, by the smoothness property (2) of H. Hence $H(u) \geq j+O(\log j)-k$. Thus if u is x-random, we have $L(j)-x = j+O(\log j)-x \geq j+O(\log j)-k$. Hence $x \leq O(\log j)+k$.

Remark:
Conversely, any random n-bit string can be extended by concatenating k bits to it in such a manner that the result is a random $(n+k)$-bit string. We shall not use this converse result, but it is included here for the sake of completeness.

Lemma 4. ("Random extensions of random strings")
Assume the string s is x-random. Consider a natural number k. Then there is a k-bit string e such that se is y-random, as long as k, x, and y satisfy a condition of the following form:

$$y < x + O(\log x) + O(\log k).$$

Proof:
Assume on the contrary that the x-random string s has no y-random k-bit extension and $y \geq x+O(\log x)+O(\log k)$, i.e. $x < y+O(\log y)+O(\log k)$. From this assumption we shall derive a contradiction by using the fact that most strings of any particular size are y-random, i.e. the fraction of them that are y-random is at least

$$1-2^{-y+O(1)}.$$

It follows that the fraction of $|s|$-bit strings which have no y-random k-bit extension is less than

$$2^{-y+O(1)}.$$

Since by hypothesis no k-bit extension of s is y-random, we can uniquely determine s if we are given y and k and the ordinal number of the position of s in {the set of all $|s|$-bit strings which have no y-random k-bit extension} expressed as an $(|s|-y+O(1))$-bit string. Hence $H(s)$ is less than $L(|s|-y+O(1)) + H(y) + H(k) + O(1)$. In as much as $L(n) = n + H(n) + O(1)$ and $|H(n) - H(m)| = O(\log|n-m|)$, it follows that $H(s)$ is less than $L(|s|) - [\, y + O(\log y) + O(\log k)\,]$. Since s is by assumption x-random, i.e. $H(s) \geq L(|s|)-x$, we obtain a lower bound on x of the form $y+O(\log y)+O(\log k)$, which contradicts our original assumption that $x < y+O(\log y)+O(\log k)$.

Theorem 1. ("Gas")
Suppose that the region R is an $O(\log n)$-random n-bit string. Consider $d = n/k$, where n is large, and k is fixed and greater than zero. Then

$$H(R) = n + O(\log n), \quad \text{and} \quad H_d(R) = H(R) + O(\log H(R)).$$

Proof that $H_d(R) \leq H(R) + O(\log H(R))$:
Let β be concatenation of tuples of strings, i.e.

$$\beta(\Theta_{i\leq k} R_i) = R_0 R_1 R_2 \ldots R_k.$$

Note that

$$H(\beta(\Theta_{i\leq k} R_i) \mid \Theta_{i\leq k} R_i) = O(1).$$

Divide R into k successive strings of size floor$(|R|/k)$, with one (possibly null) string of size less than k left over at the end. Taking this choice of partition ΘR_i in the definition of $H_d(R)$, and using the fact that $H(s) \leq |s| + O(\log|s|)$, we see that

$$H_d(R) \le O(1) + H(k+1) + \sum_{i \le k} \{ |R_i| + O(\log |R_i|)\}$$

$$\le O(1) + n + (k+2)O(\log n) = n + O(\log n).$$

Proof that $H_d(R) \ge H(R) + O(\log H(R))$:
This follows immediately from the fact that $H_{|R|}(R) = H(R) + O(1)$ and $H_d(R)$ increases monotonically as d decreases.

Theorem 2. ("Crystal")
Suppose that the region R is an n-bit string consisting entirely of 1's, and that the base-two numeral for n is $O(\log\log n)$-random. Consider $d = n/k$, where n is large, and k is fixed and greater than zero. Then

$$H(R) = \log n + O(\log \log n), \quad \text{and} \quad H_d(R) = H(R) + O(\log H(R)).$$

Proof that $H_d(R) \le H(R) + O(\log H(R))$:
If one considers using the concatenation function β for assembly as was done in the proof of Theorem 1, and notes that $H(1^n) = H(n) + O(1)$, one sees that it is sufficient to partition the natural number n into $O(k)$ summands none of which is greater than n/k in such a manner that $H(n)+O(\log\log n)$ upper bounds the sum of the complexities of the summands. Division into equal size pieces will not do, because $H(\text{floor}(n/k)) = H(n) + O(1)$, and one only gets an upper bound of $kH(n) + O(1)$. It is necessary to proceed as follows: Let m be the greatest natural number such that $2^m \le n/k$. And let p be the smallest natural number such that $2^p > n$. By converting n to base-two notation, one can express n as the sum of $\le p$ distinct non-negative powers of two. Divide all these powers of two into two groups: those that are less than 2^m and those that are greater than or equal to 2^m. Let f be the sum of all the powers in the first group. f is $< 2^m \le n/k$. Let s be the sum of all the powers in the second group. s is a multiple of 2^m; in fact, it is of the form $t2^m$ with $t = O(k)$. Thus $n = f + s = f + t2^m$, where $f \le n/k$, $2^m \le n/k$, and $t = O(k)$. The complexity of 2^m is $H(m)+O(1) = O(\log m) = O(\log\log n)$. Thus the sum of the complexities of the t summands 2^m is also $O(\log\log n)$. Moreover, f when expressed in base-two notation has $\log k+O(1)$ fewer bit positions on the left than n does. Hence the complexity of f is $H(n)+O(1)$. In summary, we have $O(k)$ quantities n_i with the following properties:

$$n = \sum n_i, \quad n_i \le n/k, \quad \sum H(n_i) \le H(n) + O(\log \log n).$$

Thus $H_d(R) \le H(R) + O(\log H(R))$.

Proof that $H_d(R) \ge H(R) + O(\log H(R))$:
This follows immediately from the fact that $H_{|R|}(R) = H(R) + O(1)$ and $H_d(R)$ increases monotonically as d decreases.

Theorem 3. ("Twins")
For convenience assume n is even. Suppose that the region R consists of two repetitions of an $O(\log n)$-random $n/2$-bit string u. Consider $d = n/k$, where n is large, and k is fixed and greater than unity. Then

$$H(R) = n/2 + O(\log n), \quad \text{and} \quad H_d(R) = 2H(R) + O(\log H(R)).$$

Proof that $H_d(R) \leq 2H(R) + O(\log H(R))$:
The reasoning is the same as in the case of the "gas" (Theorem 1). Partition R into k successive strings of size floor($|R|/k$), with one (possibly null) string of size less than k left over at the end.

Proof that $H_d(R) \geq 2H(R) + O(\log H(R))$:
By the definition of $H_d(R)$, there is a partition $\Theta\, R_i$ of R into nonoverlapping regions which has the property that

$$H_d(R) = \alpha + \sum H(R_i), \quad \alpha = H(R \mid \Theta\, R_i) + H(k), \quad |R_i| \leq d.$$

Classify the non-null R_i into three mutually exclusive sets A, B, and C: A is the set of all non-null R_i which come from the left half of R ("the first twin"), B is the (empty or singleton) set of all non-null R_i which come from both halves of R ("straddles the twins"), and C is the set of all non-null R_i which come from the right half of R ("the second twin"). Let $\dot{A}$, $\dot{B}$, and $\dot{C}$ be the sets of indices i of the regions R_i in A, B, and C, respectively. And let $\ddot{A}$, $\ddot{B}$, and $\ddot{C}$ be the three portions of R which contained the pieces in A, B, and C, respectively. Using the idea of Lemma 1, one sees that

$$H(\ddot{A}) \leq O(1) + H(\#(A)) + \sum_{i \in \dot{A}} H(R_i), \quad H(\ddot{B}) \leq O(1) + H(\#(B)) + \sum_{i \in \dot{B}} H(R_i),$$

$$H(\ddot{C}) \leq O(1) + H(\#(C)) + \sum_{i \in \dot{C}} H(R_i).$$

Here # denotes the cardinality of a set. Now $\ddot{A}$, $\ddot{B}$, and $\ddot{C}$ are each a substring of an $O(\log n)$-random $n/2$-bit string. This assertion holds for $\ddot{B}$ for the following two reasons: the $n/2$-bit string is considered to be a loop, and $|\ddot{B}| \leq d = n/k \leq n/2$ since k is assumed to be greater than 1. Hence, applying Lemma 2, one obtains the following inequalities:

$$|\ddot{A}| + O(\log n) \leq H(\ddot{A}), \quad |\ddot{B}| + O(\log n) \leq H(\ddot{B}), \quad |\ddot{C}| + O(\log n) \leq H(\ddot{C}).$$

Adding both of the above sets of three inequalities and using the facts that

$$|\ddot{A}| + |\ddot{B}| + |\ddot{C}| = |R| = n, \quad \#(A) \leq n/2, \quad \#(B) \leq 1, \quad \#(C) \leq n/2,$$

and that $H(m) = O(\log m)$, one sees that

$$n + O(\log n) \leq H(\ddot{A}) + H(\ddot{B}) + H(\ddot{C})$$

$$\leq O(1) + H(\#(A)) + H(\#(B)) + H(\#(C)) + \sum \{H(R_i) : i \in \dot{A} \cup \dot{B} \cup \dot{C}\}$$

$\leq O(\log n) + \sum H(R_i)$.

Hence

$H_d(R) \geq \sum H(R_i) \geq n + O(\log n) = 2H(R) + O(\log H(R))$.

Theorem 4. ("Bilateral Symmetry")
For convenience assume n is even. Suppose that the region R consists of an $O(\log n)$-random $n/2$-bit string u concatenated with its reversal. Consider $d = n/k$, where n is large, and k is fixed and greater than zero. Then

$$H(R) = n/2 + O(\log n), \quad \text{and} \quad H_d(R) = (2-k^{-1})H(R) + O(\log H(R)).$$

Proof:
The proof is along the lines of that of Theorem 3, with one new idea. In the previous proof we considered $\ddot{B}$, which is the region R_i in the partition of R that straddles R's midpoint. Before $\ddot{B}$ was $O(\log|R|)$-random, but now it can be compressed into a program about half its size, i.e. about $|\ddot{B}|/2$ bits long. Hence the maximum departure from randomness for $\ddot{B}$ is for it to only be $O(\log|R|)+(|R|/2k)$-random, and this is attained by making $\ddot{B}$ as large as possible and having its midpoint coincide with that of R.

Theorem 5. ("Hierarchy")
For convenience assume n is a power of two. Suppose that the region R is constructed in the following fashion. Consider an $O(1)$-random $\log n$-bit string s. Start with the one-bit string 1, and successively concatenate the string with itself or with its bit by bit complement, so that its size doubles at each stage. At the ith stage, the string or its complement is chosen depending on whether the ith bit of s is a 0 or a 1, respectively. Consider the resulting n-bit string R and $d = n/k$, where n is large, and k is fixed and greater than zero. Then

$$H(R) = \log n + O(\log \log n), \quad \text{and} \quad H_d(R) = kH(R) + O(\log H(R)).$$

Proof that $H_d(R) \leq kH(R) + O(\log H(R))$:
The reasoning is similar to the case of the upper bounds on $H_d(R)$ in Theorems 1 and 3. Partition R into k successive strings of size floor$(|R|/k)$, with one (possibly null) string of size less than k left over at the end.

Proof that $H_d(R) \geq kH(R) + O(\log H(R))$:
Proceeding as in the proof of Theorem 3, one considers a partition ΘR_i of R that realizes $H_d(R)$. Using Lemma 3, one can easily see that the following lower bound holds for any substring R_i of R:

$$H(R_i) \geq \max\{1, \log|R_i| - c \log\log|R_i|\}.$$

The max {1,...} is because H is always greater than or equal to unity; otherwise U would have only a single output. Hence the following expression is a lower bound on $H_d(R)$:

$$\Sigma\Phi(|R_i|), \tag{7}$$

where

$$\Phi(x) = \max\{1, \log x - c\log\log x\}, \quad \Sigma\ |R_i| = |R| = n, \quad |R_i| \leq d.$$

It follows that one obtains a lower bound on (7) and thus on $H_d(R)$ by solving the following minimization problem: Minimize

$$\Sigma\ \Phi(n_i) \tag{8}$$

subject to the following constraints:

$$\Sigma\ n_i = n, \quad n_i \leq n/k, \quad n \text{ large}, \quad k \text{ fixed}.$$

Now to do the minimization. Note that as x goes to infinity, $\Phi(x)/x$ goes to the limit zero. Furthermore, the limit is never attained, i.e. $\Phi(x)/x$ is never equal to zero. Moreover, for x and y sufficiently large and x less than y, $\Phi(x)/x$ is greater than $\Phi(y)/y$. It follows that a sum of the form (8) with the n_i constrained as indicated is minimized by making the n_i as large as possible. Clearly this is achieved by taking all but one of the n_i equal to floor(n/k), with the last n_i equal to remainder(n/k). For this choice of n_i the value of (8) is

$$k[\log n + O(\log\log n)] + \Phi(\text{remainder}(n/k))$$

$$= k\log n + O(\log\log n) = kH(R) + O(\log H(R)).$$

Theorem 6.

For convenience assume n is a perfect square. Suppose that the region R is an n-bit string consisting of $\sqrt{n}$ repetitions of an $O(\log n)$-random $\sqrt{n}$ bit string u. Consider $d = n/k$, where n is large, and k is fixed and greater than zero. Then

$$H(R) = \sqrt{n} + O(\log n), \quad \text{and} \quad H_d(R) = kH(R) + O(\log H(R)).$$

Proof that $H_d(R) \leq kH(R) + O(\log H(R))$:

The reasoning is identical to the case of the upper bound on $H_d(R)$ in Theorem 5.

Proof that $H_d(R) \geq kH(R) + O(\log H(R))$:

Proceeding as in the proof of Theorem 5, one considers a partition $\Theta\ R_i$ of R that realizes $H_d(R)$. Using Lemma 2, one can easily see that the following lower bound holds for any substring R_i of R:

$$H(R_i) \geq \max\{1, -c\log n + \min\{\sqrt{n}, |R_i|\}\}.$$

Hence the following expression is a lower bound on $H_d(R)$:

$$\sum \Phi_n(|R_i|), \tag{9}$$

where

$$\Phi_n(x) = \max\{1, -c \log n + \min\{\sqrt{n}, x\}\}, \quad \sum |R_i| = |R| = n, \quad |R_i| \leq d.$$

It follows that one obtains a lower bound on (9) and thus on $H_d(R)$ by solving the following minimization problem: Minimize

$$\sum \Phi_n(n_i) \tag{10}$$

subject to the following constraints:

$$\sum n_i = n, \quad n_i \leq n/k, \quad n \text{ large}, \quad k \text{ fixed}.$$

Now to do the minimization. Consider $\Phi_n(x)/x$ as x goes from 1 to n. It is easy to see that this ratio is much smaller, on the order of $1/\sqrt{n}$, for x near to n than it is for x anywhere else in the interval from 1 to n. Also, for x and y both greater than $\sqrt{n}$ and x less than y, $\Phi_n(x)/x$ is greater than $\Phi_n(y)/y$. It follows that a sum of the form (10) with the n_i constrained as indicated is minimized by making the n_i as large as possible. Clearly this is achieved by taking all but one of the n_i equal to floor(n/k), with the last n_i equal to remainder(n/k). For this choice of n_i the value of (10) is

$$k[\sqrt{n} + O(\log n)] + \Phi_n(\text{remainder}(n/k))$$

$$= k\sqrt{n} + O(\log n) = kH(R) + O(\log H(R)).$$

6. Determining Boundaries of Geometrical Patterns

What happens to the structures of Theorems 3 to 6 if they are imbedded in a gas or crystal, i.e. in a random or constant 0 background? And what about scenes with several independent structures imbedded in them – do their degrees of organization sum together? Is our definition sufficiently robust to work properly in these circumstances?

This raises the issue of determining the boundaries of structures. It is easy to pick out the hierarchy of Theorem 5 from an unstructured background. Any two "spheres" of diameter δ in the scene will have high mutual information given δ^* if and only if they are both in the hierarchy instead of in the background. Here we are using the notion of the mutual information of X and Y given Z, which is denoted $H(X{:}Y|Z)$, and is defined to be $H(X|Z) + H(Y|Z) - H(\langle X,Y\rangle|Z)$. The special case of this concept that we are interested in, however, can be expressed more simply: for if X and Y are both strings of length n, then it can be shown that $H(X{:}Y|n^*) = H(X{:}Y) - H(n)$. This is done using the decomposition (4) and the fact that since X and Y are of length n, $H(\langle n,X\rangle) = H(X) + O(1)$, $H(\langle n,Y\rangle) = H(Y) + O(1)$, and $H(\langle n,\langle X,Y\rangle\rangle) = H(\langle X,Y\rangle) + O(1)$, and thus

$H(X \mid n^*) = H(X) - H(n) + O(1)$, $H(Y \mid n^*) = H(Y) - H(n) + O(1)$, $H(<X,Y> \mid n^*) = H(<X,Y>) - H(n) + O(1)$.

How can one dissect a structure from a comparatively unorganized background in the other cases, the structures of Theorems 3, 4, and 6? The following definition is an attempt to provide a tool for doing this: An ε,δ-pattern R is a maximal region ("maximal" means not extensible, not contained in a bigger region $\hat{R}$ which is also an ε,δ-pattern) with the property that for any δ-diameter sphere R_1 in R there is a disjoint δ-diameter sphere R_2 in R such that

$$H(R_1 : R_2 \mid \delta^*) > \varepsilon.$$

The following questions immediately arise: What is the probability of having an ε,δ-pattern in an n-bit string, i.e. what proportion of the n-bit strings contain an ε,δ-pattern? This is similar to asking what is the probability that an n-bit string s satisifes

$$H_{n/k}(s) - H(s) > x.$$

A small upper bound on the latter probability can be derived from Theorem 1.

7. Two and Higher Dimension Geometrical Patterns

We make a few brief remarks.

In the general case, to say that a geometrical object O is "random" means $H(O \mid \text{shape}(O)^*) \approx \text{volume}(O)$, or $H(O) \approx \text{volume}(O) + H(\text{shape}(O))$. Here shape($O$) denotes the object O with all the 1's that it contains in its unit cubes changed to 0's. Here are some examples: A random n by n square has complexity

$$n^2 + H(n) + O(1).$$

A random n by m rectangle doesn't have complexity $nm + H(n) + H(m) + O(1)$, for if $m = n$ this states that a random n by n square has complexity

$$n^2 + 2H(n) + O(1),$$

which is false. Instead a random n by m rectangle has complexity $nm + H(<n,m>) + O(1) = nm + H(n) + H(m \mid n^*) + O(1)$, which gives the right answer for $m = n$, since $H(n \mid n^*) = O(1)$. One can show that most n by m rectangles have complexity $nm + H(<n,m>) + O(1)$, and less than two raised to the $nm - k + O(1)$ have complexity less than $nm + H(<n,m>) - k$.

Here is a two-dimensional version of Lemma 2: Any large chunk of a random square which has a shape that is easy to describe, must itself be random.

8. Common Information

We should mention some new concepts that are closely related to the notion of mutual information. They are called measures of common information. Here are three different expressions defining the common information content of two strings

X and Y. In them the parameter ε denotes a small tolerance, and as before $H(X{:}Y|Z)$ denotes $H(X|Z) + H(Y|Z) - H(<X,Y>|Z)$.

$$\max \{H(Z) \ : \ H(Z|X^*)<\varepsilon \ \& \ H(Z|Y^*)<\varepsilon\}$$

$$\min \{H(<X,Y>{:}Z) \ : \ H(X{:}Y|Z^*)<\varepsilon\}$$

$$\min \{H(Z) \ : \ H(X{:}Y|Z^*)<\varepsilon\}$$

Thus the first expression for the common information of two strings defines it to be the maximum information content of a string that can be extracted easily from both, the second defines it to be the minimum of the mutual information of the given strings and any string in the light of which the given strings look nearly independent, and the third defines it to be the minimum information content of a string in the light of which the given strings appear nearly independent. Essentially these definitions of common information are given in [17-19]. [17] considers an algorithmic formulation of its common information measure, while [18] and [19] deal exclusively with the classical ensemble setting.

Appendix 1: Errors in [5]

There are two errors in Section 5 of [5]. In this appendix we explain them using the notation of [5].

The definition of the d-diameter complexity given in [5] has a basic flaw which invalidates the entries for $R = R_2$, R_3, and R_4 and $d = n/k$ in the table in [5]: It is insensitive to changes in the diameter d. We show that for any n-bit string R, and for k fixed and n large,

$$C(n/k,R) = C_S(R) + O(1).$$

The $O(1)$-bit string A is a program for concatenating arbitrarily many strings. The $O(C_S(k))$ bit string B knows that P_i is merely the base-two numeral for i (for each i from 1 to $k+1$), that A is of fixed size, and that the remainder, the bulk of P, is the $C_S(R) + O(1)$ bit string C. The computer C_2 described by C operates as follows: Given a program which is the base-two numeral for i, it produces as output the ith piece of R when R is divided into k pieces of size floor(n/k) with a piece of size less than k possibly left over. Hence

$$C(n/k,R) \leq C_S(R) + O(k \log k).$$

Thus

$$C(n/k,R) = C_S(R) + O(k \log k),$$

since obviously

$$C_S(R) \leq C(n/k,R) + O(1).$$

Therefore all entries in the $d = n/k$ row of the table in [5] should be the same as the corresponding entries in the $d = n$ row. Thus, without having to pay a penalty, any region can be expressed as the "sum" of parts less than a kth the size of the region. The definition given in [5] fails to capture the intuitive notion that was intended.

There is also another error in the table in [5], even if we forget the flaw in the definition of the d-diameter complexity. The entry for the crystal is wrong, and should read $\log n$ rather than $k \log n$ (see Theorem 2 in Section 5 of this paper).

Appendix 2. An Information-Theoretic Proof That There Are Infinitely Many Primes

It is of methodological interest to use widely differing techniques in elementary proofs of Euclid's theorem that there are infinitely many primes. For example, see Chapter II of Hardy and Wright [20], and also [21-23]. Recently Billingsley [24] has given an information-theoretic proof of Euclid's theorem. The purpose of this appendix is to point out that there is an information-theoretic proof of Euclid's theorem that utilizes ideas from algorithmic information theory instead of the classical measure-theoretic setting employed by Billingsley. We consider the algorithmic entropy $H(n)$, which applies to individual natural numbers n instead of to ensembles.

The proof is by *reductio ad absurdum.* Suppose on the contrary that there are only finitely many primes $p_1,\ldots,p_k$. Then one way to specify algorithmically an arbitrary natural number

$$n = \prod p_i^{e_i}$$

is by giving the k-tuple $<e_1,\ldots,e_k>$ of exponents in any of its prime factorizations (we pretend not to know that the prime factorization is unique). Thus we have

$$H(n) \leq H(<e_1,\ldots,e_k>) + O(1).$$

By the subadditivity of algorithmic entropy we have

$$H(n) \leq \sum H(e_i) + O(1).$$

Let us examine this inequality. Most n are algorithmically random and so the left-hand side is usually $\log n + O(\log \log n)$. As for the right-hand side, since

$$n \geq p_i^{e_i} \geq 2^{e_i},$$

each e_i is $\leq \log n$. Thus $H(e_i) \leq \log \log n + O(\log \log \log n)$. So for random n we have

$$\log n + O(\log \log n) \leq k[\log \log n + O(\log \log \log n)],$$

where k is the assumed finite number of primes. This last inequality is false for large n, as it assuredly is not the case that $\log n = O(\log \log n)$. Thus our initial assumption that there are only k primes is refuted, and there must in fact be infinitely many primes.

This proof is merely a formalization of the observation that if there were only finitely many primes, the prime factorization of a number would usually be a much more compact representation for it than its base-two numeral, which is absurd. This proof appears, formulated as a counting argument, in Section 2.6 of the 1938 edition of Hardy and Wright [20]; we believe that it is also quite natural to present it in an information-theoretic setting.

Bibliography

[1]. L. E. Orgel, *The Origins of Life: Molecules and Natural Selection,* Wiley, New York, 1973, pp. 187-197.

[2]. P. H. A. Sneath, *Planets and Life,* Funk and Wagnalls, New York, 1970, pp. 54-71.

[3]. G. J. Chaitin, "Information-Theoretic Computational Complexity," *IEEE Trans. Info. Theor.* IT-20 (1974), pp. 10-15.

[4]. G. J. Chaitin, "Randomness and Mathematical Proof," *Sci. Amer.* 232, No. 5 (May 1975), pp. 47-52.

[5]. G. J. Chaitin, "To a Mathematical Definition of 'Life'," *ACM SICACT News* 4 (Jan. 1970), pp. 12-18.

[6]. J. von Neumann, "The General and Logical Theory of Automata," *John von Neumann – Collected Works, Volume V,* A. H. Taub (ed.), Macmillan, New York, 1963, pp. 288-328.

[7]. J. von Neumann, *Theory of Self-Reproducing Automata,* Univ. Illinois Press, Urbana, 1966, pp. 74-87; edited and completed by A. W. Burks.

[8]. R. J. Solomonoff, "A Formal Theory of Inductive Inference," *Info. & Contr.* 7 (1964), pp. 1-22, 224-254.

[9]. G. J. Chaitin and J. T. Schwartz, "A Note on Monte Carlo Primality Tests and Algorithmic Information Theory," *Comm. Pure & Appl. Math.,* to appear.

[10]. C. E. Shannon and W. Weaver, *The Mathematical Theory of Communication,* Univ. Illinois Press, Urbana, 1949.

[11]. H. A. Simon, *The Sciences of the Artificial,* MIT Press, Cambridge, MA, 1969, pp. 90-97, 114-117.

[12]. J. von Neumann, *The Computer and the Brain,* Silliman Lectures Series, Yale Univ. Press, New Haven, CT, 1958.

[13]. C. Sagan, *The Dragons of Eden – Speculations on the Evolution of Human Intelligence,* Random House, New York, 1977, pp. 19-47.

[14]. G. J. Chaitin, "A Theory of Program Size Formally Identical to Information Theory," *J. ACM* 22 (1975), pp. 329-340.

[15]. G. J. Chaitin, "Algorithmic Information Theory," *IBM J. Res. Develop.* 21 (1977), pp. 350-359, 496.

[16]. R. M. Solovay, "On Random R.E. Sets," *Non-Classical Logics, Model Theory, and Computability,* A. I. Arruda, N. C. A. da Costa, and R. Chuaqui (eds.), North-Holland, Amsterdam, 1977, pp. 283-307.

[17]. P. Gács and J. Körner, "Common Information Is Far Less Than Mutual Information," *Prob. Contr. & Info. Theor.* 2, No. 2 (1973), pp. 149-162.

[18]. A. D. Wyner, "The Common Information of Two Dependent Random Variables," *IEEE Trans. Info. Theor.* IT-21 (1975), pp. 163-179.

[19]. H. S. Witsenhausen, "Values and Bounds for the Common Information of Two Discrete Random Variables," *SIAM J. Appl. Math.* 31 (1976), pp. 313-333.

[20]. G. H. Hardy and E. M. Wright, *An Introduction to the Theory of Numbers,* Clarendon Press, Oxford, 1962.

[21]. G. H. Hardy, *A Mathematician's Apology,* Cambridge University Press, 1967.

[22]. G. H. Hardy, *Ramanujan – Twelve Lectures on Subjects Suggested by His Life and Work,* Chelsea, New York, 1959.

[23]. H. Rademacher and O. Toeplitz, *The Enjoyment of Mathematics,* Princeton University Press, 1957.

[24]. P. Billingsley, "The Probability Theory of Additive Arithmetic Functions," *Ann. of Prob.* 2 (1974), pp. 749-791.

[25]. A. W. Burks (ed.), *Essays on Cellular Automata,* Univ. Illinois Press, Urbana, 1970.

[26]. M. Eigen, "The Origin of Biological Information," *The Physicist's Conception of Nature,* J. Mehra (ed.), D. Reidel Publishing Co., Dordrecht-Holland, 1973, pp. 594-632.

[27]. R. Landauer, "Fundamental Limitations in the Computational Process," *Ber. Bunsenges. Physik. Chem.* 80 (1976), pp. 1048-1059.

[28]. H. P. Yockey, "A Calculation of the Probability of Spontaneous Biogenesis by Information Theory," *J. Theor. Biol.* 67 (1977), pp. 377-398.